NORTH CAROLINA
STATE BOARD OF COMMUNITY COLLEGES
LIBRARIES
ASHEVILLE-BUNCOMBE TECHNICAL COMMUNITY COLLEGE

Discarded
Date JUN 25 2025

TELEOPERATED ROBOTICS
IN HOSTILE ENVIRONMENTS

H. Lee Martin
Editor

Daniel P. Kuban
Editor

Rachel Subrin
Publications Administrator

Published by
Robotics International of SME
Publications Development Department
Marketing Services Division
One SME Drive
P.O. Box 930
Dearborn, Michigan 48121

TELEOPERATED ROBOTICS
IN HOSTILE ENVIRONMENTS

Copyright © 1985
Society of Manufacturing Engineers
Dearborn, Michigan 48121

First Edition

First Printing

All rights reserved including those of translation. This book, or parts thereof, may not be reproduced in any form, including machine-readable abstract, without permission of the copyright owners. The Society does not, by publication of data in this book, ensure to anyone the use of such data against liability of any kind, including infringement of any patent. Publication of any data in this book does not constitute a recommendation of any patent or proprietary right that may be involved or provide an endorsement of products or services discussed in this book.

Library of Congress Catalog Card Number: 85-50385

International Standard Book Number: 0-87263-185-0

Manufactured in the United States of America

SME wishes to acknowledge and express its appreciation to the following contributors for supplying the various articles reprinted within the contents of this book. Appreciation is also extended to the authors of papers presented at SME conferences or programs as well as to the authors who generously allowed publication of their private work.

American Nuclear Society
555 North Kensington Avenue
La Grange Park, Illinois 60525

Argonne National Laboratory
9700 South Cass
Argonne, Illinois 60439

I.F.S. Publications Ltd.
35-39 High Street
Kempston, Bedford
MK42 7BT
United Kingdom

Japan Industrial Robot Association
c/o Kikaishinko Building
3-5-8, Shibakoen
Minato-ku
Tokyo, Japan

Marine Technology
The Society of Naval Architects
  and Marine Engineers
One World Trade Center
Suite 1369
New York, New York 10048

Mechanical Engineering
The American Society of Mechanical Engineers
345 East 47th Street
New York, New York 10017

Mechanism and Machine Theory
Pergamon Press Incorporated
Maxwell House
Fairview Park
Elmsford, New York 10523

Cover photo courtesy of Oak Ridge National Laboratory

PREFACE

As man expands his domain to new horizons and seeks to control hazardous materials in these environments, the need and demand for remote manipulation capabilities increases. This collection of papers and articles provides a sampling of the equipment, applications, and developments of remote manipulators for performing a variety of tasks in hostile environments.

The advent of nuclear energy brought the initial development of mechanical master/slave manipulators as a means of reducing human exposure to radioactive materials. As larger facilities were built, the need for large-volume manipulation led to the development of electromechanical manipulators in the 1950s. The pioneering work in this field was performed at the Argonne National Laboratory under the direction of Ray Goertz. At the same time, the initial research in robotic systems was beginning at other institutions.

These two related fields are now converging and accomplishing tasks their originators only imagined. Access to undersea resources and exploration of outer space are two significant manipulative challenges. The handling of radioactive materials, munitions, and chemical wastes continues to pose unique problems that justify a wide variety of applications for remote manipulation. As resource demands increase and environmental constraints tighten, the capabilities of this type of equipment will continue to be expanded and proven.

To meet these challenges, teleoperator performance is being improved by the implementation of more powerful, yet lower cost computing hardware. The combination of robotic (autonomous) and teleoperated (manual) manipulator technology—the telerobot—is now made possible by recent developments in computers, artificial intelligence, manipulators, and human-machine interfaces. The ability of master/slave manipulators to scale human forces and motions, and yet allow human intelligence to guide unstructured activities offers potential usefulness in many future applications.

This volume is intended to be a resource for those presently considering the development or use of manipulation systems in unique or unstructured environments where human access is limited or impossible. The selected works included represent only a few of the many excellent papers on the subject. Chapter One provides an overview of the development of remote manipulators and reviews their present applications and future potential. Teleoperated manipulator applications in nuclear, undersea, and space environments are detailed in Chapters Two, Three, and Four. The use of autonomous robotic systems in hostile environments is the subject of Chapter Five. Chapter Six reviews several developments in mobility methods used to deliver manipulators to the worksite. The Appendix offers Goertz's early work, "Philosophy and Development of Manipulators."

We would like to thank all of the companies, organizations, publishers, and authors who gave permission to have their articles reprinted in this volume. Thanks also to the Publications Development Department staff at SME for their assistance in the research and development required to making this book possible. The base of knowledge contained within the Transactions of the Remote Systems Technology Division of the American Nuclear Society has been instrumental in the preparation of this collection.

H. Lee Martin, P.E.
Oak Ridge National Laboratory

Daniel P. Kuban, P.E.
Oak Ridge National Laboratory

ABOUT THE EDITORS

H. Lee Martin is a Development Engineer in the Instrumentation and Controls Division of the Oak Ridge National Laboratory which is operated by Martin Marietta Energy Systems. He has been involved with the development of digital controls for force-reflecting manipulators to be used in future nuclear fuel reprocessing facilities. He assisted in the development of the first fully digital master/slave servomanipulator, which was awarded an IR 100 award by *Research and Development* Magazine.

Mr. Martin holds a BSME from the University of Tennessee and an MSME from Purdue University. He is presently a doctoral candidate at the University of Tennessee. He has organized and led tutorials on teleoperated manipulators and has authored more than 10 papers on manipulators. As a consultant for Remotec Corporation, he developed a master/slave manipulator control system for a high lift-to-weight ratio manipulator system.

In addition to being the founding chairman of the Knoxville/Oak Ridge Subchapter of Robotics International, Mr. Martin is also a member of the Remote Systems Division of Robotics International, a member of the American Society of Mechanical Engineers and a registered Professional Engineer in Tennessee. He holds one patent.

Daniel P. Kuban heads the Equipment Development Engineering Section in the Engineering Division of Martin Marietta Energy Systems. He is actively involved in design, fabrication, and application of teleoperators, servomanipulators, advanced robotics, and specialized remote equipment. He has published several papers and has taught professional development courses in these areas. For the past three years, he has concentrated on the design and analysis of precision servomechanisms.

Mr. Kuban received a BSME from the University of Illinois and an MSME from the University of Tennessee; he has also completed course work toward an MBA at the University of Tennessee. He holds three patents, is a registered Professional Engineer in Tennessee, and holds membership in the American Society of Mechanical Engineers, Robotics International, and several honor societies.

RI/SME

Robotics International of SME (RI/SME) was founded in 1980 as an applications and research-oriented organization covering all phases of research, design, installation, operation, and maintenance of industrial robots. In addition, RI/SME serves as a liaison between industry, government, and education to determine areas where technological development is needed. RI/SME is the vital link in the chain that connects the individual to the many aspects of robotic-oriented manufacturing.

Members of RI/SME are involved in these three categories:

- Planning, selecting, and applying robots.
- Research and development leading to the creation or improvement of robots.
- Other creative robotic activities in administration, education, or government.

The members of RI/SME update their knowledge and skills through educational programs, chapter meetings, conferences, expositions, and publications including *Robotics Today*. This book is one example of the organization's wide-ranging activities.

The largest American scientific and educational association exploring the expanding use of robots, RI/SME is an official association of the Society of Manufacturing Engineers and is headquartered in Dearborn, Michigan. RI/SME addresses the informational needs of the engineer involved in the application of robots and provides a forum for discussion among concerned individuals. RI/SME is a partner to industry and a major force in an engineer's professional advancement.

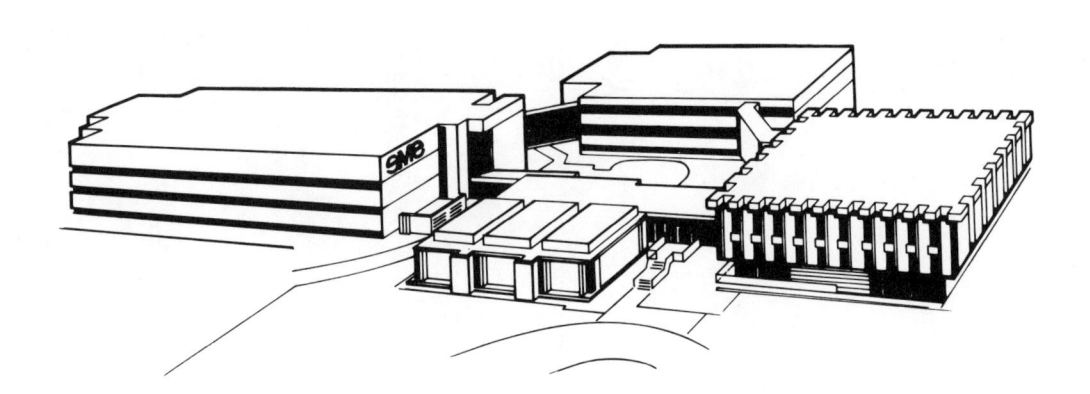

MANUFACTURING UPDATE SERIES

Published by the Society of Manufacturing Engineers and its affiliated societies, the Manufacturing Update Series provides significant up-to-date information on a variety of topics relating to the manufacturing industry. This series is intended for engineers working in the field, technical and research libraries, and also as reference material for educational institutions.

The information contained in this volume doesn't stop at merely providing the basic data to solve practical shop problems. It also can provide the fundamental concepts for engineers who are reviewing a subject for the first time to discover the state of the art before undertaking new research or applications. Each volume of this series is a gathering of journal articles, technical papers and reports that have been reprinted with expressed permission from the various authors, publishers, or companies identified within the book. Educators, engineers, and managers working within industry are responsible for the selection of material in this series.

We sincerely hope that the information collected in this publication will be of value to you and your company. If you feel there is a shortage of technical information on a specific manufacturing area, please let us know. Send your thoughts to the Manager, Publications Development, Marketing Division at SME. Your request will be considered for possible publication by SME or its affiliated societies.

TABLE OF CONTENTS

CHAPTERS

1 OVERVIEW OF REMOTE APPLICATIONS

Joining Teleoperation with Robotics for Remote Manipulation in Hostile Environments
By *H.L. Martin* and *W.R. Hamel*
Presented at the RI/SME Robots 8 Conference, June 1984 3

The Manipulator as a Means of Extending our Dexterous Capabilities to Larger and Smaller Scales
By *Carl R. Flatau*
Presented at the American Nuclear Society 21st Conference on Remote Systems Technology, November 1973 19

Design and Application of Remote Manipulator Systems
By *Carl Witham*, *Anthony Fabert*, and *A.L. Foote*
Presented at the American Nuclear Society 26th Conference on Remote Systems Technology, 1978 23

Robots and Manipulators
By *Ewald Heer*
Reprinted from *Mechanical Engineering*, November 1981,
Copyright © The American Society of Mechanical Engineers 30

2 NUCLEAR APPLICATIONS

The MA 23 Bilateral Servomanipulator System
By *Jean Vertut, Paul Marchal, Guy Debrie, Michel Petit, Danier Francois* and *Philippe Coiffet*
Presented at the American Nuclear Society 24th Conference on Remote Systems Technology, 1976 41

SM-229—A New Compact Servo Master-Slave Manipulator
By *Carl R. Flatau*
Presented at the American Nuclear Society 25th Conference on Remote Systems Technology, 1977 54

The State-of-the-Art Model M-2 Maintenance System
By *J.N. Herndon, D.G. Jelatis, C.E. Jennrich, H.L. Martin* and *P.E. Satterlee, Jr.*
Reprinted from the Proceedings of the 1984 National Topical Meeting on Robotics and Remote Handling in Hostile Environments, American Nuclear Society, pages 147-154 59

MINIMAC—The Remote-Controlled Manipulator with Stereo Television Viewing at the SIN Accelerator Facility
By *Eyke Wagner* and *Albin Janett*
Presented at the American Nuclear Society 26th Conference on Remote Systems Technology, 1978 67

An Advanced Remotely Maintainable Force-Reflecting Servomanipulator Concept
By *D.P. Kuban* and *H.L. Martin*
Reprinted from the Proceedings of the 1984 National Topical Meeting on Robotics and Remote Handling in Hostile Environments, American Nuclear Society, pages 407-415 70

3 UNDERSEA APPLICATIONS

Robotics Undersea
By *Robert L. Wernli*
Reprinted from *Mechanical Engineering*, August 1982,
Copyright © The American Society of Mechanical Engineers 81

Cable Controlled Deep Submergence Teleoperator System
By *J. Charles* and *J. Vertut*
Reprinted with permission from *Mechanism and Machine Theory*, Volume 12, J. Charles, J. Vertut, "Cable Controlled Deep Submergence Teleoperator System," Copyright © 1977, Pergamon Press, Ltd. 89

Experience with an Unmanned Vehicle-Based Recovery System
By *Robert L. Wernli*
Reprinted from *Marine Technology*, January 1983 101

Development of "Underwater Robot" Cleaner for Marine Live Growth in Power Station
By *Kyosuke Edahiro*
Presented at the 1983 International Conference on Advanced Robotics.
Copyright © Japan Industrial Robot Association 108

4 SPACE APPLICATIONS

Synergy in Space—Man-Robot Cooperation
By *Sam Walters*
Reprinted from *Mechanical Engineering*, January 1985,
Copyright © The American Society of Mechanical Engineers 119

Mars Viking Surface Sampler Subsystem
By *Donald S. Crouch*
Presented at the American Nuclear Society 25th Conference on Remote Systems Technology, 1977 131

A Large-Scale Manipulator for Space Shuttle Payload Handling—The Shuttle Remote Manipulator System
By *H.J. Taylor*
Presented at the American Nuclear Society 26th Conference on Remote Systems Technology, 1978 142

5 ROBOTIC ADAPTATIONS TO HOSTILE ENVIRONMENTS

Application of High Performance Heavy Duty Industrial Manipulators
By *Henry G. Krsnak* and *Michael J. Howe*
Presented at the SME 5th International Symposium on Industrial Robots, September 1975 153

Remote Demilitarization of Chemical Munitions
By *Robert H. Leary*, *James M. McNair*, and *John F. Follin*
Presented at the RI/SME Robots 8 Conference, June 1984 163

Applying Robots in Nuclear Applications
By *John J. Fisher*
Presented at the RI/SME Robots 9 Conference, June 1985.. 175

6 MOBILE SYSTEMS

Manipulator Vehicles of the Nuclear Emergency Brigade in the Federal Republic of Germany
By *Gerhard Wolfgang Kohler*, *Manfred Selig*, and *Manfred Salaske*
Presented at the American Nuclear Society 24th Conference on Remote Systems Technology, 1976 197

Teleoperators in the Nuclear Industry
By *J.A. Constant* and *K.J. Hill*
Reprinted with permission of the authors and IFS Publications, Ltd. Presented at the 3rd Conference on Industrial Robot Technology & 6th International Symposium on Industrial Robots, March 1976 220

Technology for Mobile Robotics in Nonmanufacturing Applications
By *Thomas G. Bartholet*
Presented at the RI/SME Robots West Conference, November 1984 230

VIRGULE Variable-Geometry Wheeled Teleoperator
By *Jean Vertut*, *Jean-Pierre Guilbaud*, *Guy Debrie*, *Jean-Claude Germond* and *Francois Riche*
Presented at the American Nuclear Society 20th Conference on Remote Systems Technology, 1972 240

Walking Robot for Underwater Construction
By *Yoshitane Ishino*, *Toshihisa Naruse*, *Toshiyuki Sawano* and *Norikazu Honma*
Presented at the 1983 International Conference on Advanced Robotics.
Copyright © Japan Industrial Robot Association 247

APPENDIX

Philosophy and Development of Manipulators
By *R.C. Goertz*
Reprinted with permission from Argonne National Laboratory, Copyright © 1951 257

INDEX 263

CHAPTER 1
OVERVIEW OF REMOTE APPLICATIONS

Presented at the RI/SME Robots 8 Conference, June 1984

Joining Teleoperation with Robotics for Remote Manipulation in Hostile Environments

by H.L. Martin
and
W.R. Hamel
Union Carbide

Teleoperators have been used by the nuclear industry for nearly three decades to perform manipulative tasks within hostile environments.[1] This technology has been implemented and improved by applications in space, under water, and in laboratory development environments. Heightened interest in robotics is presently occurring in our society. It is therefore appropriate to review teleoperator applications as an important past and future foundation of robotic developments and research. Such a perspective will highlight the similar research goals but contrasting approaches in these two fields. Technology exchange between robotics and teleoperator developers should increase as these systems approach common goals.

The ultimate manipulation system might be described as being totally adaptable yet fully automated. Teleoperators offer adaptability due to their man-in-the-loop control schemes, whereas robots are normally operated in an autonomous mode to reduce labor costs. A middle-ground class of systems that function either autonomously or with real-time human interaction is envisioned for future systems. Developments in hardware and software for both robotic and teleoperated systems will make this goal a reality.

This paper presents the background of teleoperator development and principles. It is intended to emphasize the operational similarities and differences between teleoperators and industrial robots. Major milestones in the historical progression of teleoperator technology are given and compared to the history of robotic activity. A review of national and international applications of teleoperators is also provided. These activities range from outer space to undersea manipulators, and span projects from France to Japan. Research directions are reviewed, concentrating on the convergence of teleoperator and robot technology. Common goals pursued through different techniques are observed in present research. Improvements in mechanisms, modularity, kinematics, and man-machine interfacing techniques will have useful transfer from the teleoperator realm to the robotic domain. The expanse of research activity related to robotics will result in better motors, electronics, software, and sensors to improve the performance of both system types. We begin our review by discussing the fundamentals of teleoperation as background.

II. FUNDAMENTALS OF TELEOPERATION

A teleoperator system is a general-purpose, dexterous man-machine system that augments man by projecting his manipulation capabilities, often across distance and through physical barriers into hostile environments.[2] The most famous example of a teleoperator system is the space shuttle's remote manipulator system (RMS), used to deploy and retrieve satellites, but several other applications exist and are planned. For example, the handling of radioactive materials with teleoperators has been ongoing for 35 years, and undersea exploration and exploitation are being pursued by both the public and private sectors.

---

*Research sponsored by the Office of Spent Fuel Management and Reprocessing Systems, U.S. Department of Energy under Contract No. W-7405-eng-26 with Union Carbide Corporation.

By acceptance of this article, the Publisher and/or recipient acknowledges the U.S. Government's right to retain a nonexclusive, royalty-free license in and to any copyright covering this paper.

A teleoperator system is not a robot. Most teleoperator systems consist of a manipulator that is similar to a robot in many aspects. What sets teleoperators apart is the man-machine interface which allows realtime interaction between the human operator and the mechanical manipulator. The most common form of teleoperator man-machine interface is the replica master arm. This interface allows the user to operate the slave throughout the workspace simply by moving the master manipulator arm, which is normally a kinematic replica of the slave. Many forms of master controllers exist:

1. <u>Replica Master</u>. Kinematically identical to the slave, this method of control usually provides force reflection to the operator. A high degree of proprioceptive feedback is present due to the similarities between master and slave. Normally the systems are 1:1 in geometric proportion, but scaling factors as great as 3:1 have been used by TeleOperator Systems (TOS).

2. <u>Switch Box Control</u>. Similar to robotic pendant controllers, these control structures were used for the first teleoperator systems because of their ease of implementation. Operator efficiency and dexterity is lacking in this method due to the absence of force reflection and proprioception.

3. <u>Joystick Control</u>. This is superior to switch box control because coordinated motions can be made using computer guidance to accomplish movement. True force reflection through joystick control has not yet been accomplished commercially, and bilateral force-reflecting replica masters still hold certain advantages in efficiency.

Table 1 gives a comparison of the relative task efficiency using various controllers and manipulator systems. The emphasis on the man-machine interface is made to point out the major difference between teleoperators and robots. The primary operating mode for a teleoperator is with man as the decision maker in the control loop. This allows teleoperators to successfully accomplish tasks and conquer unexpected situations within unstructured environments. Man's ability to reason combined with the machine's attributes of strength and resistance to hostile environments produces a beneficial relationship. In contrast, the robot control system is designed to be autonomous. All but the most sophisticated robot applications are in structured surroundings with a limited set of operational requirements. The need for human intervention is limited to start-up programming with a pendant or by lead-through teaching. As sensor technology and artificial intelligence methods improve, the adaptability of robots will approach teleoperators, but these two systems presently find application in different environments.

TABLE 1
TASK EFFICIENCY FOR VARIOUS TELEOPERATOR CONTROL METHODS AND MANIPULATION TYPES

Manipulation/control	Task completion time (direct human 1:1)
Crane - Impact wrench with switch control	500-800:1
Unilateral (no force reflection) with switch control	500:1
Unilateral with joystick control	60-80:1
Force-reflecting servomanipulator with 1 arm	16:1
Force-reflecting servomanipulator with 2 arms	8:1
Suited human	8:1

The foundation of teleoperators is in the nuclear industry,[3] where the requirement to handle radioactively hazardous material led to the development of the first mechanical master-slave manipulator. These manipulators, such as the one shown in Figure 1, were able to perform dexterous tasks with force reflection through stainless steel tapes connecting master to slave (similar to a pantograph in concept).[4] Visual feedback was through shielded windows. Mechanical master-slave manipulators have one significant shortcoming: they have a small volumetric coverage limited by a fixed pivot point. To expand this coverage required elimination of the mechanical connection between master and slave. Two drive methods have been explored through the years: electromechanical and hydraulic. Hydraulics are most favorably applied in underwater or extremely heavy duty applications. Hydraulic systems offer a high power-to-weight ratio if the sump tank is not considered, and they function well under the high static pressure environments of deep-sea applications. The electromechanical systems offer 1% of peak load force sensitivity through backdrivable gear trains. Most modern teleoperators are of the electromechanical manipulator type.

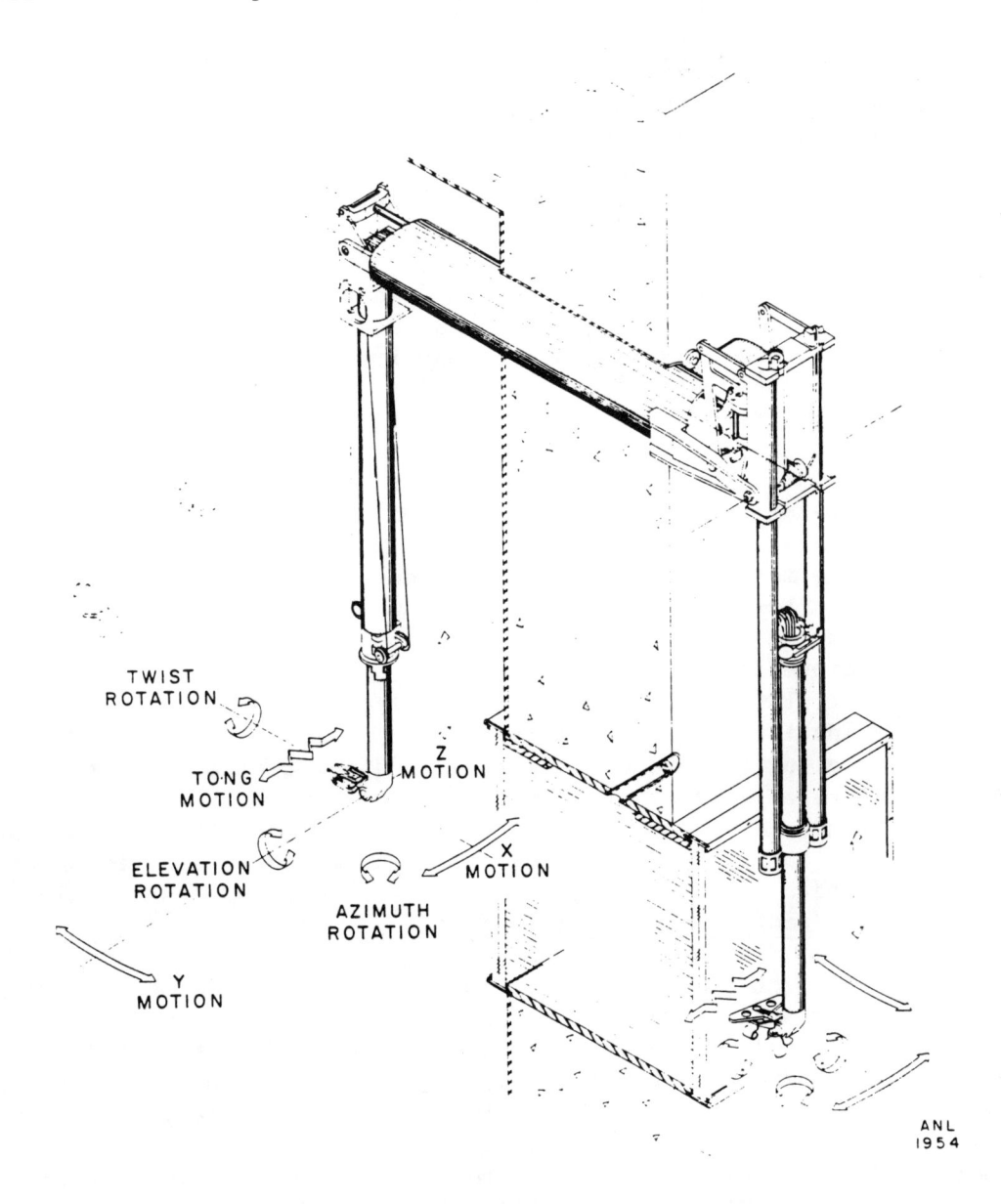

Figure 1 Mechanical master/slave manipulation system.

Early development efforts by Argonne National Laboratory and General Mills (later Programmed and Remote Systems and now GCA) addressed unilateral concepts--that is, they provided no force reflection and were operated from a simple on/off switch box. Later implementations included force reflection through bilateral servo loops with low friction, high-efficiency torque transmission methods. The original force-reflecting servomanipulator development was performed by the Remote Control Division of the Argonne National Laboratory under the direction of Ray Goertz. Their research in mechanics and controls laid the foundation for development which continues today in both teleoperators and robotics. Table 2 gives a brief chronology of the major milestones of teleoperator developments.[5]

TABLE 2
TELEOPERATOR DEVELOPMENT HISTORY

Year	Milestone
1948	Goertz and Argonne National Laboratory (ANL) developed first bilateral mechanical master-slave manipulator.
1948	General Mills produced a unilateral manipulator based on electric motors with switch control. Used for high capacity, large volume tasks.
1954	Goertz built an electric master-slave manipulator incorporating servos and force reflection. This was the first bilateral force-reflecting servomanipulator.
1958	General Electric built the Handyman electrohydraulic manipulator incorporating force feedback, articulated fingers, and an exoskeletal master control.
1961	The General Mills Model 150 manipulator was fitted for manned deep-sea operation.
1963	The U.S. Navy began deep-submergence projects which included developing underwater manipulators.
1965	ANL combined manipulation with head controlled TV camera and receiver.
1970's	NASA sent teleoperators into space: unmanned soil samplers went to the Moon and Mars.
1970's	NASA began development of a space shuttle manipulator in cooperation with SPAR, a Canadian firm.
1970's	The nuclear community reinstituted efforts to develop improved teleoperators for facility maintenance.
1980	Supervisory control techniques in hardware were demonstrated by Brooks at Massachusetts Institute of Technology.
1980	Universal controller techniques were developed and refined by the Jet Propulsion Laboratory and Stanford University.
1982	Oak Ridge National Laboratory and Central Research Laboratories designed and fabricated the first fully distributed, digitally controlled servomanipulator.
1983	ODETICS developed a tetherless electromechanical walking functionoid with a lift-to-weight ratio greater than one.

As implied thus far, teleoperators and robots share many common subsystems. An understanding of the fundamental control structure of a force-reflecting teleoperator is necessary to fully appreciate operational differences. Figure 2 shows the basic block diagram of a one-degree-of-freedom servo loop for a teleoperator. Motor types, encoders, amplifiers, and even digital control electronics are very similar between a teleoperator and a robot. It is the operational control of a teleoperator through real-time human interaction via a master controller that results in significant differences in control philosophy and methods. Force is reflected in a bilateral force-reflecting teleoperator by backdriving the motor to create a position offset (this can also be accomplished by direct force measurement, i.e., motor current or strain gages). A stiff position loop causes the slave to follow the master quickly until an obstacle is encountered, which usually causes deceleration of the slave.

This in turn results in a force generated at the master due to the increased positional error. The velocity loops are for stabilization and inertia compensation. A typical robot control system will appear similar to the bottom half of Figure 2 as the master motion is simulated by computer playback of commands previously taught. Robotic control loops will also include integral gain to eliminate steady state errors and some form of preprocessing to generate the desired path to follow.[6]

III. THE EVOLUTION OF ROBOTICS AND TELEOPERATED SYSTEMS

To consider the potential future relationship between robotics and teleoperators, it is desirable to review the evolution of these two classes of manipulator applications. Robot and teleoperator manipulators, although quite similar in basic mechanical concepts, have followed very different evolutionary paths.

Teleoperator systems[4] were developed with the basic objective of accurately projecting a human operator's motor capabilities into a remote environment. In these systems, from the purely mechanical designs to the later servomanipulator designs, the fidelity accomplished in replicating human functions was the principle performance criterion.

Industrial robotics appears to have followed a different path, directed toward different objectives. The first industrial robots were, in essence, programmable

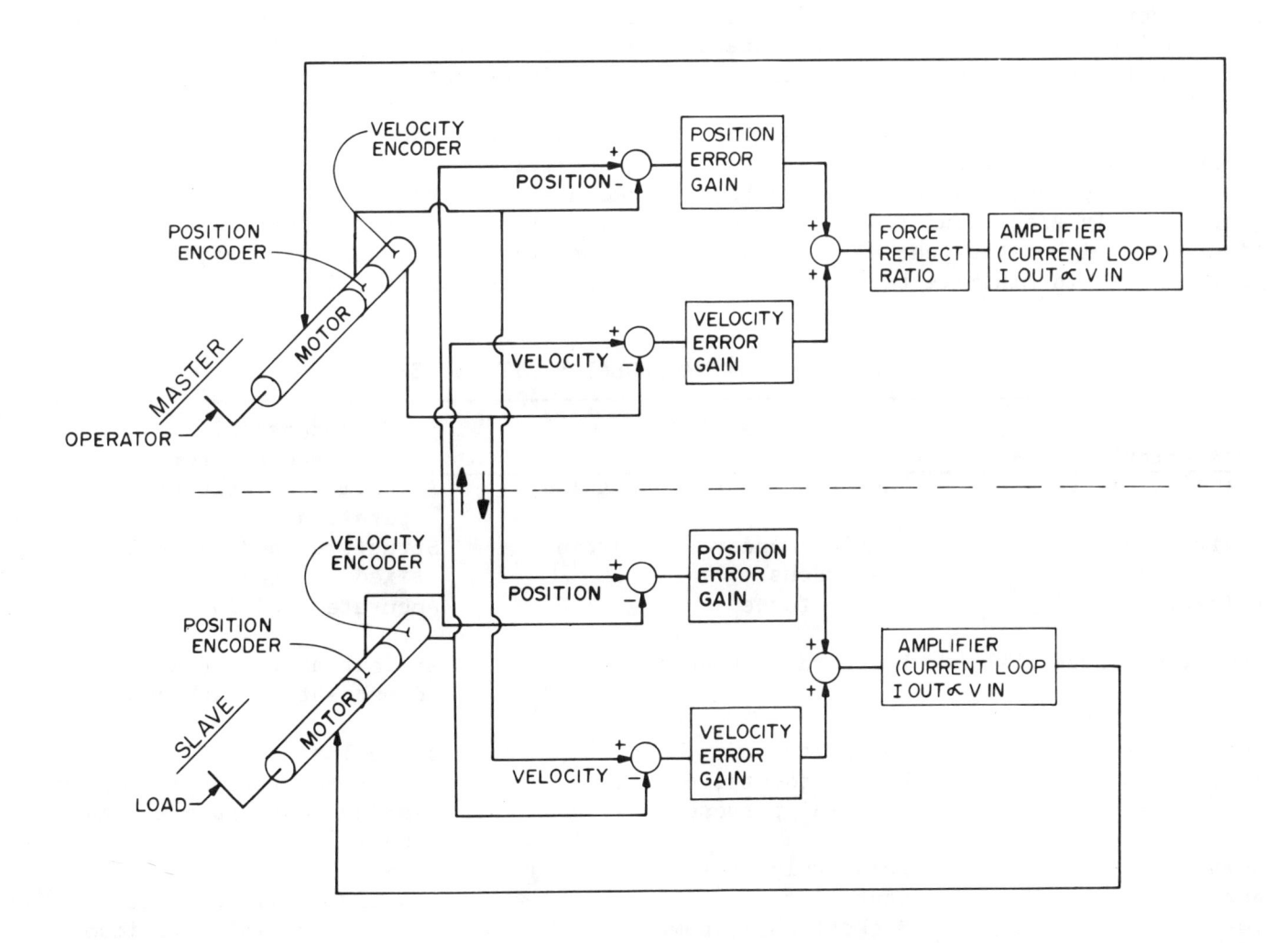

Figure 2 Block diagram of bilateral servomechanism concept.

parts-handling systems capable of fast operation, large load capacity, operation in harsh environments, and precise positioning. For the most part, the first robots were replacements for human production workers with the advantages of greater stamina, faster operation, and less cost. They were, however, limited to simple and relatively structured tasks.

The differences in development objectives between the two classes manifest themselves in the basic design parameters associated with the systems. Reviewing some of these parameters provides a good comparative framework for understanding the differences. Table 3 lists various design features (or objectives) and corresponding technical attributes (or implications) for manipulators in teleoperation and industrial robotics. In the case of human operation, it is critically important to provide the operator with a sense of feel and speed compatibility. This in turn requires that the manipulator be designed with minimum inertia and be operable in the human dynamic range. As depicated in Table 3, this form of low friction and inertia design involves the use of centralized actuators (located to reduce arm-link moments of inertia and to reduce motor size) and high efficiency power transmissions such as backdrivable gear trains and cable or metal-band pulley drives. Because these designs achieve low friction and are backdrivable, actuator drive torque (drive current in the case of dc servomotors) is an accurate indication of applied load (within 1% of maximum torque).

In contrast, industrial robots were developed at least initially to increase the ratio of production output to effective labor cost.[7] The effective labor cost of a robot is the complete operating and capital recovery cost of the system for its planned life. With emphasis on manufacturing productivity, industrial robots are designed to create economic advantage by trading off operating speed and precision against capital cost. Controlling capital cost translates into minimizing the purchase price of the robot, which in turn leads to design tradeoff decisions that reduce recurring manufacturing costs. The two design objectives of good position control and minimum production cost seem to have pushed most industrial robot designs (particularly electrical robots) in the direction of distributed actuator configurations with high-ratio gear drives (for torque amplification), utilizing adjustable gear centers to minimize backlash at the expense of meshing friction. In this class

TABLE 3
COMPARISON OF MANIPULATOR ATTRIBUTES

Attribute	Teleoperated Servomanipulators	Industrial Robot Manipulators
Principal function	Master/slave teleoperation	Autonomous, repititious operation
Environment	Complex, uncertain, often hazardous	Structured and generally fixed
Primary control parameter	Output force	Accurate position
Operating speed	Human range 0-40 in./s	As fast as possible, dependent on task and design
Load range	20-50 lbs	5-200 lbs
Kinematics	Rotary joints, 6 dof, general purpose tong	Rotary and prismatic joints, various dof, specialized end effectors
Compliance	Relatively flexible	Usually stiff
Actuators	Centralized	Usually local to joint
Torque transmission	Backdrivable, some backlash	Non-backdrivable, minimum backlash

of manipulators production cost is comparatively low. Robotic manipulators are capable of achieving relatively high accuracy and repeatability by virtue of their high stiffness and low backlash, but they are non-backdrivable and have relatively large friction thresholds.

Perhaps one would not be surprised by the fundamental differences between these two classes of manipulators since they were in essence addressing fundamentally different design objectives. The most fundamental factor relating these classes may be the issue of force modulation and reaction in the work task environment. In the material handling applications associated with industrial robots, force interactions are accommodated through position precision, jigging/fixture design, and special end-effector design. It has only been with the more sophisticated assembly applications that the issues of force interaction have become of interest. Assembly,[8] especially for high precision components, involves force interaction to facilitate part fit-up or insertion. In such applications the robot cannot rely completely upon position control but must also control applied forces to achieve the task.

The prevalent research approaches to achieving force control capability include augmenting the conventional designs with torque or force transducers. The installation of a 6-degree-of-freedom force transducer[9] at the robot wrist is a common approach. This type of transducer is used to resolve the three-dimensional forces and moments at the wrist. Appropriate transformations and algorithms are then used to close force control servo loops at the joint axes. This, of course, does not alleviate the problem of the large friction/non-backdrivability nonlinearities which are inherently present.

Manipulator development was initially dominated by applications in nuclear technology development.[3] In these systems a great deal of emphasis has been placed upon the engineering issues of force interaction in complex work environments. Microelectronics advances made the implication of George C. Deval's 1954 programmable automated robot patent realizable.[10] Since the first developments focused on pick-and-place parts handling, a class of industrial robot manipulators evolved emphasizing cost, speed, and position control. The number and type of industrial robotics applications have increased steadily. Today, many applications have lead developers into the realm of force control. It is believed that these types of applications will produce a continuing evolution which will decrease the fundamental differences we now see between industrial robotic manipulators and the electronic servomanipulators used in teleoperation.

IV. OVERVIEW OF TELEOPERATOR SYSTEMS APPLICATION AND DEVELOPMENT

Although many research efforts have been expended to develop teleoperators, only a few functioning systems exist. These systems invariably find application in hostile environments of one form or another. With the advent of low-cost digital controls, improved control techniques, and improved subsystems (motors, encoders, software, etc.) it is conceivable that cost reduction and improved versatility will justify increased use of teleoperator systems. A review of present application and development thrusts should give insight into the future direction of teleoperator technology.

Only two American companies currently market complete force-reflecting master-slave teleoperator systems commercially: Central Research Laboratories (Figure 3) and TeleOperator Systems (Figure 4). Both of these companies have been making master-slave teleoperators for some time and have units in operation around the country. Due to the small production volume, these systems remain labor intensive and are priced accordingly. Up to now these systems have been used primarily in hazardous radiation environments, but teleoperators are being developed for other areas. Four areas of

Figure 3 Model M2 maintenance system by Central Research Laboratories.

application are reviewed here: nuclear, space, military, and industry. The first three are primarily government based, but private industry is also entering the field.

In the nuclear field the use of teleoperators is related to equipment maintenance and material handling in radioactive environments. Successful implementation of teleoperators has been ongoing in target maintenance for several years at particle accelerator facilities at Fermi Laboratory and Los Alamos National Laboratory.[11,12] In these applications, the teleoperator is required to replace target materials and fixtures between tests. Remote maintenance systems for nuclear fuel reprocessing facilities have used teleoperated systems, but only recently have research efforts been renewed to address large-volume dexterous coverage and remote maintainability of the remote slave manipulator system.[13] Other applications in the nuclear environment include systems to perform maintenance on radioactive heat exchangers in commercial reactors. Westinghouse has developed a modular manipulator which can be assembled quickly *in situ* and allows the remote teleoperation of tooling to check and replace failed tubes in a steam generator.[14] General purpose teleoperators for inspection and maintenance in nuclear power reactors have been constructed, but their access is limited by facility geometry and tethered operation. Fusion research facilities plan to use teleoperators to maintain tiles and shields and to set up experiments within Tokamak reactors. Mobility is a problem in this application also, as it will be extremely difficult to develop transporter systems that can insert the manipulator system into and maneuver about the extremely complex internal structures of various fusion reactor concepts (e.g., Tokamak, Torus).[15]

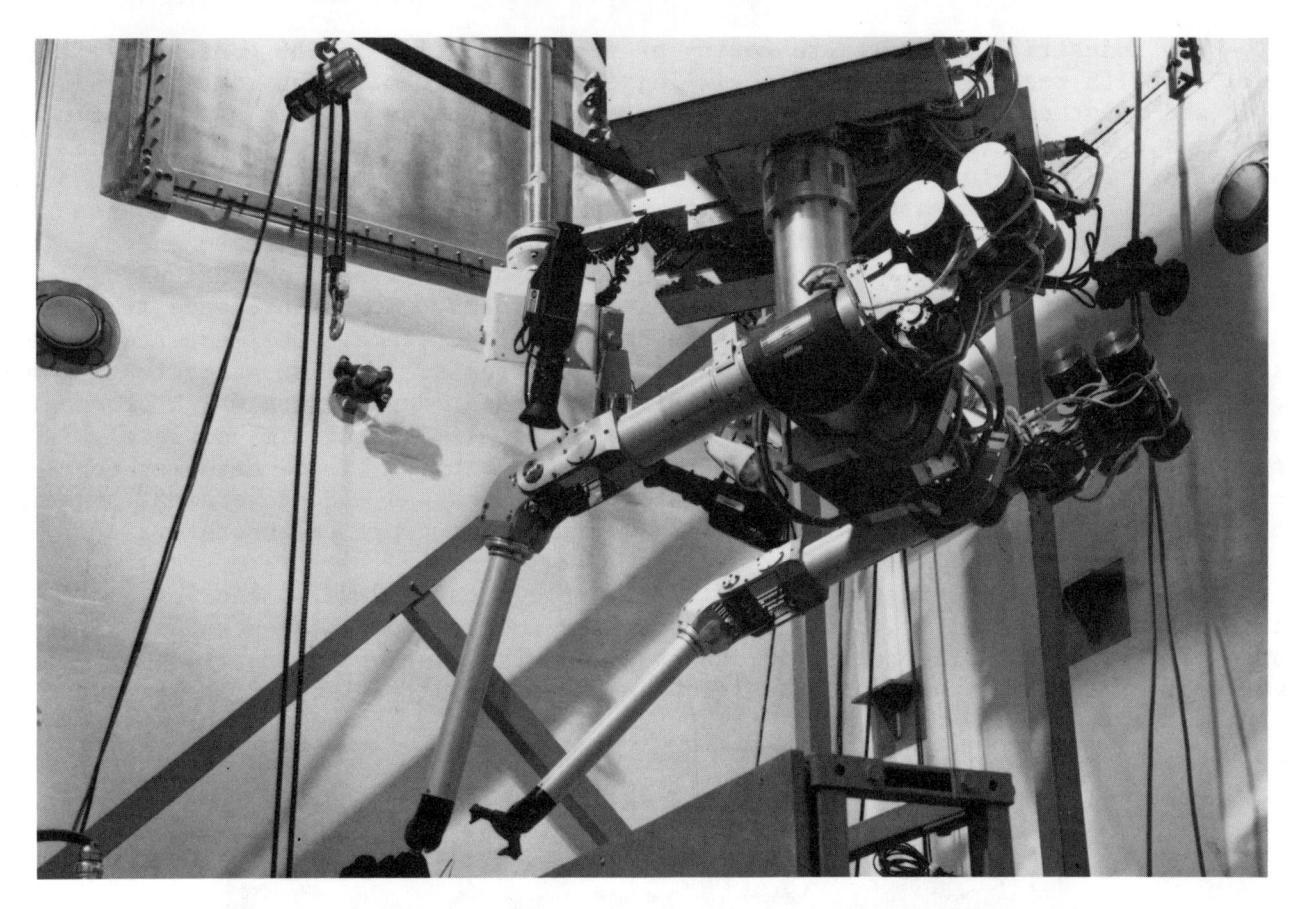

Figure 4 SM-229 servomanipulators by TeleOperator Systems.

Manipulator technology has been applied in the space arena in four instances. Manipulators were used to collect and analyze soil samples as part of the U.S. Lunar Surveyor Program and the USSR Lunar Exploration Program in the late 1960s. The Surveyor manipulator was used to dig a trench and perform other simple functions under remote control. The USSR manipulators acquired soil samples and deposited them in a return capsule. The third application was in the U.S. Mars Viking Program in 1976, when a computer-controlled manipulator helped to perform experiments on the Martian surface. In 1982 the U.S. space shuttle employed a Canadian-built 48-foot-long attached manipulator to unload payloads from its cargo bay and handle other objects in space.[16] Although only the space shuttle RMS is classified as a real-time teleoperator, all are examples of using machines to extend man's capabilities. Plans to develop sensory feedback on future space manipulators will make them more similar to earth-based teleoperator systems.

The U.S. Navy appears to be the leader in military teleoperator system development. The Navy's interest stems from the need to perform work tasks under water (especially in the hostile environment of extreme depths). The Navy has developed and operated a family of remotely operated vehicles (ROV) for nearly two decades. They usually have an umbilical cable for power and control and use a manipulator to retrieve submerged items.[17] Two non-routine recoveries have been performed by Navy ROVs: a nuclear weapon off Palomaries, Spain, in 1965 and the manned submersible Pisces III off Cork, Ireland, in 1975, just prior to the exhaustion of the pilot's air supply. The Navy continues significant efforts to develop tetherless vehicles and improved control techniques under the auspices of the Naval Ocean Systems Center. More recently the Army has shown interest in applying robotics in the battlefield to augment manpower requirements.[18] As this effort takes shape, it will be interesting to observe the tradeoffs made between the autonomous robot and the man-aided teleoperator.

A few industrial projects are worthy of special mention in the context of this paper. ODETICS recently announced a six-legged walking robot which combines teleoperator and robot control features. The outstanding features of this system include tetherless operation (self-contained power supply and radio wave control transmission), and a payload-to-weight ratio greater than one. Odex 1, shown in Figure 5, is targeted for sentry duty applications with possible military uses.[19] Exxon Research and Production has developed a submersible petroleum production platform and a maintenance teleoperator system. The objective is to perform well-head operations on the ocean floor rather than build large, expensive platforms. The maintenance teleoperator is then used for upkeep of the submerged platform items. One tethered, teleoperated manipulator would service many submerged platforms. Shallow-water testing has already been performed successfully. A third commercial application which shows great promise is that of mining. Remoting the operator or eliminating him completely is the thrust of British efforts through the National Coal Board.[20] These examples represent the diversity of teleoperator activities within industry and U.S. government organizations, but many groups worldwide are also working to improve teleoperators.

Several foreign countries are involved in teleoperator development for hazardous material handling, primarily in nuclear applications. In West Germany, mobile platforms and force-reflecting manipulators for emergency situations have been developed.[21] The MF3 manipulator vehicle is a mobile platform supported by four

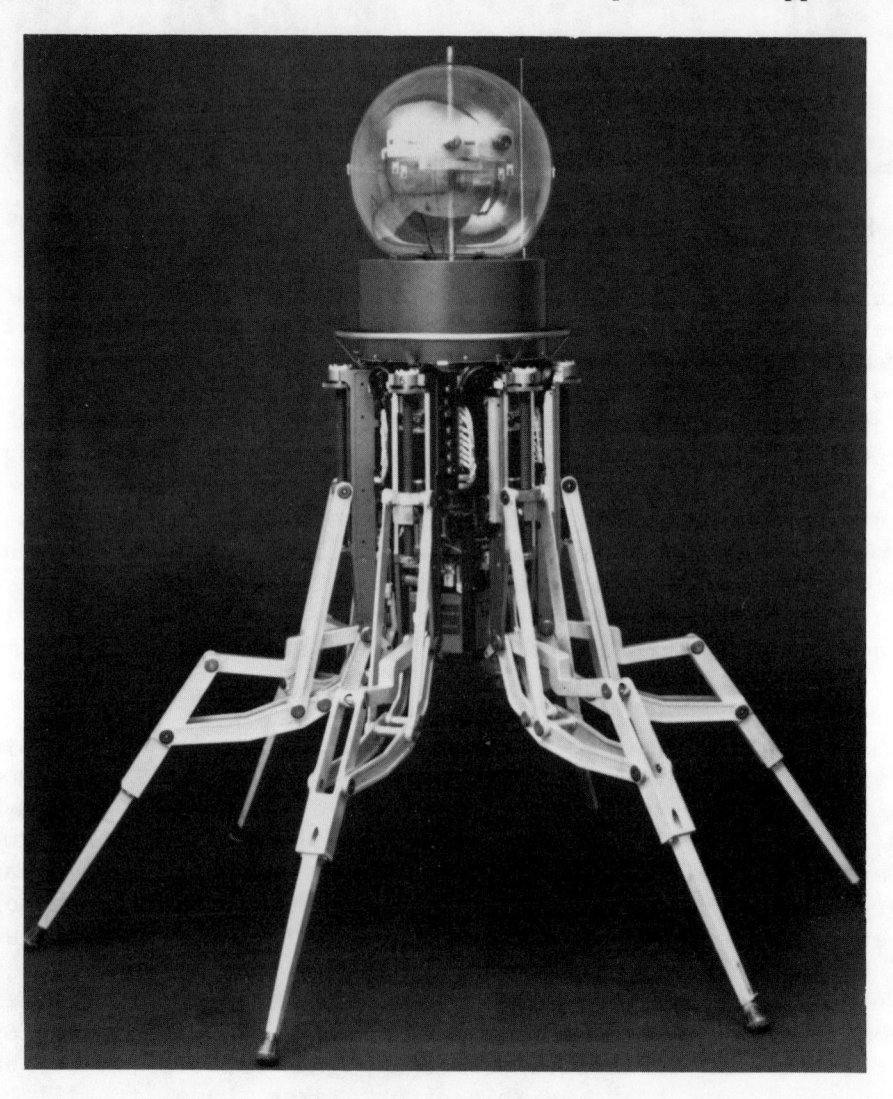

Figure 5 Walking robot, Odex I by ODETICS.

independent crawler tracks (two per side). These tracks can be articulated automatically to maintain a level platform surface. The vehicle can climb a 45 incline or stairs, can climb over obstacles 3 feet in height, and can bridge gaps up to 3 feet. Several West German police forces have used the vehicle, which is produced by Blocher Motor Company, for bomb disarming and disposal. This vehicle has a 500-pound capacity and a control tether of 300 feet. Their servomanipulator system, the EMSM I, was the first to fully implement electronic counterbalancing. The 9-degree-of-freedom arm has a continuous capacity of 25 pounds, with somewhat compliant force reflection. This system also makes use of torque tubes for drive power transmission. The 9-degree-of-freedom kinematics allows the arm to take on many possible configurations for a given end-effector orientation and position. The redundant kinematics offers obstacle avoidance advantages, but it causes some operator interaction difficulties unless special control functions are implemented.

The French atomic energy agency, CEA, has supported an active manipulator development program for more than a decade. The MA-23 servomanipulator is the last major output of this program.[22] This system has been improved for remote glove box maintenance,[23] and similar systems have been outfitted for end-effector camera tracking, compliance correction, and insertion force control.[24] The French have two companies involved in manipulator manufacturing: (1) La Calhene, licensee for the MA-23 servomanipulator, and (2) Ateliens et Chantiers de Bretagne (ACB), manufacturer of power arms and overhead handling equipment. An MA-23 servomanipulator has been used with a large shielded cask and telescoping tube to produce a hot cell intervention system. The tube and manipulator fully retract into the cask, which is then set into position over a man-hole penetration for access to the cell.

Several other nuclear-related remote operations developments are being pursued in Europe. The Joint European Torus (JET) Fusion Program is incorporating remote maintenance for application using the Italian MASCOT servomanipulator designs. As an offshoot of this work, the Italian atomic energy agency is considering rejuvenation of its manipulator development to upgrade the MASCOT to today's electronics technology.[25]

The Japanese have been noted for their ability to develop, manufacture, and apply robotic technology. Their original development efforts in robotics and teleoperator systems, however, have been limited. Recently they have increased their efforts to develop teleoperator systems[26] for nuclear applications. The BILARM-83A is a cross between power manipulators and servomanipulators. High inertia resulting from large gear reductions requires that strain gages be used for force sensing. This particular force-reflection implementation results in slow system response and a lack of good operator control, even though an anthropomorphic master controller is used. Another Japanese teleoperator development uses distributed actuators, microprocessor-based controls, and electronic counterbalancing.[27] Crawling and walking vehicles are also being researched in Japan.[28,29] The major research centers are Meidensha Electric Manufacturing Co., Hitachi Energy Research Laboratory, and Toshiba. With such diversity of activity, the Japanese effort should be observed closely in the coming years.

The SPAR Corporation of Canada has developed and fabricated the teleoperator system for the U.S. space shuttle.[30] Its remote manipulator system has special considerations for operating in outer space, including: (1) distributed brushless actuators for operation in an oxygen-free environment, (2) joystick control to minimize operator station size, and (3) special docking sensors developed with the Jet Propulsion Laboratory.

Remote technology development has been a continuing effort in programs of the United Kingdom Atomic Energy Authority.[31] Because hot cell volumetric sizes have remained relatively small, and as a matter of philosophy, mechanical master/slave manipulators and unilateral power manipulators have been the basic tools of remote maintenance, development activities have focused on reliability refinements of basic manipulator designs. No major activity has been assigned to the development of servomanipulators.

V. FUTURE RESEARCH DIRECTIONS

The ultimate goal of robot and teleoperator systems is to provide unlimited flexibility and a high level of autonomy. The two system concepts are approaching this goal by different evolutionary pathways as discussed earlier. Today's industrial robot is perceived as quite autonomous, since once programmed further intervention is not normally required. Because of sensory limitations, these systems have limited flexibility in adapting to work environment changes, which is the motivation of computer vision research. The teleoperator is very adaptable, but it relies on human control to accomplish its work. Figure 6 graphically shows the relationship of these systems to the ultimate goal and gives insight into the focus of present research activities. Advanced robotic systems are striving for task adaptability through enhanced sensory feedback. Research areas concentrating on artificial intelligence, sensor integration, computer vision, and off-line CAD/CAM robot programming will make robots more universal and economical. Teleoperator systems are moving toward autonomous operation as an enhancement to human control. Research in supervisory control, man-machine interface methods to reduce operator burden, and computer data base management is intended to improve operator efficiency. Many research activities are common to both systems and are aimed toward reducing implementation cost and expanding the realm of application. These include improved communications methods, advanced

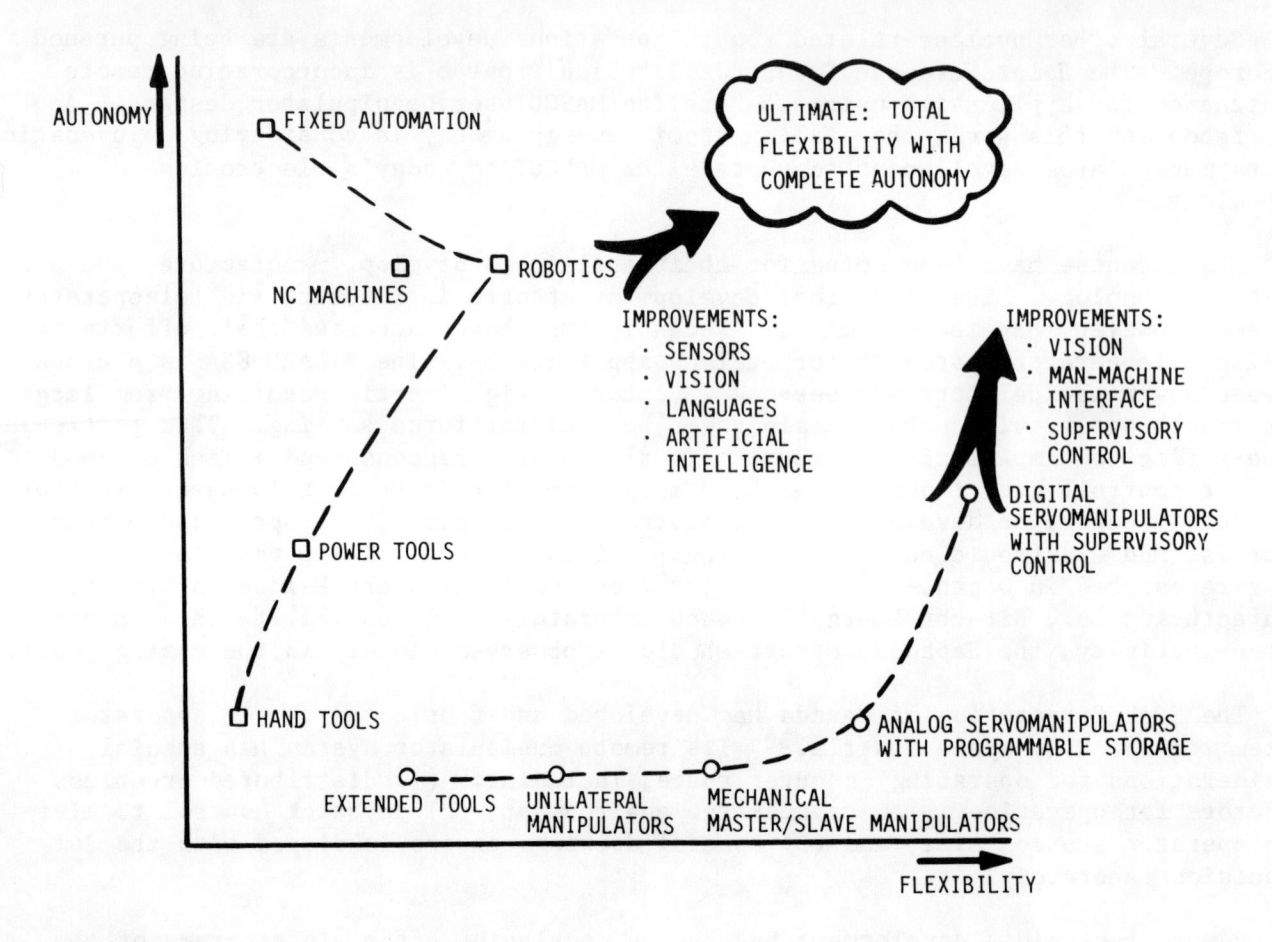

Figure 6 Automation and versatility of various manipulation methods (adapted from conversations with Jean Vertut).

digital control techniques, basic sensor development (force, tactile, vision, etc.), mobility, modularity, and subsystem components (actuators, amplifiers, materials, etc). A review of teleoperator research activities will show possible relationships to robotic developments.

The Jet Propulsion Laboratory has developed a universal manipulator controller.[9] The fundamental concept is to develop a master arm that can be used to operate any slave arm system by means of real-time transformation of the kinematic dissimilarities. Such a method could be applied in the future as a programming device for industrial robots. One would use a single controller and different transformation software to teach the desired motions to various commercial robot configurations. This concept would also facilitate single-operator supervision of several robots to provide remote recovery from fault conditions. A related effort is also ongoing at the University of Florida.

Distributed digital control techniques have recently been applied to teleoperator systems to improve their operational flexibility.[32] The advent of easily reprogrammable microprocessor control systems has led to improved diagnostic methods, control implementation, and reliability in teleoperator systems. Digital controls also allow sophisticated compensation algorithms for inertia, friction, counterbalancing, and other nonlinear effects which were difficult or impossible to accomplish accurately with analog circuitry. Servomanipulators tend to be mechanically compliant due to the emphasis on low inertia (small structure) and centralized actuators (long force transmission path). Current intelligent digital systems will allow such compliance to be corrected by control adjustments and will improve the positional accuracy of teleoperators.[33]

Mechanical modularity is another area in which both teleoperator and robot manipulators can be improved. A modularly constructed robot can be reconfigured for heavier loads, greater reach, or different kinematic constraints. American Robot Corporation describes its latest system as a Modular Expandable Robot Line (Merlin).[34] Modular construction should also provide increased availability by facilitating faster *in situ* repair of failed systems. It is this aspect that is receiving emphasis in the latest teleoperator developments.* Modularity will also enhance mobility, in that the ability to reduce a manipulator to small, easily assembled parts should improve its transportability to various work sites.

Kinematic improvements in teleoperators have resulted from the development of manipulators which operate in the anthropomorphic (man-like) stance. This elbows-down configuration required development of unique wrist mechanisms to provide yaw, pitch, roll, and grip actuation while avoiding mechanical singularities. A joint effort between ORNL and TeleOperator Systems resulted in a mechanism similar to the Cincinnati Milacron three-roll wrist, but without mid-range singularities. The anthropomorphic stance reduces manipulator obstruction of viewing the work site, increases operator comfort at the master controller, and increases horizontal reach dexterity.

Improvements in man-machine implementation methods are aimed at increasing the operational efficiency of teleoperators. Graphic display techniques provide the operator with information concerning everything from system fault conditions to

---

*A major activity at the Oak Ridge National Laboratory is addressing the development of advanced remote teleoperator concepts, including manipulator modularization. This work is sponsored by the Department of Energy through the Consolidated Fuel Reprocessing Program, and publication and foreign disclosure of some of this information is restricted by law. Interested parties may contact the authors concerning additional information.

obstacle avoidance in a dynamic environment. The problem of acquiring data, decomposing it, and presenting it to an operator in a logical sequence is very similar to sensor integration for a computer system; artificial intelligence is required to condense available inputs into useful forms. Efforts in improved graphic presentation methods are common within the process control industry, whereas intelligent action based on input data is being extensively pursued by the defense community.

Underlying the operation of successful multifunctional systems is the ability to successfully partition actions into a series of defined acts. Such partitioning leads to a structured organization of programming tasks. The concept of hierarchical systems applies equally well to both teleoperator and robot systems.[35] Both systems can be decomposed into similar subsystems (servo control, sensor data acquisition, kinematic transformation determination, etc.) with congruent goals (manipulate, orient, enunciate, etc.). The efforts by the National Bureau of Standards[35] to develop interfacing criteria for robotic systems will also be beneficial to the teleoperator industry.

Vision research for teleoperators has emphasized the human aspects: multiple views, optimal perspective, stereo versus monocular, color versus black and white, and optimal lighting. The goal is to optimize the human perception of the work environment. As supervisory automation is pursued, the use of end-effector cameras and pattern image processing techniques and other carry-overs from robotic developments will be evident. Teleoperation research into environmental coloring to aid operator recognition and multiple viewing to improve depth perception may also find application in advanced computer vision.

Research activities in teleoperators and robotics are strongly complementary. The major motivation in robotic research is economics, but the different perspective offered by teleoperator development could open new areas useful to both. As the complexity of industrial robotics systems and applications increases, it is expected that these two related technologies will continue to converge.

VI. SUMMARY

Teleoperator systems comprising man-controlled remote manipulators have undergone extensive development in nuclear applications over the past three decades. Today, teleoperator research has expanded and diffused into other fields such as space, military, and industrial applications. Modern teleoperator systems utilize the latest advances in materials, electronics, and microcomputer technology, all of which are very closely related to industrial robotics systems. Developmental similarities in kinematics, controls, mechanisms, modularization, mobilization, and man-machine interfacing are expected to increase.

Future trends will lead to increased automation (robot operating modes) in teleoperation and autonomous intelligent operation in robotics. These trends will reduce the apparent distinction between the two classes of systems. It is expected that force-reflection techniques in teleoperated servomanipulators will contribute to robot force control applications in assembly. Research in robot sensors, especially computer vision, will surely lead to extensions in teleoperation.

BIBLIOGRAPHY

1. Goertz, R., "Some Work on Manipulator Systems at ANL; Past, Present, and a Look at the Future," *Proc. 1964 Seminars on Remotely Operated Special Equipment*, USAEC Report CONF-641120, May 1964.

2. Corliss, W. R., and Johnsen, E. G., *Teleoperator Controls - An AEC-NASA Technology Review*, NASA Report NASA SP-5070, December 1968.

3. Bejczy, A. K., "Remote Manipulator Systems, Technology Review and Planetary Operations Requirements," Jet Propulsion Laboratory Report 760-77, July 1972.

4. Goertz, R. C., et al., "ANL Mark E4A Electric Master-Slave Manipulator," *Proc. 14th Conf. Remote Syst. Technol.*, American Nuclear Society, Hinsdale, Illinois, pp. 115-123, 1966.

5. Johnsen, E. G., and Corliss, W. R., *Teleoperators and Human Augmentation*, NASA Report NASA SP-5047, 1967.

6. Luh, J. Y. S., "Conventional Controller Design for Industrial Robots - A Tutorial," *IEEE Trans. Syst., Man, Cybern.*, Vol. SMC-13, No. 3, pp. 298-316, May-June 1983.

7. Engeberger, J. F., *Robotics in Practice*, AMACOM, 1980.

8. Hogan, N. and Cotter, S. L., "Cortesian Impedance Control of a Nonlinear Manipulator," *Robotics Research and Advanced Applications*, ASME, New York, pp. 121-128, 1982.

9. Bejczy, A. K., and Salisbury, J. K., "Controlling Remote Manipulators Through Kinesthetic Coupling," *Computers in Mechanical Engineering*, pp. 48-60, July 1983.

10. Ayres, R. and Miller, S., "Industrial Robots on the Line," *Technol. Rev.*, pp. 35-46, May-June 1982.

11. Grisham, D. L., "Monitors 1980: Now There are Two," *Proc. 28th Conf. Remote Syst. Technol.*, Vol. 2, American Nuclear Society, p. 83, 1980.

12. Grimson, J., "Design of the Fermilab Neutrino Remote Target Maintenance System," *Proc. 23rd Conf. Remote Syst. Technol.*, American Nuclear Society, p. 32, 1975.

13. Feldman, M. J., "REMOTEX - A New Concept for Efficient Remote Operations and Maintenance in Nuclear Fuel Reprocessing," *Proc. 28th Conf. Remote Syst. Technol.*, Vol. 2, American Nuclear Society, p. 3, 1980.

14. Donnelly, T. J., "The Service Robot - Myths, Benefits, and Needs," *NSF Workshop on the Impact on the Academic Community for Research in Generalized Robotic Manipulators*, University of Florida, pp. 8-10, 1978.

15. Flanagan, C. A., et al., "Fusion Engineering Device Design Description," Oak Ridge National Laboratory Report ORNL/TM-7948, Vol. 2, pp. 8-55 thru 8-67, 1981.

16. Heer, E., "Robots in Space," Decade of Robotics, *Ind. Robot*, pp. 104-106, 1983.

17. Wernli, R., "The Silent World of the Undersea Robot," Decade of Robotics, *Ind. Robot*, pp. 100-102, 1983.

18. Brown, D. R., et al., "R&D Plan for Army Applications of AI/Robotics," SRI International Report ETL-0296, May 1982.

19. "Multifunctional Six-Legged Robot is Walking Wonder," *Design News*, July 23, p. 20, 1983.

20. Foster, Prof., "Can the Robot Be the Miner's Friend?," Decade of Robotics, *Ind. Robot*, pp. 102-103, 1983.

21. Kohler, G. W., et al., "Manipulators and Vehicles of the Nuclear Emergency Brigade in the Federal Republic of Germany," *Proc. 24th Conf. Remote Syst. Technol.*, American Nuclear Society, pp. 196-218, 1976.

22. Vertut, J., et al., "The MA-23 Bilateral Servomanipulator System," *Proc. 24th Conf. Remote Syst. Technol.*, American Nuclear Society, pp. 175-187, 1976.

23. Vertut, J., et al., "Protection, Transfer, and Maintenance of the MA-23 Bilateral Servomanipulator," *Proc. 26th Conf. Remote Syst. Technol.*, American Nuclear Society, pp. 125-133, 1978.

24. Liegiois, A., et al., "Learning and Control for a Compliant Computer-Controlled Manipulator," *IEEE Trans. Auto. Control*, Vol. AC-25, No. 6, pp. 1097-2003, December 1980.

25. Hamel, W. R., personal communications with Mr. Carlos Mancini of the ENEA of Italy, January 1983.

26. Yamamoto, M., et al., "Remote Maintenance Equipment for Hot Cell Facilities," *Trans. Am. Nucl. Soc.*, Vol. 43, pp. 755-756, 1982.

27. Suzuki, M., et al., "A Bilateral Servo-Manipulator for Remote Maintenance in Nuclear Facilities," *Trans. Am. Nucl. Soc.*, Vol. 43, pp. 756-757, 1982.

28. Heki, H., et al., "Crawler Driven Robotic Vehicle with Steering Mechanisms for Nuclear Power Plants," *Trans. Am. Nucl. Soc.*, Vol. 43, pp. 757-758, 1982.

29. Ichikawa, Y., et al., "Five-Legged Vehicle for Remote Systems in Nuclear Power Plants," *Trans. Am. Nucl. Soc.*, Vol. 43, pp. 758-759, 1982.

30. SPAR Corporation informal presentation, 1981.

31. Peagram, G. W., "Remote Handling and Viewing for the Future," *Post Irradiation Examination*, British Nuclear Energy Society, London, 1981.

32. Martin, H. L., et al., "Distributed Digital Processing for Servomanipulator Control," *Trans. Am. Nucl. Soc.*, Vol. 43, pp. 752-753, 1982.

33. Book, W. J., "Analysis of Massless Elastic Chains with Servo Controlled Joints," *ASME J. Dynamic Syst. Meas. Controls*, Vol. 101, pp. 187-192, September 1979.

34. Stefanides, E. J., "Electric Robot Meets Range of Needs," *Design News*, pp. 70-72, July 23, 1983.

35. Albus, J. S., "Hierarchical Control of Robots in an Automated Factory," 13th International Symposium on Industrial Robots and Robots 7, Conf. Proc., Robotics International of SME, Vol 2, pp. 13-29 to 13-43, April 1983.

Presented at the American Nuclear Society 21st Conference
on Remote Systems Technology, November 1973

THE MANIPULATOR AS A MEANS OF EXTENDING OUR DEXTEROUS CAPABILITIES TO LARGER AND SMALLER SCALES

CARL R. FLATAU, *TeleOperator Systems*
Shoreham, New York 11786

KEYWORDS: *manipulator, design-equipment, remote handling, servo system, computer control*

ABSTRACT

It has become apparent that manipulators are a viable means of extending our dexterous capabilities to large as well as small working scales. Detailed studies of the requirements and design principles for a 40-ft-long, bilateral force-reflecting (BFR) manipulator were contrasted to those for an 8-in.-long, force-sensing, mini-robot arm. In both cases, extensions and scaling of the technology of BFR manipulators were applied.

INTRODUCTION

Manipulators were first used to extend our dexterous capabilities to isolated environments. Initially the hazardous conditions of the radioactive laboratory motivated their use, and later they were applied in other hazardous environments. New application areas involve extension into activities requiring working scales substantially larger or smaller than unity, since these are circumstances in which our dexterous capability performs poorly when applied directly.

The author has studied scale extensions to both very large manipulator sizes and rather small ones. In each case the study involved manipulators of the force-sensing bilateral variety.

To describe scaling of manipulators, one should define a unity-scale device. Guided by human size statistics, we arbitrarily chose to define a unity-scale arm as one having a total arm length of 24 in. and a handling capacity of 30 lb. These basic parameters are rather similar to those of the Brookhaven Arm[1] and our model SM-2090, now under development.

THE ORBITING SHUTTLE ATTACHED MANIPULATOR

One of the first studies about the manipulator in space was performed by Goertz et al. at ANL.[2] This study showed dramatically how BFR manipulators could effectively do docking and cargo handling operations under inertial (zero-gravity) conditions, and how to absorb kinetic energy of delicate objects by catching them with BFR manipulators. After the study and during the early stages of American manned-space operation, a number of people knowledgeable in the art of manipulator design emphasized how useful BFR manipulators would be for performing the difficult docking operations, the necessary cargo handling, and maintenance routines of future space explorations.

The next docking and cargo handling problems planned concern the missions of the orbital shuttle with 65,000-lb cargo capacity, and a capability to build a space station out of 65,000-lb modules to which eventually the 320,000-lb shuttle might be docked with manipulators. By the time shuttle manipulator arms became "baseline equipment" (which means an essential requirement), reach requirements had been determined to be about 40 ft (12 m), but stowage space was available for only a 12-in.-diam, 40-ft-long tube. Initial requirements are summarized in Table I.

UPSCALING OF MANIPULATORS

It is obvious that the Shuttle Attached Manipulator (SAM) requirements cannot be satisfied by scaling from any known manipulator components.

The available storage space alone imposes an aspect ratio significantly larger than anything existing. The absence of gravity in its operational environment, as well as the stringent safety requirements, prohibit complete scaling of the arm. However, an exercise in scaling tells us which of the familiar techniques are attractive for use when one upscales a manipulator.

Several thousand pages of design documentation have been accumulated to date on the subject.[3-5] Only the briefest description can be given here.

The extreme slenderness of the arm and the absence of gravity require and allow one to work with rather modest forces and very low accelerations. Therefore, tip capacity can be as low as 30 lb, and accelerations with maximum cargo are down to about 0.18 in./sec^2 or 4.6×10^{-4} G. The above factors are the ones that most influence the design.

The SAM control system is of great interest. Since the prime function is the acceleration, translation, and deceleration of very large masses with application of small forces, the resulting visual feedback cues have values below operator acceptance range. From simulation one knows that, without explicit acceleration display, the SAM operator will not know whether the cargo is accelerating or decelerating. Since the SAM's joint drives must be easily back-drivable, the need for a BFR control system is strongly indicated. It becomes obvious that, in the absence of gravity, all forces are inertial, and normal BFR control mode is, in effect, acceleration feedback. By displaying acceleration as a force in the control input device (CID), or master, acceleration feedback is quite efficient.

The normal value for the CID tip speed at no load is about 3 ft/sec. With replica control, the velocity capability requirement at the tip of the SAM would be 50 to 60 ft/sec. This would lead to dangerously high kinetic energy and severe velocity resolution requirements. To prevent this, a multimode control system was invented.

In the capture mode, we matched linear velocities between the CID and the manipulator, which results in CID motions similar to those used in unity-scale BFR arms. For acceleration, deceleration, and translation, linear velocities are approximately in a 20:1 ratio, which means angular velocities are matched and manipulator tip velocities are limited to ~2.5 ft/sec. As a result, the CID behaves like a quasi-isometric force controller. One lets the cargo inertia drive the CID, and exerts forces on it only for acceleration and deceleration operations and velocity vector modification. However, work in the first mode requires a transformation between input and output coordinates, which no longer bear a joint-to-joint relationship. The resulting system is called an "$\overline{X}$ Reference Controller."

Since "$\overline{X}$ Reference Control" is capable of continuously transforming motion from one coordinate system to another, it is no longer necessary to maintain geometric similarity between the CID and the manipulator.

The "one-G" simulator is a SAM version that can operate in normal gravitational environment. Due to low accelerations involved in the SAM, inertia criteria are of much less importance; thus, all motors and gear trains can be mounted joint-proximate, allowing free arrangement of the degree-of-freedom (DOF). To balance all joints with minimum deviation from space hardware configuration, one must counterbalance wrist motion without addition of extraneous weight to reduce cascading of total counterbalance requirements. One also must find a new torque transmission mechanism capable of transmitting lower-arm unbalance motion to an articulated counter-balance mass.

At present, final detail design and construction of the "one-G" simulator is ready to start.

THE OTHER END OF THE SCALE

The SAM is a case wherein BFR manipulator techniques are utilized to solve handling problems brought about primarily by large object size. As a contrast, one might examine the extension of our dexterous capabilities to rather small scales. The deterioration of capability, which accompanies any attempt to work in small objects, is quite familiar. It is basically caused by the demagnification of motion which we automatically attempt in order to match motion rates to small scales, and tends to magnify, rather than demagnify, the output forces at the small end. This results in a loss of the accustomed force-feedback and a deterioration of dexterity. However, if one used a BFR servo manipulator which is scaled down to match the visual magnification required, but magnifies the forces encountered before feeding them back to the operator, there is no difficulty in scaling both linear motions and forces. In this way, demagnification of both forces and motion can be accom-

TABLE I

Maximum cargo mass	65,000 lbs
Maximum cargo size	15-ft-diam × 60 ft long
Maximum docking mass	320,000 lb
Reach requirements	30 to 40 ft
Maximum Δv at capture	0.4 ft/sec
Available storage space	12-in. diam × 40 ft long
Safety requirements	fail safe, fail operational

plished. The author studied this problem while working with the MIT Artificial Intelligence (AI) Laboratory.[a]

The present program at the AI Laboratory calls for development of a desktop computer-controlled-arm system that allows AI robot work to go on in an average-size office. Furthermore, the Laboratory would like to study a number of small-scale operations with computer-controlled arms. An outline of this mini-manipulator is shown in Fig. 1, superimposed, for contrast, on the motor chosen to drive the SAM joints. The mini-arm has an upper and lower arm length of 4 in. each, which is $\frac{1}{60}$ of the total arm length of the SAM.

The mini-arm has been designed as a computer-controlled device. However, the design approach utilized a considerable amount of scaling from larger arms, and the mini-arm was driven from the CID of a full-size BFR manipulator.

---

[a] Work reported herein was conducted for the Artificial Intelligence Laboratory, a Massachusetts Institute of Technology research program supported in part by the Advanced Research Projects Agency of the Department of Defense and monitored by the Office of Naval Research under contract number N00014-70-A-0362-0005.

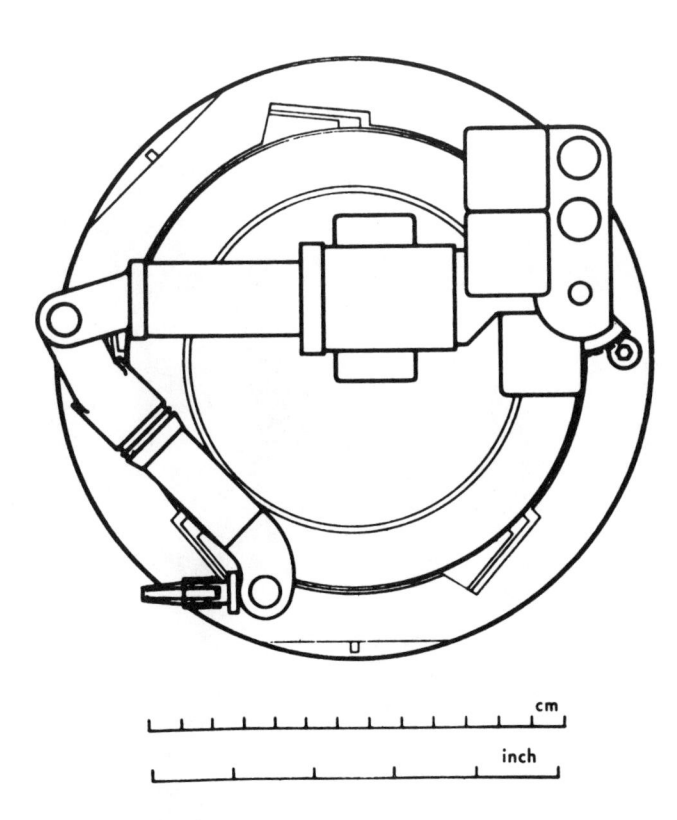

Fig. 1. Mini-robot arm superimposed on shuttle manipulator motor.

SCALING OF MANIPULATOR COMPONENTS

When one wants to scale a manipulator, it is useful to have some understanding of the scaling characteristics of components contained in it. First, one must choose the scaling rules for the manipulator performance parameters. One must decide how to scale lifting capacity, velocities, and accelerations of the arm.

If one linearly scales objects to be handled by a parameter L, then the weight of these objects scales as L^3. On the other hand, operations involving fitting objects together call for L^2 scaling.

For the design of the mini-manipulator, we chose to use time-invariant scaling. This means that the arm is designed to traverse similar, but scaled distances in the same time-independent scale. We also chose L^3 scaling for capacity. Then sealing rules require that velocity and acceleration scale as L and torques as L^4. Table II shows the appropriate scaling exponent for each parameter and compares it with the required scaling exponent for favorable downscaling. For upscaling, the inequality signs should be reversed.

The scaling behavior in Table II has been derived from first principles and, in the more complex cases, is verified by comparing behavior to families of commercial components.

As can be seen from Table II, most components downscale favorably except for starting friction of ball bearings and permanent magnet motors.

TABLE II

Exponent of Scaling Parameter and Required Scaling Exponent for L^3 Scaling of Capacity

Parameters and Components	Actual Exponent	Required Exponent
Structural parts		
Stress	1	≥ 0
Deflection	2	≥ 1
Gears		
Stress	1	≥ 0
Deflection	2	≥ 1
Limit torque on wear	3	≤ 4
Tapes and Cables		
Tensile stress	1	≥ 0
Bending stress	0	≥ 0
Elongation	2	≥ 1
Ball Bearings		
Capacity	2	≤ 3
Running friction	4	≥ 4
Starting friction	1	≥ 4
Permanent magnet motors		
Torque	4.1	≤ 4
Friction torque	2.8	≥ 4
Power rate	3.2	≤ 3

For modest downscaling, the one parameter not in good form is the frictional behavior of the motor used, but, as is well known, the unfavorable frictional behavior of motors can be compensated by use of force transducers. These must be provided for the computer-controlled arm so programming techniques can be extended beyond geometrical exercises.

Table II also shows that, for L^3 upscaling of manipulators, one must contend with structural difficulties. However, L^2 force scaling for larger manipulators seems a definite possibility. An arm scaled by $L = 10$ (20 ft long and 3000-lb capacity) seems definitely a possibility. Whether there is a true requirement for such an arm (considering its probable cost) remains to be seen.

On the other hand, downscaled manipulators do not have this large cost penalty and probably a larger number of applications can utilize a true BFR mini- or micro-manipulator with either manual or computer control.

CONCLUSION

Motivations exist for using manipulators, other than for hazardous environments. The motivation of handling scaled-up or scaled-down objects has been described and examples for very large and very small manipulator designs have been shown. With proper adjustment to the requirements of scale, man's dexterity can very likely be extended to both very large and quite small scales. In either case, high dexterity can be assured by adjusting the necessary sensory inputs to an optimal human sensing level.

REFERENCES

1. C. R. FLATAU, "Compact Servo Master-Slave Manipulator with Optimized Communication Links," *Proc. 17th Conf. Remote Syst. Technol.*, 154 (1969).

2. "Manipulator Systems for Space Applications," Technical Report prepared by ANL Remote Control Engineering Division for George C. Marshall Space Flight Center (April 1967).

3. "Preliminary Design of Shuttle Docking and Cargo Handling System," Final report, MSC, 05218, Martin-Marietta Corporation (December 1971).

4. "A Shuttle and Space Station Manipulator for Assembly, Docking, Maintenance, Cargo Handling and Spacecraft Retrieval," Final report, MSC, 05219, MBAssociates (January 1972).

5. "Attached Manipulator System Design and Concept Verification for Zero-G Simulation," JFC-08021, Martin-Marietta Corporation (June 1973).

Presented at the American Nuclear Society 26th Conference
on Remote Systems Technology, 1978

DESIGN AND APPLICATION OF REMOTE MANIPULATOR SYSTEMS

CARL WITHAM, ANTHONY FABERT, and A. L. FOOTE
MBAssociates, P.O. Box 196, San Ramon, California 94583

KEYWORDS: *remote vehicle, manipulators, underwater*

ABSTRACT

Manipulator systems have been used in locations that are either too dangerous, or for tasks that are too fatiguing for man. Several remote manipulation and viewing systems have been developed recently to remove man from hostile environments. General design guidelines for such systems should include man-machine interface, overall weight, power, and control considerations. System design is discussed for systems that have been developed for use in undersea, underground mining, explosives handling, and nuclear maintenance applications.

Remote manipulation equipment has been used for many years to augment man's manipulative capabilities in various environments. This augmentation has taken the form of teleoperator systems that have grown more dexterous and, therefore, more complex in the past few years. Recent technological advancements have allowed computer aid of teleoperator performance and will ultimately allow more intelligent autonomous manipulator control. The applications for remote manipulators in the past and in the near future generally consist of aiding man in tasks that are: (a) too dangerous, (b) too fatiguing, or (c) beyond the physical capacities of man. This paper discusses a few systems designed to provide an aid for all three areas with an emphasis on removing man from hostile environments.

Some of the hostile environments that man must currently deal with are high nuclear radiation, high-power explosives, toxic substances, high pressure underwater, underground mining, and extraterrestrial space. These environments pose a large number of varying constraints on systems and equipment designed for them. These varying constraints require varied designs to approach an optimal system configuration. This is contrary to the industrial robotics market where environmental performance requirements are usually not as stringent. There are, however, a few design criteria that can help guide the design of systems in the general case.

GENERAL DESIGN CRITERIA

One of the major areas that must be addressed in developing a usable manipulator system is the man-machine interface. Optimal, human-operated controls can vary greatly and are directly dependent on task type and load. Prior to control conceptualization, a careful analysis must be made of the operator's working area, his visual capabilities (including any proposed visual aid systems), the speed required to perform tasks, task repetitiveness (is it mentally fatiguing?), the dexterity required for the tasks, absolute accuracy requirements, task requirements for kinesthetic and tactile feeling, and varying operator conditions. The results of this analysis not only will affect controller design, but also will usually guide slave design as well. The general trend should be to provide a man-machine interface that is simple from a human-factors standpoint. It should provide a system that is easy to understand, relatively idiot-proof, and requires a minimum of operator training time.

Unfortunately, controls that are simpler to operate and understand are usually more complex and, hence, more expensive. Any cost analysis of a system must, however, take into account operational costs over the life of the manipulator system. Dramatic cost savings can come about from more rapid operator training and from increased efficiency and speed during normal operations. A more complex system will often be the least costly in the long run.

Another criterion that should be addressed is overall manipulator system weight. In many cases, there are weight constraints placed on the design of a manipulator system. However, weight is often overlooked because the system is carried by a crane or attached to the floor or wall. Weight of arm elements at the tip are cascaded as one moves toward the base. For example, a heavier actuator on the wrist will apply a larger load on an elbow actuator and its structure. This larger load will require higher drive system capacity, which is equivalent to a larger weight cascaded toward the base. Drive actuator, structural stiffness, and bearing requirements can become difficult to satisfy at the base of the manipulator if weights are not minimized. For this reason, compact, possibly tubular, sections fabricated from a material with a high stiffness-to-density ratio should be used for arm sections. Actuators should be as lightweight as possible. (Sometimes operating speed can be sacrificed to achieve lighter weight). MBAssociates (MBA) has developed a set of very lightweight hydraulic actuators specifically to achieve these design goals.

One criterion that is often overlooked is the desire for a "clean" arrangement of manipulator elements and electrical, pneumatic, or hydraulic lines. Slave-manipulator environments are usually complex, with many appendages to tear cabling from a manipulator. One trend at MBA has been an attempt to route all required cabling (those that operate the arm as well as those for tools and end effectors) down the center of arm sections and joints. In this way, the arm protects its vital cabling from external damage.

A choice that must be made early in the development of any manipulator system is which powering technique to use. Available approaches include electric, hydraulic, pneumatic, and mechanical. Each mode has its advantages and an analysis must be made of the environment, tasks, and other design constraints to arrive at a decision. Hydraulic power is generally superior where a fluid system can be used because of the high power per added arm weight ratio. Sometimes it can be used in difficult locations through the judicious choice of operating fluid (example: water glycol mixtures).

Careful consideration should also be made of the manipulator arm kinematic configuration. The arrangement of joints and link lengths affects the zones of reach and orientation of the end effectors within space. Some techniques for finding configurations have been developed recently, including computer operation of these techniques.[1] Ultimately a model or mockup of the configuration should be made to study general dexterity and operational modes. If the manipulator is to be computer operated or computer augmented, kinematics can become very important, dictating the real-time computational effort. Sometimes minor joint ordering changes can bring about a significant reduction in computations.

Many other design considerations must be taken into account to develop a usable system. These considerations are a result of specific system requirements; designs must be optimized according to these custom requirements. The manipulation systems described below have been developed recently by MBA to satisfy the need to remove man from a dangerous environment (or an environment impossible for man to enter).

UNDERSEA MANIPULATOR SYSTEM

Figure 1 shows a manipulator system designed for use with the Submersible Craft Assisting Repair and Burial (SCARAB), which is used for undersea telephone cable repair and surveillance. The manipulator arms mounted to the support vehicle are shown in Fig. 2. The system is designed to locate, unbury, attach, cut, recover, and rebury a malfunctioning cable in a minimum amount of time at depths to 1829 m.

The manipulator and end effectors for this system are of special-purpose design to perform these specific tasks. Kinematically, the manipulator consists of four degrees of freedom; base azimuth, elevation, roll, and a linear extension joint. The tool package can be placed and oriented at any point in the space between two hemispheres of 1.5- and 1.0-m radii. The end effector tool package consists of a selection of tools for dredging, gripping, and cutting.

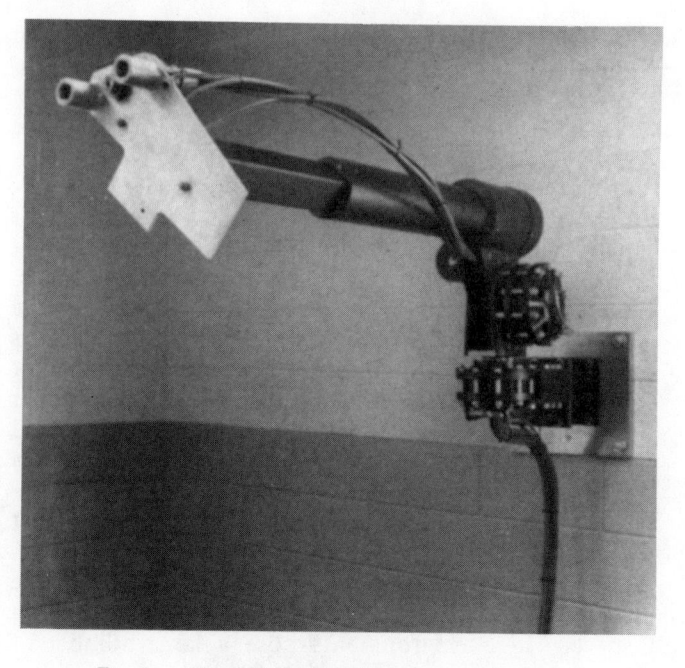

Fig. 1. The SCARAB underwater manipulator.

Fig. 2. The SCARAB vehicle with manipulators mounted.

The manipulator is hydraulically powered using a fluid that has low viscosity change with temperature. Therefore, it is compatible with the low-temperature and high-absolute-pressure environment. Corrosion is another adverse effect of this environment. To counteract this, materials of construction and coatings were carefully chosen. These consist of hard anodized aluminum, stainless steel, electroless nickel plating, and epoxy coatings.

Two types of teleoperator controls have been used with this system. The first type uses variable rate control. This involves individual joint commands that control the joints for a motion in either direction, or control the joint for zero motion. One problem that arises under a rate control of this type is the fact that joint angles can change rates or drift under varying loads. This type of control system is less expensive than a position servocontrol, but is a poor man-machine interface because of the drift. The other control system that is used is a position-position servocontrol that uses individual joint commands. With this system, variability problems are relieved as the slave automatically locks to the position commanded. This control system represents an adequate control where extremely fine positioning, manipulative tasks, and rapid, precise motions are not required.

Much attention was given to weight constraints within the design, as added weight greatly impacts support vehicle performance requirements. Manipulator actuators are hydraulically powered rotary vane and linear piston actuators specially designed for low weight. The arm and end effector mounting plate weighs ~66.0 kg in air. Handling capacities are 45.0-kg minimum reaction force with a 22.6-kg tool load (in air). This affords an excellent lift to a weight ratio of >1:1.

Extensive system testing has been performed in undersea environments with a high success rate.[2-4] The vehicle and arms have been operated to a depth of 1835 m, which is a world record for a commercial system of this type. Human control has been shown to be excellent for the required tasks. Dexterity and use have been shown to surpass original task requirements with the use of the manipulators to recover previously expended tools from the ocean floor.

UNDERGROUND MANIPULATOR SYSTEM

Coal mining in the world today represents a large-volume production operation within a hostile environment. Many of the coal mines currently in production are of floor-to-ceiling seam heights as low as 122 cm. These conditions cause coal mining to be one of the most unsafe occupations for workers in the U.S. Over 25% of all underground accidents occur while the victim is handling materials. This brings about a requirement for a general materials handling aid.

Figure 3 shows a manipulator and vehicle system designed under the constraints that materials used within a mine must be handled without repackaging into more easily handled units. To replace the manual handling presently done in underground mines, it was necessary to develop a relatively dexterous, heavy-duty, mobile manipulator system. It must be capable of reaching into cramped areas to pick up or deposit a wide variety of supplies and materials.

Safety and size considerations dictated the development of an hydraulic rather than an electric power system for the manipulator. A 208-cm reach, 118-kg lift capacity is used here to satisfy the functional requirements. To satisfy these requirements and to provide durability for underground service, a relatively heavy (364-kg) manipulator was required.

Manipulator kinematics involve a 6-deg-of-freedom anthropomorphic design. Joint configuration includes shoulder azimuth, shoulder elevation, elbow, forearm roll, wrist flexure, and wrist roll. This configuration was chosen over other designs because of its increased work area and better articulation flexibility for handling awkward unit loads.

The variability of manipulative tasks also dictated a position servocontrolled system utilizing a replica master controller. All joints are position-position servoed with the exception of the elevation, which is rate controlled through a lever located on the replica master. Figure 4 shows the replica controller that is mounted to the support vehicle in a position easily accessible to the operator. A replica master allows the operator simultaneous control of all joints and removes the requirement for mental correlation from individual master control to slave joint (a very fatiguing process under rapid, precise tasks). The operator performs his work under direct visual contact, which enhances the man-machine interface.

The end effector shown in Fig. 3 is a general-purpose device, designed to handle a large range of varying materials. Other types of grips could be designed for special item handling in a more efficient manner. The grip is controlled through the use of a foot pedal. This technique works well for use with a replica master where the joints do not follow the joint motions of the operator's arm. Also note that the operator's position with respect to the pedal is relatively constant, which makes this a usable solution.

Special note should be made of cable routing. Due to the variability of the underground environment and the type of service encountered, it is essential that the cabling be protected from impact

Fig. 3. Mine materials manipulator.

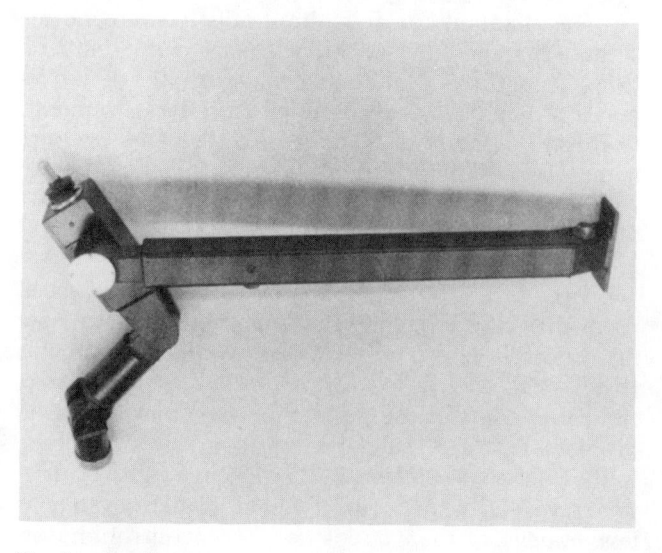

Fig. 4. Replica master control for mine materials manipulator.

or tear-off damage. One technique used with this design to simplify the routing is to place the electrohydraulic servocontrol valves at various protected locations on the arm, feeding directly into the joint actuators. This reduces the number of hydraulic flow lines traversing the whole arm to two: pressure and return. This technique generally increases the overall weight of an arm due to the weight-cascading effect mentioned above. However, in this case, a high weight was dictated by service durability requirements, and the servovalve weight does not impact it greatly.

The power source for this system should also be noted, as it applies to other mobile systems. An hydraulic pump is driven by a battery-operated electric motor. All of these components are located in the rear of the vehicle. The battery is a rechargeable type and is designed for simple changeout placement so that a worn battery can be replaced by a fresh battery, while the worn one is charged during continuing operation.

System testing has indicated a high degree of controllability. The manipulator has been used to handle large heavy items, as well as smaller, lightweight items requiring more precise positioning and manipulation. Improved safety in handling existing unit items of supply has been shown through use of this system.

EXPLOSIVES-HANDLING MANIPULATOR SYSTEM

Explosive and/or toxic materials handling can represent difficult and dangerous tasks for man. The manipulator and vehicle system shown in Fig. 5 is designed for use with an automated/remotely manned system that downloads defective 105-mm ammunition. The total system included equipment to depalletize boxes of shells, remove the shells from wooden boxes, and remove the defective fuses. A remotely driven forklift conveys pallets of shells to the automatic machinery. This forklift also acts as a support vehicle for a detachable manipulator system.

The manipulator arm is provided to clear ammunition or components jammed in the various pieces of automated machinery. Pieces removed by the manipulator system are then placed in a static location to allow entry of maintenance personnel into the facility. The arm is also capable of introducing ammunition into the machinery for disassembly, or disposing of components.

The manipulator is a 6-deg-of-freedom unit with the following motion capabilities: base azimuth, elevation, upper arm roll, linear extension, wrist flexure, and wrist roll. These motions are necessary to reach within various pieces of machinery and allow enough flexibility to reach components while avoiding obstacles. This requires a relatively high degree of dexterity because of the potentially hazardous nature of the handled components. Note, however, that speed requirements are low; in fact, maximum speeds are limited.

The manipulator is hydraulically powered through rotary vane and linear piston actuators. The manipulator mounting base also contains the hydraulic power unit, which receives its electrical drive power from the forklift. Manipulator control signals also come from the forklift at a remotely detachable interface. The manipulator weighs ~100 kg and has a 68-kg lift capacity. Maximum reach distance is 211 cm, with positioning and orientation ability at most of the points within a hemisphere of this radius.

Control of this arm is very critical because of the explosive loads it handles. The operator controls all joints simultaneously through the use of a miniature replica master arm. The operator views his work through a stereo-foveal head-aimed television system (HATS) mounted to the forklift support vehicle.[5] All joints (including grip) are position-position servoed to provide a high degree of accuracy and controllability in its teleoperator mode. Operator training time required for adequate operations and safety is much shorter than for less precise systems (rate-controlled and/or switchbox types). This reduction in time is necessary since the manipulator is not constantly used and learned, but operator proficiency is an absolute requirement when the manipulator is needed.

The U.S. army is currently using this system to defuse 81 000 defective 105-mm shells.

Fig. 5. Remote-controlled forklift and manipulator system for munitions handling.

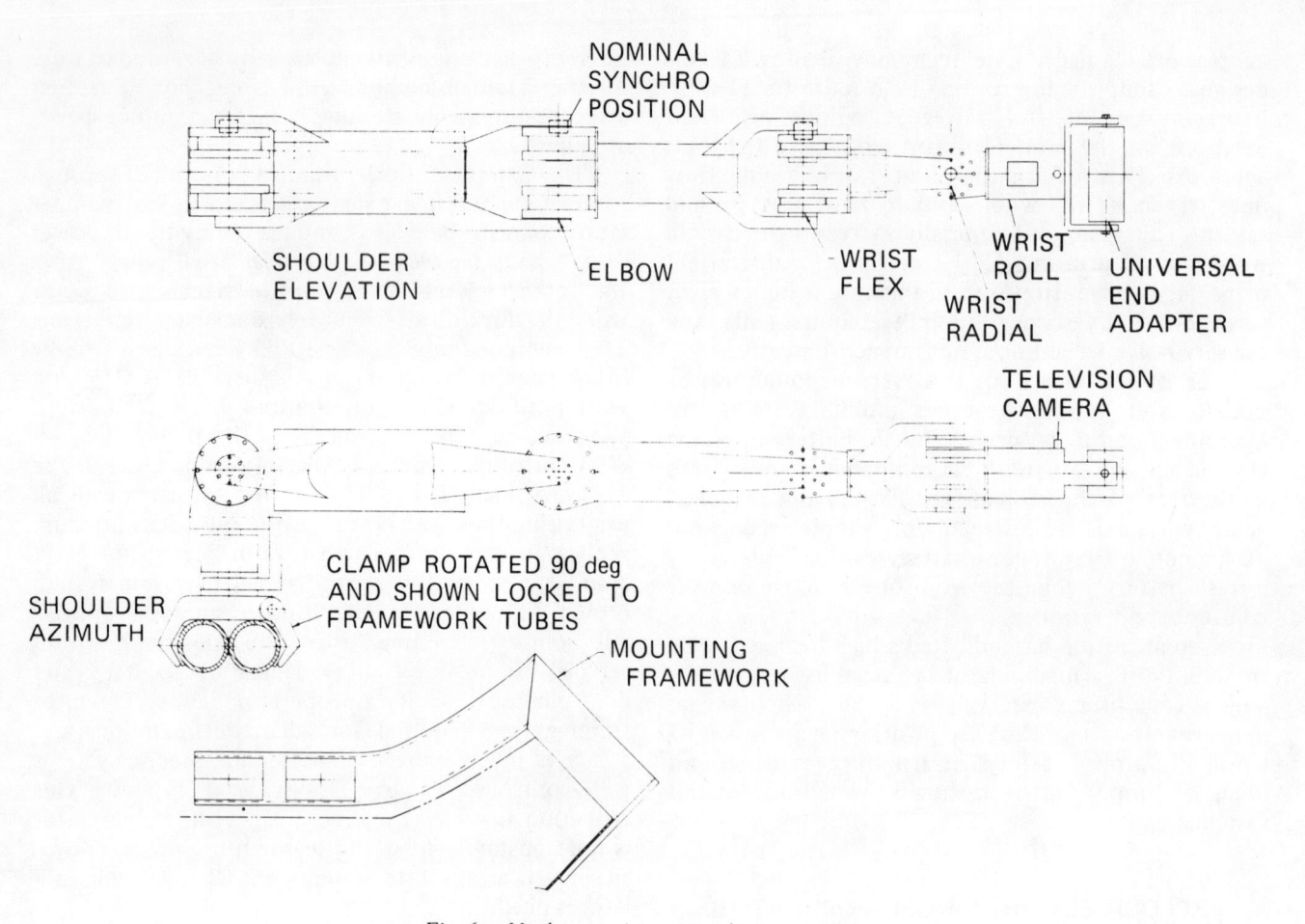

Fig. 6. Nuclear service manipulator concept.

NUCLEAR MAINTENANCE MANIPULATOR SYSTEM

The nuclear industry presents an environment that is an important market for manipulator systems designed to remove man from hostile environments. The system described here has been recently designed (conceptually) to replace the manned access requirement for inspection and repair of components within a reactor containment building.

Several constraints are placed on a system of this sort. These include: high reliability, radiation compatibility, light weight, capability for rapid assembly and disassembly for hand transport, high reaction load capability (for machining operations), high absolute accuracy (for absolute location of components within the work area), flexibility and dexterity (for work within confined locations), automated manipulator compensation for varying conditions, and precise operator controls and visual systems. These constraints have led to an advanced, high-accuracy, high-load capacity, lightweight manipulator system with its associated controls, visual systems, and ancillary equipment.

The manipulator conceived consists of 6 deg of freedom to reach points within its operating space with any end effector angular orientation (see Fig. 6). Several tools can be used at the end of the arm; these can be changed while the system is still in place at its work location. Utilities and cabling for the arm functions and end effectors are routed totally internal to the manipulator structure where they are protected from accidental damage.

The high-accuracy needs require a complex control system with computer monitoring and compensation of both slave and master. Use of the system in a complex surveillance mode requires an automated autonomous computer control to step from inspection point to inspection point rapidly and accurately. Manual operations must be very exact, and thus a bilateral force-feedback minimaster control system is proposed. Another control mode that can be made available is a simple rate control. This mode could be used as a control for checking out manipulator installation prior to full control station setup. The rate control could also be used in the event of failure of other modes. These considerations lead to the possible use of a hybrid analog-digital-control system for maximum reliability and utility.

Computer augmentation and autonomous control bring about requirements for computer positional analysis, trajectory computation planning and optimization, and collision avoidance system considerations. Positional analysis, the basis for the other features, can be implemented in the computer by using system kinematic equations that do not require iterative solutions. The trajectory planning and optimization routines use these position solutions to determine an optimal path to a target, while avoiding obstacles. The obstacles are modeled in the computer through the use of spatial zones that are set up around the obstacle. The computer must then ensure that arm position solutions do not intersect these zones.

The objectives of such a system are to reduce the total accumulated radiation exposure of a service team in performing a nominal service call and to reduce plant downtime. This system has been shown to significantly reduce radiation exposure and downtime over currently designed systems and maintenance techniques.

CONCLUSION

Remote manipulator technology has been advancing in the past few years and represents a viable technique to remove man from environments that are dangerous to him, but within which he must work. Manipulation systems have been shown to be very applicable for various tasks within highly diversified environments. Most special applications of this sort can gain greatly through the use of a system that has been customized and optimized for the tasks that will be encountered. These gains include overall cost reduction over the lifetime usage of the system (due to reduced operator training times and more rapid operations reducing downtimes) and higher quality work performed.

REFERENCES

1. J. DUFFY, *Mechanisms and Robots,* Edward Arnold and Company (1978).

2. "Record-Setting Vehicle," *Ocean Industry,* p. 79 (Apr. 1978).

3. "Industrial Research Salutes the IR-100 Winners," *Ind. Res.,* **19**, *10*, 39 (Oct. 1977).

4. "SCARAB Built for Cable Burial, Recovery," *Ocean Industry,* **II**, *10*, 76 (Oct. 1976).

5. A. L. FOOTE and R. J. DOMPE, "Application for Head-Aimed Television Systems," *Proc. 26th Conf. Remote Syst. Technol.,* 83 (1978).

Robots and Manipulators

EWALD HEER

If, as many believe, the 1960s were the decade when numerical control (NC) came to maturity, and the 1970s were the decade of the computer, and, in particular, computerized numerical control (CNC), the 1980s promise to be the decade in which the robot achieves maturity and full acceptance in industry, and takes over many boring and dangerous jobs now done by humans. The basic reason is its cost effectiveness. Typically, one robot can displace three or four workers, and reduce labor costs by as much as 80 percent. Over 30,000 are already at work in many industrial and nonindustrial applications.

What are the attributes that make a machine a robot irrespective of its particular application? The Robot Institute of America defines a robot as "a programmable, multi-function manipulator designed to move material, parts, tools, or specialized devices through variable programmed motions for the performance of a variety of tasks." The emphasis here is on doing a job that requires the handling of physical objects.

This definition appears to be somewhat at variance with the concept of a robot (spacecraft) in the space program. Here, information gathering spacecraft, such as Earth orbiters and planetary spacecraft, are considered to be robots, and, similarly, unmanned undersea vehicles used only for information gathering may also be called robots. Instead of concentrating on handling physical objects, they handle data and information. However, during this process, they must move to and position themselves at the appropriate location to obtain the desired data. They themselves are the "tools and specialized devices" to accomplish the data gathering tasks and, thus, they can be considered robots within the definition of the Robot Institute of America. In addition, their information gathering function implies that they are equipped with appropriate sensory devices.

Such spacecraft and undersea vehicles are really remotely controlled systems with many of their functions preprogrammed. Other functions may be programmed or reprogrammed into new action sequences during the progress of the mission. The planning and structuring of these action sequences are done by humans at a proximally located control station, and are then communicated to the robot. These kinds of supervisory modes of control actions of unmanned systems (robots) are generally designated *teleoperation*. Teleoperated robots are being developed for many special applications other than in space or undersea. Usually they can do work in inaccessible and dangerous areas, e.g., for fire fighting, rescue in mines, or repair work in nuclear hot cells.

While mobility units are important elements of such systems in order to navigate along desired trajectories, and to maintain prescribed orientations, they are of little significance for industrial robots. Most industrial robots today are stationary. Their main actuator systems are the general purpose, programmable manipulators, which determine their essential characteristics. Most work on robotics to date, both in research and in applications, has been done in the context of industrial manipulator requirements.

The technical sophistication of manipulators and robots is indicated mainly by their ability to make their own decisions during operation. This ability requires that the robot device be equipped with a sensor system that receives information about the surrounding objects, the environment, and its own state. Such systems may include vision, ranging, force-torque, touch, proximity, and other sensors. In addition, it requires a control computer that processes the sensory information, subject to given objectives and constraints, and develops action decisions for the robot actuators.

Industrial Systems

Frequently, remotely operated manipulators, such as the master/slave manipulators used to avoid radiation hazards in the nuclear energy field, or automated machines that resemble a manipulator, are called robots. However, none of these devices can truly be called robots if they are not

---

The author: Program manager, autonomous systems and space mechanics, Jet Propulsion Laboratory, Pasadena, Calif.

programmable. Such misnomers often lead to misleading statements about the number of robots in use. Japan, for example, rates first in the world with about 30,000 robots in operation. But this number also includes simple devices similar to the Seiko Model 700 shown in Fig. 1, which have been in use for almost 40 years. If the Seiko Model 700 is still classified as a robot, it is at the lowest end of the scale of robot sophistication, performing simple pick and place functions.

Industrial robots were put into practice during the early 1970s. However, initially there was a comparatively slow rate of penetration into U.S. manufacturing plants. Only during the last three to four years have U.S. companies begun to make wider use of robots in manufacturing, largely due to the pressures of strong competition from abroad. Today, robots already do work previously done by humans in machining centers, and take their place, e.g., as components of flexible manufacturing systems (Fig. 2). They perform complex transfer, placement, and positioning operations, and blend naturally into modern concepts of customized product and small batch manufacturing.

All major U.S. auto makers— General Motors Corp., Ford Motors Co., Chrysler Corp., American Motors Corp., Volkswagen of America, Inc.— are using about 5000 to 7000 robots for welding, painting, loading, machining, etc. These numbers are rapidly increasing and, as technology advances, robots are also finding acceptance in areas with highly complex requirements. For example, General Electric announced recently the establishment of a new division to serve worldwide markets with advanced automation systems, including robots. General Electric expects robots to take over 15,000 of the jobs now performed by its 37,000 blue-collar workers. IBM and Westinghouse created new divisions to produce and use robots internally. Westinghouse installed 50 robots last year and plans to add 200 more this year. By 1983, General Motors wants to add 2000 robots to its present 300, and for 1990, the planned number is 14,000. RCA uses robots in the manufacture of picture tubes and has established a robotics research laboratory. Texas Instruments uses vision-assisted robots for the assembly of small parts, and a new company, Automatix, is expected to produce robot systems that include charge-coupled device cameras.

Robot Types. Today we subdivide industrial robots and manipulators into special-purpose and general-purpose robots. These *special-purpose robots* are designed and produced for a limited range of specific tasks. They are used in approximate order of frequency for welding, machine loading, painting, casting-forging, assembly, machining, and material handling. They usually employ manipulators with limited degrees of freedom. Each of the discrete movements must be recorded in the robots' memory in sequence. These special-purpose robots include electrically and pneumatically activated arms and are generally less costly than the multipurpose systems. Some can be bought for less than $5000.

The *general purpose robots*, or universal robots, are designed and produced to perform a large variety of tasks. In order of increasing sophistication, they fall into non-servo controlled, servo controlled, and sensory robots.

The simplest, the *non-servo controlled robots*, move their arms in an open-loop fashion between exact endpoint positions on each axis, or along precisely predetermined trajectories in accordance with fixed sequences. They can operate over an infinite number of points that describe the envelope of operation. In the first case, the *point-to-point control*, the start and end points that must be passed on each axis, are given. During motion, there is then no prescribed functional connection between the various axes, and no defined trajectory is maintained by the end effector (hand). In the second case, the *trajectory control*, all points between the start and end points are also predetermined, thus enabling motions along straight lines, circles, spirals, or arbitrary curves. Because of their inherent limited control sequence characteristics, non-servo controlled robots are usually used for special purpose functions.

The *servo-controlled* robots have, in addition to the above, also internal sensors, i.e., position, velocity, acceleration, force and torque sensors, which measure the internal states of the manipulator. The states are compared with the predetermined operational parameters of the control program, and the deviations are transmitted to the servo system for corrective action. Depending on the quality of the internal sensors, the controllers, and the type of motors, these systems can be operated with high precision (0.1- to 1-mm repetition of position) and high speed. However, already for relatively low operational precision and speed, these systems tend to become involved and expensive, while their flexibility and adaptability to the manufacturing process remains still relatively limited.

Most UNIMATION robots are typical of the simplest servo-controlled robots (Fig. 3). They have a single simple arm with six degrees of freedom and a high lifting capacity between 30 and 400 Kg. They are programmed by leading the robot through the desired operations, called *teach-in* programming. These robots can be applied to spot or arc welding, machine loading, material handling, or practically any other task that can be predictably automated.

A more sophisticated type of servo-controlled robot is the articulated robot, which has increased flexibility as compared with the simpler types, and is capable of approximating the motions of a human arm. The Cincinnati Milacron T-3 robot (Fig. 4) is an example of this type. The T-3 is mounted on a pedestal 1.5-m high, which can swivel the arm through 240 deg. The arm can swivel vertically at the shoulder through 90 deg. The robot can carry heavy loads of up to 80 Kg at full speed and up to 135 Kg at reduced speed. It can calculate automatically the shortest route between any two points, can call any of a number of programs from its memory, and is capable of continuous as well as point-to-point motion.

Sensory Robots. These are the most sophisticated type of universal robots. In addition to the internal control sensors (potentiometers, straingages, etc.), they usually have a set of external sensors consisting of television cameras, pressure detectors, magnetic sensors, laser range finders, force-torque sensors, and the like. These sensors serve to determine the robot's relation to its environment and, in particular, to determine the location of parts to be handled. The sensory data are fed back to the control computer, which can effect appropriate program modifications to cope with changing requirements and unpredictable situations, such as a random collection of parts. The important characteristic of adaptable sensory robots is that they have the ability to recognize through "intelligent sensors" changes or events in their environment and make the corresponding changes in the control program in order to perform reliably in the new situation. Such robots may be considered the first level of intelligent machines.

One of the most advanced robots of this type is produced by Hitachi, Japan. It has two arms, each with eight degrees of freedom. Using seven television cameras, it is capable of

Fig. 1 Seiko Model 700 is a four-axis, air operated, simple pick and place manipulator.

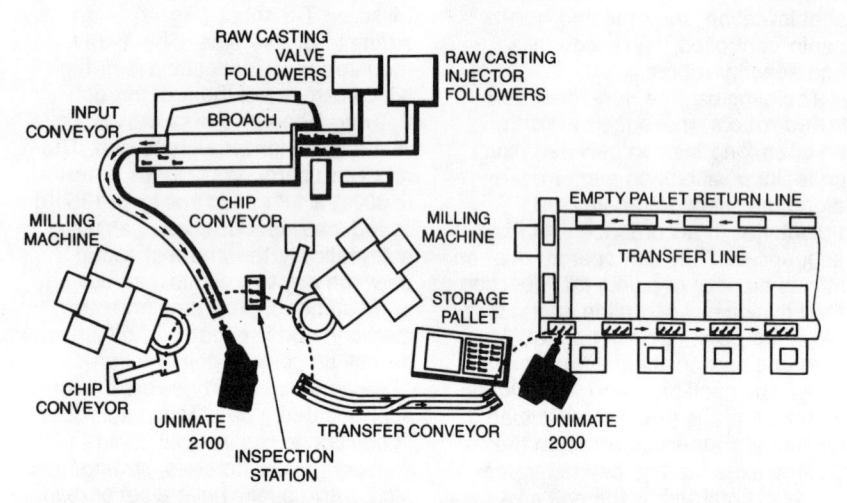

Fig. 2 Auto-handling and transfer system. Unimate Series 200 robots are in operation and integrated into a manufacturing system. The robots are equipped with random program selection for changing to an alternate program if desired, or for aborting a program to unload parts without reloading if there is a malfunction. (Courtesy UNIMATION Inc.)

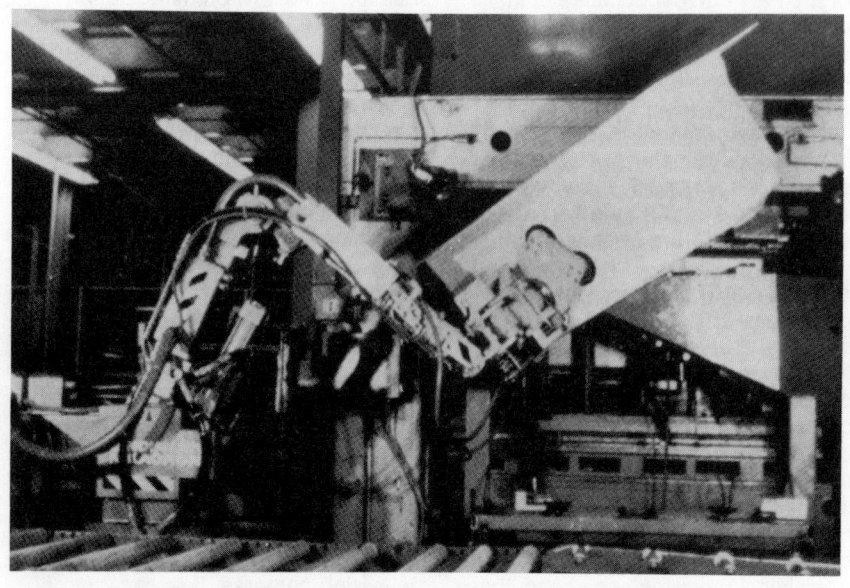

Fig. 3 A UNIMATE robot in operation. (Courtesy UNIMATE Inc.)

identifying a given object in a random collection of objects and assembling it with other parts into a complex device, such as a vacuum cleaner, in a few minutes.

The Soviet built LPI-1 robot has similar capabilities, which are provided by two electromechanical arms equipped with tactile and distance sensors, a hot diode sensing disk, and a television camera. This robot works automatically or under the direction of a human through a voice control unit.

A third sensory robot is produced by Auto Place. This robot, the Opto-Sense, is capable of recognizing a part in 0.02 s, then picking it up regardless of orientation, inspecting it, and accepting or rejecting it for future processing. Its camera, the TN-2000, scans row by row, and a microprocessor calculates the area of the observed object. The rate of change of the area as the scan proceeds, as well as the overall area, is used by the minicomputer to recognize the object and determine its orientation.

Vision is considered one of the most important keys to broadening the base of robot applications in industry. A lot of work is being done in this area. For example, Automatix, Inc., introduced its Autovision system for high-speed inspection of up to 300 parts per min, for sorting parts, and for automatic assembly. Together with a PUMA robot (Fig. 5), UNIMATION introduced the UNIVISION robot system which uses a Machine Intelligence Corp. VS-100 vision system. This system incorporates pattern recognition principles by selecting distinctive parts of the object's silhouette, and operates on its information using advanced software algorithms.

Another area of potential is the development of a general-purpose end effector: a universal hand. Proximity, tactile, and force-torque sensors are examples of components that can provide general-purpose capabilities to the end effector when appropriately integrated. Proximity sensing refers to the robot's ability to sense without touching when it is close to another object, while tactile sensing allows the robot to grasp gently. Force-torque sensors, usually in the manipulator's wrist, assure that the robot will not overexert itself or cause damage to other objects.

Most sensory robots have not yet progressed much beyond the laboratory state. The reason is that the technology is not sufficiently matured. Most robots still work "blind." Also, there are still questions as to whether or not the frequency of requirements for their sophisticated capabilities will justify their relatively high cost, even when they can be produced in large quantities.

The prime characteristics of robots in the manufacturing area are their programmability and their general applicability in handling. This implies that a gripper or a tool can be moved to any point through any trajectory within the workspace, and at any orientation within certain physical constraints, and that the gripper can grasp almost any objects or tools. Based on these characteristics and the availability of computer logic and memory, complete manufacturing process sequences for general purpose handling can be combined and implemented. Such sequences can generally be combined in a computer program in any arbitrary succession subject to constraints of computer type, memory size, programming language, and the physical constraints of the manipulator geometry and workspace.

Depending on the level of development, the trajectory and stopping points of the manipulators end effector can be defined by manual programming, "playback" and "teach-in" methods, or textual programming.

In manual programming, the stopping points of each degree of freedom are fixed by physical means (hard wire, limit switches, etc.). No computer is necessary in such systems.

In playback or walkthrough programming, all motion points are stored into memory (disks, tapes, etc.), by leading the manipulator through the desired motions. In the teach-in method, on the other hand, the manipulator end effector is steered toward the required successive positions and orientations using an appropriate mechanism that has switches for various motions, joints, time intervals, and velocities.

Textual programming of robots requires that all operations be described symbolically in the correct sequence and with the required precision indicators. This requires that all manipulator and robot positions and motions be determined beforehand by elaborate measuring processes. To avoid these difficulties, techniques similar to playback or teach-in methods are often used to determine the geometrical data required in textual programming. Although textual programming requires more forethought and appears to be more elaborate and expensive, it is gaining progressively greater importance in practical applications. Some of the reasons are the greater

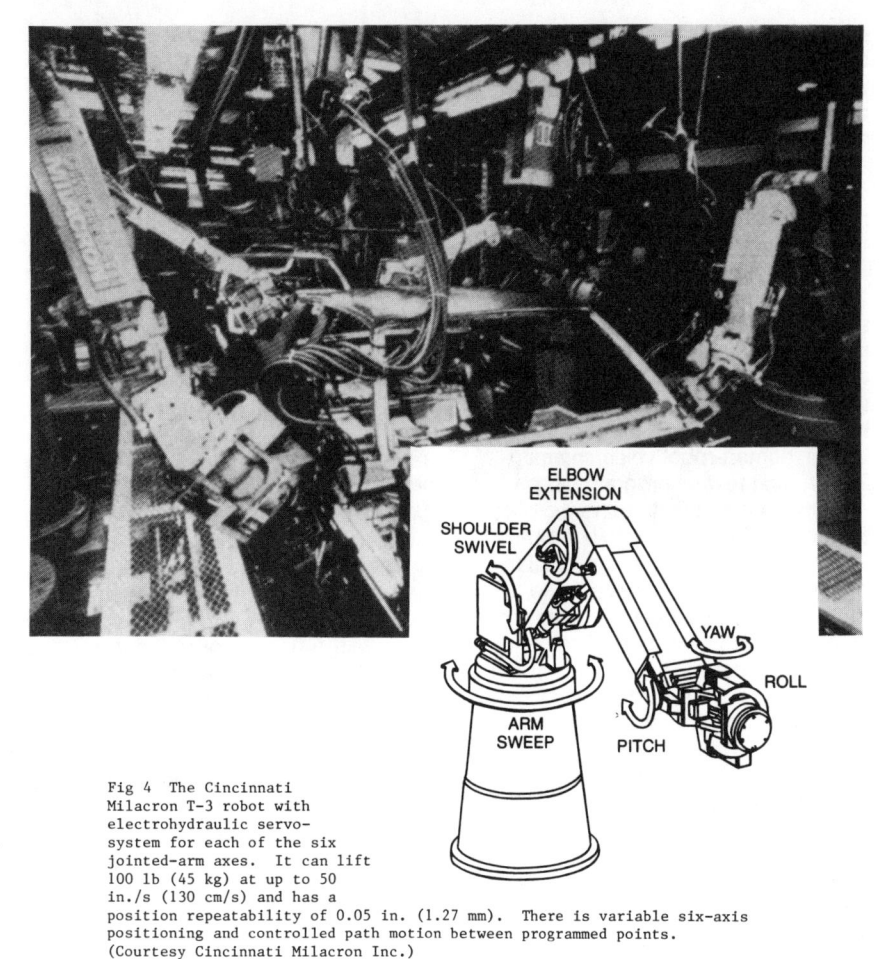

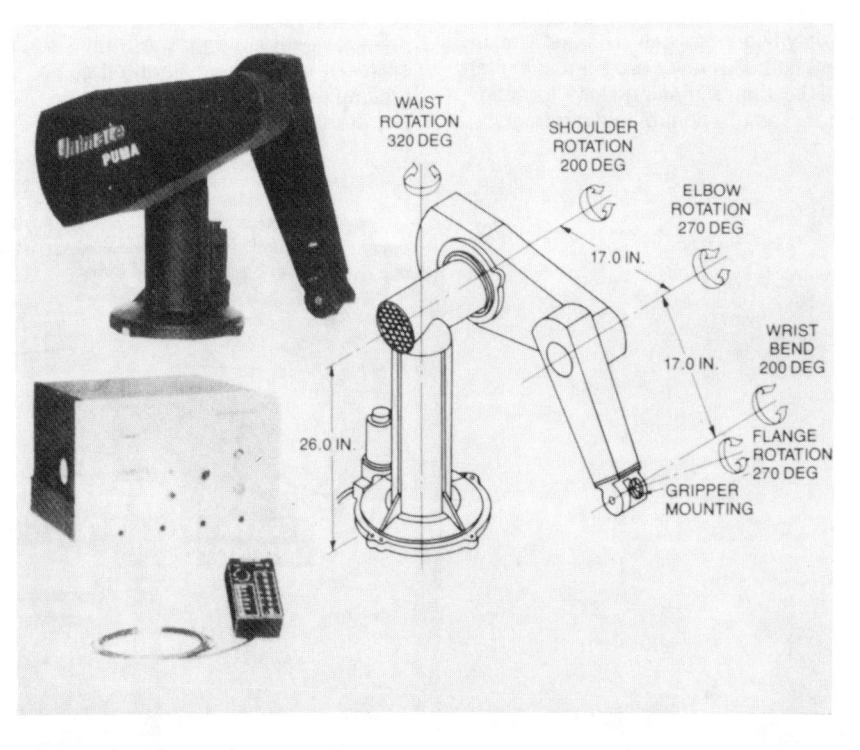

Fig 4 The Cincinnati Milacron T-3 robot with electrohydraulic servo-system for each of the six jointed-arm axes. It can lift 100 lb (45 kg) at up to 50 in./s (130 cm/s) and has a position repeatability of 0.05 in. (1.27 mm). There is variable six-axis positioning and controlled path motion between programmed points. (Courtesy Cincinnati Milacron Inc.)

Fig. 5 The PUMA 500 is a five-axis, electric powered robot manipulator with 0.004-in. (0.10-mm) position repeatability, 13.2-lb (6-kg) static load capacity, 5.0-lb (2.3-kg) load capacity at maximum tip velocity of 3.3 fps (1 m/s) and teach-in programming method. (Courtesy UNIMATION Inc.)

33

complexity of modern production systems, which include robots; the proliferation of sensor systems and associated evaluation of sensor data; the possibility to modularize the software structure and to effect changes easily; the preparation of programs off-line; and readable documentation.

Present developments suggest that the important and interesting areas of research for industrial robots and manipulators are in vision sensors, end-effector sensors, multiple manipulator coordination, computer-directed hand trajectories, general purpose hands, human-robot voice communication, and textual programming.

Perhaps the most important problem that should be solved within the next five years is the effective integration of appropriate sensor systems so that the robot can cope with changing environments in real time. Another important problem is the development of concepts for multi-processor systems with distributed and hierarchical configurations. The effective combination of computer-aided design (CAD), computer-aided manufacturing (CAM), and robots into well functioning end-to-end integrated systems still needs much research and development work and is not expected to be available within the next five years.

Space Systems

To date, manipulator technology has been applied in the space arena in only two instances. The first was in the Lunar Surveyor program in the late 1960s when a manipulator was used under remote control to dig a trench and perform other simple functions on the Moon. The second was in the Mars Viking program in 1976, when a computer controlled manipulator helped to perform a number of scientific experiments on the Martian surface. Within the next two years, it is expected that the Space Shuttle will employ a Canadian-built 48-ft (14-m) long attached manipulator to unload payloads from its cargo bay and handle other objects in space (Fig. 6). The actions of these manipulators are in the teleoperator mode or preprogrammed without sensory feedback. While the Surveyor manipulator is non-servo controlled, the other two fall into the servo-controlled category. Developments are now under way to design future space manipulators with sensory feedback, i.e., as sensory robots.

Planetary and Earth orbital spacecraft with functions of gathering scientific and utilitarian data and transmitting them to Earth are equipped with appropriate sensory devices, not only to acquire scientific information, but also to navigate, point accurately, and perform other programmed functions. These space robots also fall into the general class of sensory robots.

Future robots and manipulators in space may be put into four general application categories: (1) space exploration; (2) global services; (3) space industrialization; and (4) space transportation.

Space exploration robots may be exploring space from Earth orbit like orbiting telescopes, or they may be planetary flyby and/or orbiting spacecraft like the Mariner and Pioneer families. They may be stationary landers with or without manipulators like the Surveyor and the Viking spacecraft, or they may be wheeled like the Lunakhod and the proposed Mars rovers. Others may be penetrators, flyers, or balloons, and some may bring science samples back to Earth (Fig. 7). All have in common that they can acquire scientific and engineering data using their sensors, process the data with their computers, plan and make decisions, and send some of the data back to Earth. Some robots are, in addition, able to propel themselves safely to different places and to use actuators, manipulators, and tools to acquire samples, prepare them, and perform in-situ experiments or bring them back to Earth.

It is clear that the spacecraft autonomous capabilities, including automated decision making and problem solving by the onboard computer system, will be major elements in the operation of these systems. The robot spacecraft must be able to implement sophisticated control strategies and manage its on board resources. It must also be able to assure self-maintenance of its functions and make the appropriate decisions toward achieving the assigned tasks.

Present and projected developments in machine intelligence suggest that in the future many of the presently ground-based data processing and information extraction functions can be performed autonomously on board the robot spacecraft. Only the useful information would be sent to the

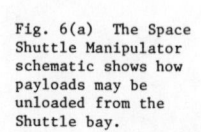

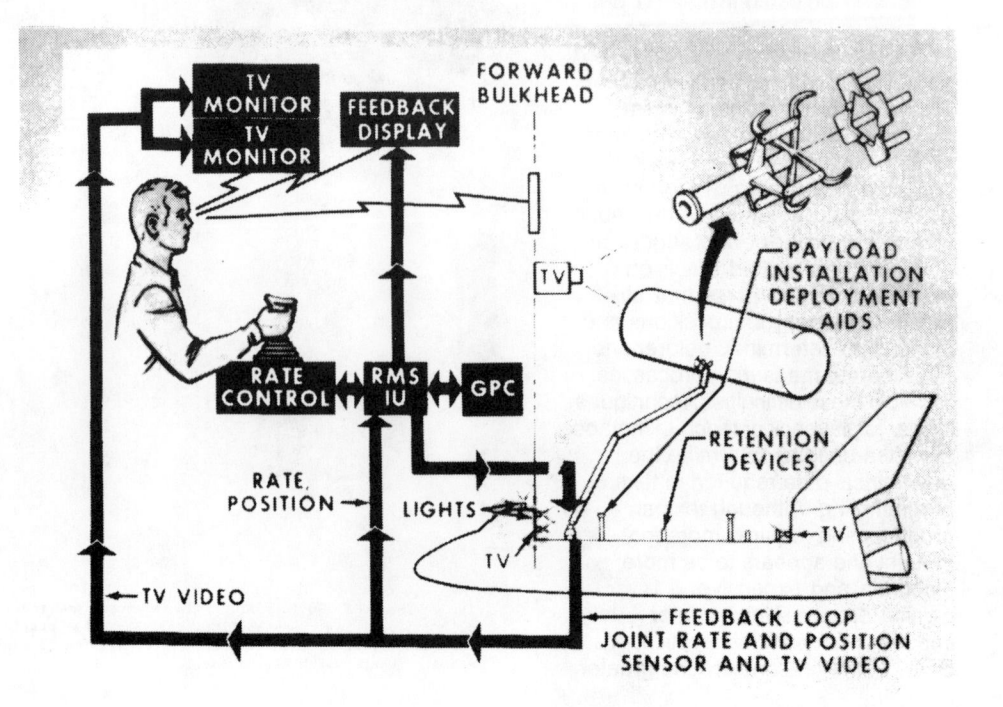

Fig. 6(a) The Space Shuttle Manipulator schematic shows how payloads may be unloaded from the Shuttle bay.

ground and distributed to the users. This would imply that the robot is able to make decisions on its own as to what data to retain and how to process them to provide the user with the desired information.

Space Industrialization Systems. Such systems, including space utilization and space transportation systems, require a broader spectrum of robotics and automation capabilities than those for space exploration and global services. The multitude of systems and widely varying activities envisioned in space until the end of this century will require the development of space robots and automation technologies on a broad scale. It is in this area that robot and manipulator technologies will have the greatest impact.

Large area systems, such as large space antennas, satellite power systems, and space stations, require large-scale and complex construction facilities in space. Relatively small systems, up to 100 m in extent, may be deployable and can be transported into orbit with one Shuttle load, and handled with the Shuttle attached manipulator. For intermediate systems of several hundred meters extension, it becomes practical to shuttle the structural elements into space and assemble them on site, using, for example, free-flying robots and teleoperated manipulators (Fig. 8).

Very large systems require heavy-lift launch vehicles, which will bring bulk material to a construction platform, where the structural components are manufactured using specialized automated machines (Fig. 9).

The structural elements will be handled by teleoperators or self-acting cranes and manipulators, which bring the components into place and join them. Free-flying robots will transport the structural entities between the Shuttle or the fabrication site and their final destination, and connect them. These operations require a sophisticated general-purpose handling capability. In addition to transporting structural elements, the robot must have manipulators to handle them, and work with them and on them. Large structural subsystems must be moved from place to place and attached to each other. This usually requires rendezvousing, stationkeeping, and docking operations at several points simultaneously, and with high precision. These robot systems could be controlled remotely like teleoperator devices, or they could be under *supervisory control* with intermittent human operator involvement.

Maintenance and Rescue. After a system has been constructed, its subsequent operation will require service functions that should be performed by free-flying robots or by robots attached to the structure. The functions that such a robot should be able to perform include calibration, checkout, data retrieval, re-supply, maintenance, repair, replacement of parts, cargo and crew transfer, and recovery of spacecraft.

During and after construction, there should be a robot on stand-by for rescue operations. An astronaut drifting into space could be brought back by a free-flying robot. Such devices could also be on stand-by alert on the ground. The delivery systems for these rescue robots need not be man-operated. They can deliver expendable life support systems or encapsulate the astronaut in a life-support environment for return to a shuttle, space station, or Earth. They could also perform first-aid functions.

Materials Processing. Utilization of space includes processing of materials in space. Space materials processing requires additional sophisticated technology, including remotely operated automated machinery and robotic devices. Basic types of processes envisioned include solidification of melts without convection or sedimentation, processing of molten samples without containers, diffusion in liquids and vapors, and electrophoretic separation of biological substances.

It is expected that techniques will also be developed for the acquisition and conversion of raw space materials to forms useful for space system components and for other space and terrestrial applications. A first step in this process calls for a lunar base. After a lunar surface survey with robot (rover) vehicles, an automated precursor processor system could be placed on the Moon. This system would collect solar energy and use it in experimental, automated physical/chemical processes for extracting volatiles, oxygen, metals, and glass from lunar soil delivered by automated rovers (Fig. 10). The products would be stored, slowly building up stockpiles in preparation for a lunar base. The lunar base would be constructed using automated

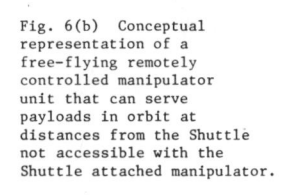

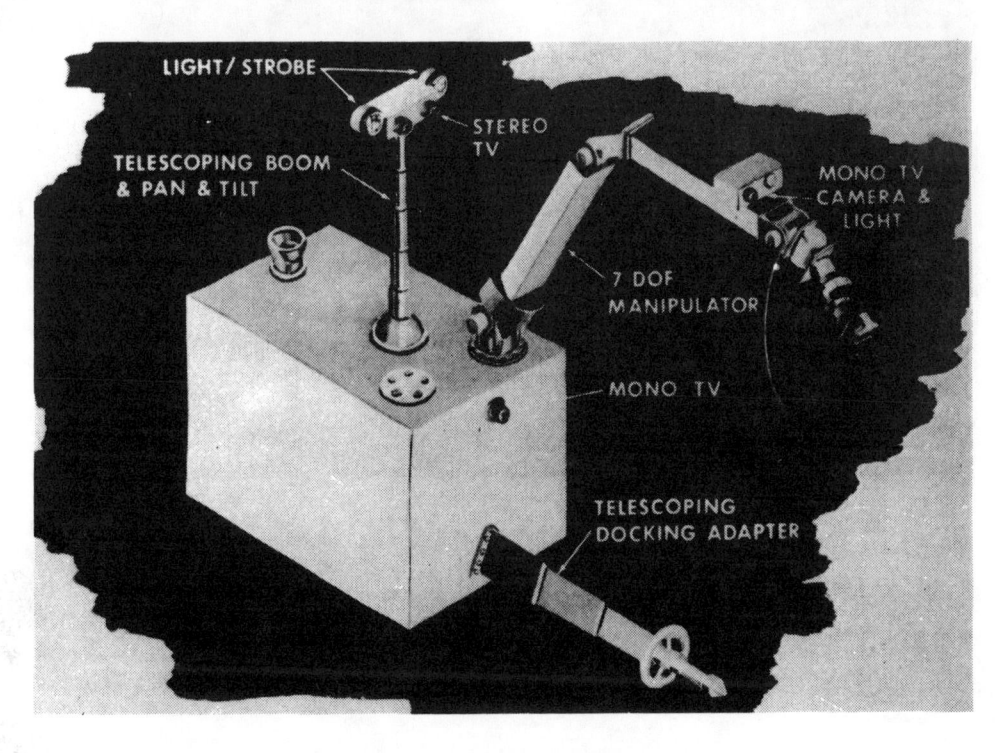

Fig. 6(b) Conceptual representation of a free-flying remotely controlled manipulator unit that can serve payloads in orbit at distances from the Shuttle not accessible with the Shuttle attached manipulator.

equipment and robots similarly as in Earth orbit. After construction, general-purpose robot devices would be necessary for maintenance and repair operations. In addition, the lunar base would use all types of industrial automation (qualified for operation in space) that are generally used on Earth for similar tasks.

Undersea Systems

Undersea manipulators have had much more frequent application than manipulators in space. The use of these machines has been varied, and each design has to cope with its own special problems. The need for increased undersea robot vehicles that contain manipulators with significant sensory feedback or computer control so that work can be performed at a wider variety of tasks and at greater depths, is pressing. Already, exploration for offshore oil is taking place in water depths of 1500 m (4920 ft), and production platforms are pumping oil from depths of 310 m (1015 ft). Oil and gas pipelines are under construction, which will be in water 914 m (3000 ft) deep, and more than 12,893 km (8000 mi) of undersea pipeline are in U.S. waters alone. In 1977, according to the Exxon Co. offshore oil accounted for 16 percent (10 million bbl daily) of the worldwide crude production. Offshore proved crude oil reserves are estimated at 26 percent (170 billion bbl) of the world total.

At a depth beyond 400-500 m there is no diving capability. The only means available are manned or remotely operated vehicles, both of which may be tethered or free swimmers. The latter has problems of acoustic communication through water not present in space (otherwise the technologies appear to be similar, at least in principle). For the untethered vehicle, this puts an extraordinary demand on on-board machine intelligence capabilities, similar to those required for planetary surface vehicles.

Probably the most systematic and generally applicable development work in the area of undersea robotics is being conducted by the U.S. Naval Research Laboratory and the Naval Ocean Systems Center. The Ocean Systems Center is presently developing a robot test-bed, i.e., an untethered submersible to allow demonstration of new, improved undersea robot system technology. The submersible, which is 2.2 m (7 ft) long, about 20 in. (50 cm) high, and 20 in. wide, has a modular construction which allows expansions to accommodate additional payload and new sensor systems as the technology for those systems becomes feasible to demonstrate. The vehicle is designed to follow a set of predetermined program tracks such as a parallel-path search or a figure-8 demonstration run. In this mode of operation, the vehicle is programmed via a computer console and an umbilical cable, which is disconnected after the initial pre-programming phase. The vehicle is then allowed to follow this course until its mission is complete. A microprocessor is used to compare programmed altitude, heading, depth, and run sequence input data with measured data coming from an on-board altimeter, gyrocompass, depth sensor, and clock, respectively. The microprocessor generates digital

Fig. 7 Artist concept of a Mars surface scientific processing and sample return facility. Airplanes transport samples into the vicinity of the processing station. Tethered small rovers then bring the samples to the station for appropriate analysis and return to Earth.

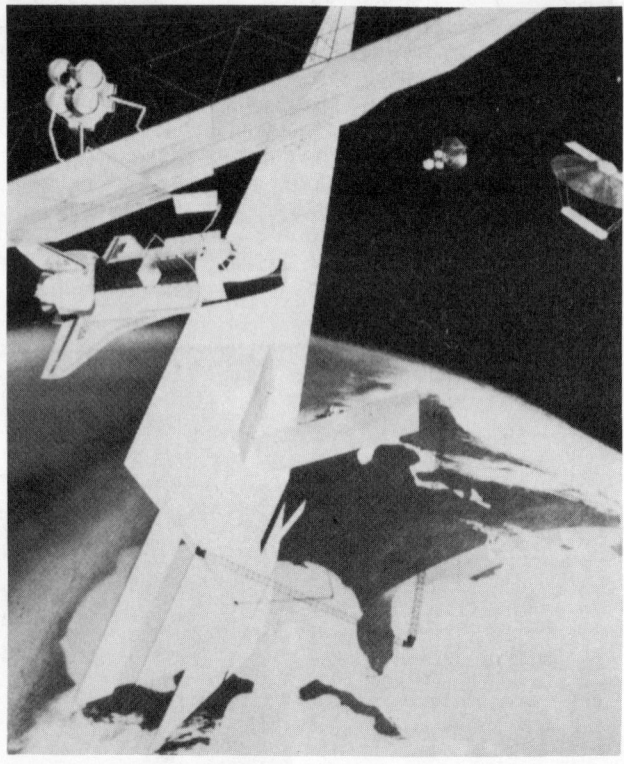

Fig. 8 Large space systems require robot and automation technology for fabrication, assembly, and construction in space.

error signals between the programmed values and the measured values, and issues error signals to the appropriate motor controllers. The motor controllers then power the d-c motors, which directly drive the propellers from a separate 24V battery supply. If an emergency arises, there are automatic procedures which allow the vehicle to turn on an emergency beacon, which shuts off all thrusters and is recovered at the surface.

Other methods of vehicle command control and communications are being added. Communication with the vehicle while it is underwater is being incorporated by means of an acoustic link from the surface. The same programmable controls used for setting up initialization of the vehicle through the hard wire link on the surface will then be incorporated into its real time control system, together with some editing commands. The vehicle will then be able to alter its preprogrammed mission sequence, and/or respond to direct control commands from the surface. The end result will be a system which is not limited by cable drag and cable handling problems. It should then autonomously perform rudimentary tasks.

Generally, however, there appears to be no U.S. program that plans to capitalize on the many and diverse programs going on in the undersea arena, by ultimately bringing the end products together into a prototype operational vehicle combining real-time, through-water television transmission, a mission-oriented operational duration, and significant operating depth. If the technological problems now being addressed are successful, other problems remain. These include vehicle command/control techniques, navigation, launch/retrieval, and integration of the individual components and subsystems into an operational unit.

Conclusions

Robots and manipulators at various stages of mechanical complexity and intellectual sophistication are rapidly becoming important elements in industry, space, and undersea. In industry, the primary driving force is increased productivity. In space and undersea operations (and other applications), robots and manipulators become necessary because of remoteness, danger, and inaccessibility to humans. While applications for robots and manipulators are varied and manifold, many fundamental research problems are very similar, centering on vision sensors, end-effector sensors, various artificial-intelligence techniques, control system architectures, and peripheral interfaces. The importance of the last point is being increasingly recognized: for an average industrial robot installation project, it will usually cost as much to engineer, tool, and install the robot as it costs to purchase the robot itself.

After a decade of uncertainty and vacillation, American industry is moving toward automation and is adopting robots and manipulators as vital elements in the production process. Within reach are computer-controlled systems of robots and other complex machines that promise unprecedented gains in productivity. A radical restructuring of work will lead to jobs for humans becoming more complex than ever.

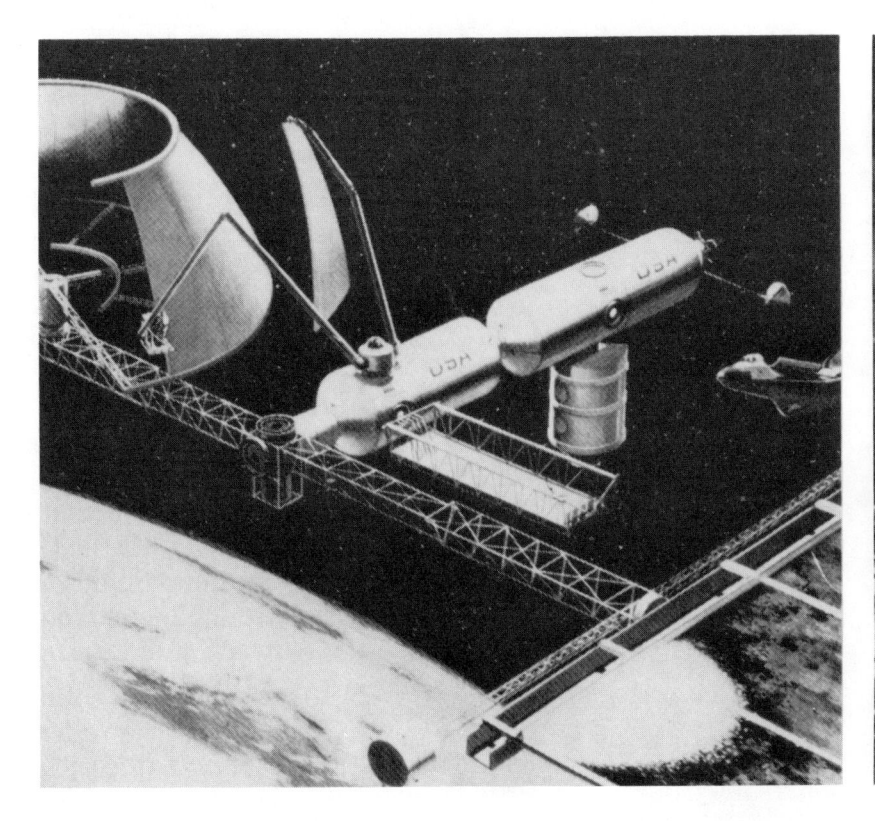

Fig. 9 Construction of a space station. Bulk material is brought by the Shuttle. Structural elements are fabricated at the construction facility and then assembled by remotely controlled manipulators.

Fig. 10 Automated material processors on the lunar surface are serviced by robot vehicles with raw lunar soil.

CHAPTER 2
NUCLEAR APPLICATIONS

THE MA 23 BILATERAL SERVOMANIPULATOR SYSTEM

JEAN VERTUT, PAUL MARCHAL, and GUY DEBRIE
Commissariat à l'Energie Atomique
91190, Gif-sur-Yvette, France

MICHEL PETIT and DANIER FRANCOIS *Société La Calhène*
5, rue Emile Zola, 95870, Bezons, France

PHILIPPE COIFFET *Laboratoire d'Automatique de Montpellier*
Université des Sciences et Techniques du Languedoc
34060, Montpellier, France

KEYWORDS: *manipulator, remote vehicle*

ABSTRACT

The MA 23 is an advanced zero-backlash arm design, with a d.c. servo system derived from a previous version, the MA 22. The modular concept of the system offers a standard master arm, normal- and heavy-duty slave, and standard electronics, including digital multiplex and tape recording, plus a developing computer-aided control. A booting assembly for the hot cell, remote welding, undersea and industrial applications (forging, welding, etc.) were developed by a group with the Comissariat à l'Energie Atomique, the University, and the navy.

INTRODUCTION

Initially, work on servomanipulators in France was started in cooperation with Carl Flatau of Tele Operator Systems, as part of the Virgule remotely operated vehicle.[1,2] In 1972, the MA 22 arm was designed.[3] In 1973, a first prototype was built and tested with servo motors and control system derived from Flatau's contribution. The first pair were completed in 1974 and were mounted on the Virgule vehicle (Fig. 1). In mid-1976, control of the entire unit via a single coaxial cable was tested; radio control is planned for mid-1977.

In 1973, the French navy became interested in the development of a deep-submergence master-slave manipulator that was controlled from the surface. From this cooperative effort, it seemed possible to extend this effort to other participants concerned with computer control and advanced industrial robot applications.

With this full range of applications in mind, the MA 23 program was started in late 1974 in close cooperation with the Commissariat à l'Energie Atomique and La Calhène, for the design and construction (mechanics and electronics), the University of Montpellier, for computer control, and the French navy, for undersea applications. The group now is working on a medical project, i.e., an advanced manipulator for quadriplegics, as well as industrial projects.

The first MA 23 arm was built in mid-1975. One 10-kg capacity and one 25-kg capacity system have been working since late 1975. One 10-kg arm and one 5-kg arm are used for computer control. A total of five arms have been built.

The MA 23 Modular System Concept

This system was designed to provide different operational modes as illustrated in Fig. 2. The basic modes are:

1. the classical master-slave mode
2. a tape recording and playback mode
3. computer control.[4]

The system provides possible combinations, like a computer supervisory control applied to a master slave, a computer-assisted programmed mode, etc. These modes are all along the path between full-time manual teleoperation and advanced automation (artificial intelligence).

The various subsystems are illustrated in Fig. 3.

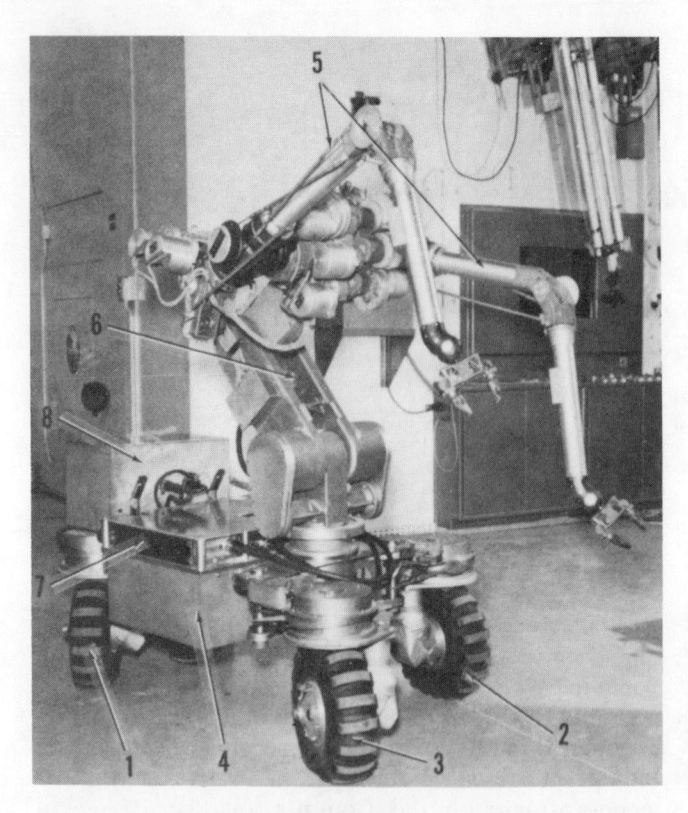

Fig. 1. The Virgule teleoperator: (1) four self-contained propulsion and steering wheels with special tread pattern for stair climbing; (2) extended front wheel (both extended give stability); (3) retracted front wheel (both retracted allow passage through a narrow door); (4) batteries; (5) pair of MA 22; (6) three-deg freedom-articulated support; (7) right arm power amplifier; (8) multiplex communication.

Different Types of Arms

The system now is based on three types of arms (Fig. 3):

1. the master arm (1)
2. the normal-duty slave arm (2)
3. the heavy-duty slave arm (4).

Other arms are under design evaluation; all are specifically designed so that they can be interchangeably coupled together.

Bilateral Servo-Electronics

The following equipment is shown in Fig. 3:

1. The power amplifier packages (5) near the master and slave, or attached to the arms [see (1), (2), and (4)].
2. The power supplies (6); one also is located near each arm.
3. The servo-pack (7)—one pack for master plus slave, located near the master. This works directly with multiwire cable communication to the slave.
4. The digital multiplexer (8)—provides master-to-slave communication on a single coaxial line or by radio link.
5. The digital tape recorder (9) is connected to the servo-pack.
6. The minicomputer control unit (11) is connected to the servo-pack and/or the tape recording; it allows control of the arm from a console.

MECHANICAL DESIGN AND PERFORMANCE

The General Arrangement of the Master and Normal-Duty Arms

The two arms are very similar, the principal difference is a lower reduction ratio on the master. Figure 4 shows a normal-duty slave arm with motors (1) on both sides of the body (2). This arm does not have the attached power amplifier package(s) as shown on arm (2) of Fig. 3.

The shoulder box is in front of the body and moves around a horizontal axis to generate the general rotation (X-motion). The arm is in two segments [(4) and (5)] articulated on a shoulder axis perpendicular to the general rotation axis, so that shoulder motion (Z-motion), elbow motion (Y-motion), and general rotation (X-motion) are the basic coordinates.

The lower arm (5) rotates around its own axis (azimuth). A standard wrist joint is mounted at the end and is fitted with a standard tong (6). Counter weights (7) balance the arm in any position of the whole manipulator. In Fig. 3, the master arm handle (13) is a new spring-balanced model for light-duty manipulators.

The coverage in this geometry is exactly the same as the mechanical MA 11 model H by CRL (Ref. 5). The use of a general rotation as X-motion improves the lateral coverage compared to the Argonne $E3$ (Ref. 6), $E4$ (Ref. 7), Mascot (Ref. 8), or MA 22 (Ref. 3).

A motion lock [Fig. 3 (14)] on the master arm enables the operator to lock the slave via the servo-loop without need for slave arm brakes. This is safer since the slave arm is free in case of any electrical failure.

General Arrangement of the Heavy-Duty Slave Arm

Figure 3 (4) shows this heavy-duty slave arm, which has the same length dimensions as the normal-duty arm. The body is designed to pass

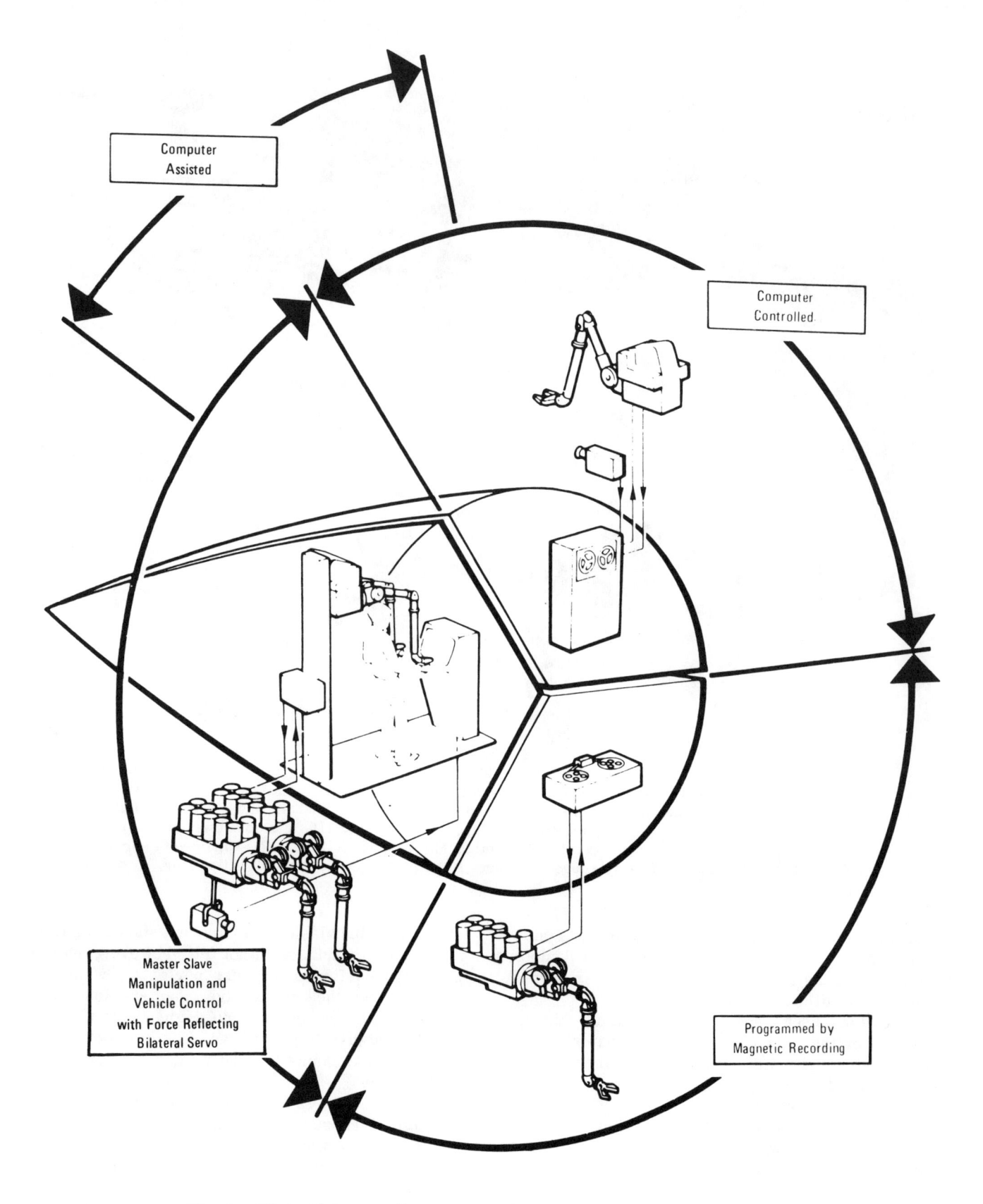

Fig. 2. Bilateral force reflecting manipulators—different modes of operation.

Fig. 3. The MA 23 system modules: (1) master arm; (2) normal-duty slave arm; (3) normal-duty slave with booting assembly (booting removed); (4) heavy-duty slave arm; (5) power amplifier pack; (6) power supply for one arm (or pair of master); (7) servo-unit and control box for master and slave; (8) multiplex unit (master); (8') multiplex unit (slave) identical; (9) tape recorder unit; (10) minicomputer interface; (11) rack for minicomputer control; (12) computer-controlled MA 23 arm; (13) master handle with spring balance; (14) master motion lock; (15) single torque-motor; (15') twin torque motor; (16) lifting ring.

through a narrow opening with all motors on top. Most motors are double, with two identical motors in a single housing (15). A single lifting ring (16) enables handling.

The undersea model is identical to this arm except for the corrosion-resistant materials used to construct it; i.e., titanium castings, special bearings, etc.

Figure 5 shows a model of the ERIC II cable-controlled remote vehicle in its working position. (NOTE: Bodies with other geometries are envisioned, including a single body for one pair of arms.)

Zero-Backlash Actuator System

The MA 22, like all existing servo-arms and like nearly all mechanical master slave manipulators, has backlash. In many movements against gravity, this backlash is not felt; however, even if the servo performance were perfect, it would be degraded by the cumulative master-plus-slave backlash. On the other hand, high-precision low-backlash high-efficiency gear reduction boxes are expensive and cannot be easily interchangeable for maintenance.

From the MA 22 experience, we optimized the mechanical characteristics by coupling a higher-torque d.c. motor to a lower-reduction transmission based on block-and-tackle mechanisms without gears.

Figure 6 is a close-up of the heavy-duty slave arm shoulder box. Two antagonist 6-tape block-and-tackle units [(1) and (2)] are attached to 3.5-mm ($\frac{1}{8}$-in.) cables [(3) and (4)] fitted to the shoulder-driving pulley (5). These are at the left

Fig. 4. Normal-duty arm with self-contained ventilated booting assembly: (1) d.c. torque-motor in sealed housing; (2) manipulator body; (3) shoulder box; (4) upper arm; (5) lower arm; (6) standard tong on standard wrist joint; (7) balancing counterweight (one on each side); (8) upper lifting and handling attachment for powered manipulator; (9) foot; (10) rear attachment for handling by the heavy-duty powered manipulator; (11) ventilation blower protected by (12); (12) inlet quick-change filter; (13) exhaust multifilter; (14) rear plate booting fitting; (15) intermediate ring; (16) booting (shown here only in outline).

side of the box. At the right side, similar 8-tape block-and-tackle units [(6) and (7)] act on the elbow pulley (8).

A lever (9), driven by a pulley (8), and a tie bar (10) drive the elbow. These form a parallelogram, which is prolonged on each side to hold the balancing counterweights (11).

By using basic torque-motors that develop 3 Nm torque (26 lb-in.), a twin motor drives a heavy-duty shoulder through a 24:1 ratio, and drives the elbow through a 32:1 ratio.

This actuator system produces zero backlash; deflection is ~25 mm (1 in.) at full load. This is illustrated in Fig. 7 for one of the X-, Y-, Z-motions of the heavy-duty slave arm.

Friction depends on the initial tensioning of the tapes, and follows the same variations as deflection. Friction due to the d.c. motors is about the same as the no-load friction of the arm.

Table I summarizes the arm performance. The following clarification of the table is necessary:

1. All characteristics are referred to the arm terminal.

2. All characteristics have been measured on arms after more than 100 hours of operation.

3. The mass capacity means the mass to be lifted against gravity.

4. The maximum force (stall load) and capacity is the minimum to be found in the worst arm

45

Fig. 5. The ERIC II cable-controlled deep-submergence teleoperator: (1) undersea heavy-duty slave arms retractable into their cavities in the (2) wet submarine remote vehicle; (3) floating titanium vessels with electronics at normal pressure; (3') floating titanium vessel movie camera; (4) television camera operated by head control; (5) two fore and aft propellers; (6) two lateral-movement propellers; (7) two vertical-movement propellers. The combination gives six-deg freedom controlled with force feedback by a pilot on the surface ship; (8) self-oriented rear propeller to automatically compensate for the main cable traction force.

configuration. Of course, combinations of motions can give at least 1.7 times more force in certain directions in particular arm configurations.[2]

5. About friction, deflection, and inertia; they all depend on the different motions. Table I gives maximum and minimum figures.

6. Friction is greatest in lateral rotation and azimuth rotation, which we now are working to improve.

7. Deflection is greatest for general rotation (X-motion) in the fully extended arm configuration.

8. Inertia is greatest in general rotation, with retracted-arm configuration, and in the shoulder motion.

All of these figures are close to those of corresponding mechanical master-slave manipulators. (NOTE: As far as friction is concerned, a new Mark II power amplifier, now under development, will bring significant improvement.)

Bilateral Servo-Loop Principle and Performance

This loop is an improvement on the Mark II loop of the MA 22 (Ref. 2). It is a position-to-position force-reflecting servo-loop, illustrated in Fig. 8. The d.c. torquers (1) have built-in potentiometers (2). Power amplifiers (3) can be placed

Fig. 6. The heavy-duty slave arm shoulder box: (1) and (2) shoulder motion block-and-tackle actuators with six tapes; (3) and (4) twin $\frac{1}{8}$-in. cable antagonists on (5) [(4) is not visible in the photo]; (5) shoulder driving pulley, (6) and (7) elbow motion block-and-tackle actuators with eight tapes; (8) elbow motion driving pulley, fitted to (9); (9) elbow motion lever; (10) tie bar; (11) balancing counterweights.

close to the motors, or at a distance if necessary. The servo-card itself processes the position error signal derived from the master and slave potentiometers. It is located at any distance from the power amplifiers, but preferably at the master control cabinet. The position error signal is fed into the two separate servo-loops controlling the master (4) and slave (5) motors. Figure 9 shows the servo-card (1) and the power amplifier (2).

Plastic potentiometers and differentiating circuits (not shown on Fig. 8) are used to derive velocity signals in place of tachometers. Optimized gains on position and velocity signals allow the attainment of static deflection identical to dynamic deflection.

Separate circuits [(6) and (7)] compensate viscous friction and part of the inertia of each loop. Force feedback can be switched from high to low ratio or to logarithmic response (8). In addition, another circuit (9) suppresses 80% of the slave coulomb friction to the force feedback. This Mark III servo-loop provides an electric deflection smaller than the mechanical deflections of the manipulator.

Performance data on the master-slave system are given in Table II, which includes load versus time and duty-cycle limitations based on static heat dissipation tests made with natural cooling. (NOTE: The master friction still is high and will be improved with new cabling and Mark II power amplifiers.)

Deflection is measured with the slave held fixed and by applying zero to maximum force on the master. The MA 23 performance is like the best

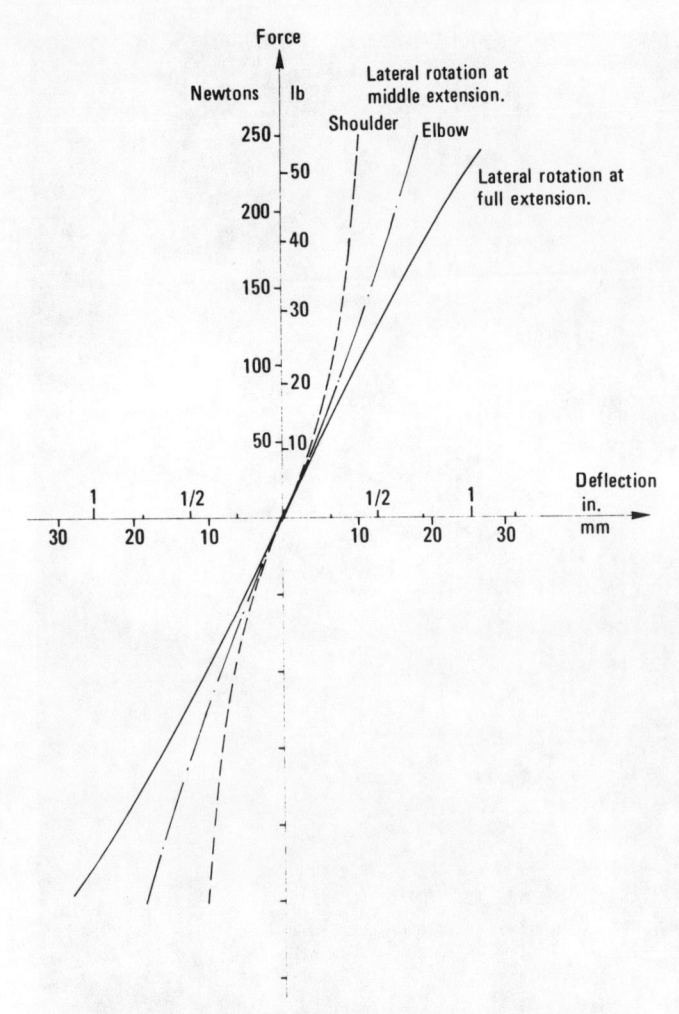

Fig. 7. Diagram of heavy-duty arm deflection.

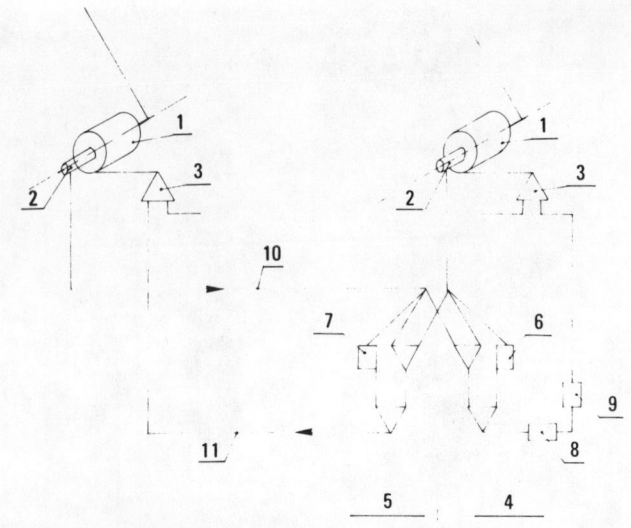

Fig. 8. Diagram of the bilateral servo-loop—master is on the right side: (1) d.c. torquers; (2) potentiometers; (3) power amplifiers; (4) master position loop; (5) slave position loop; (6) and (7) friction and inertia-correcting circuits; (8) logarithmic force feedback circuit; (9) slave coulomb friction correction circuit; (10) slave position feedback; (11) slave power amplifier input.

light-duty arm with slave capacity of standard duty and heavy duty, except the greatest friction is in different motions (in telescopic arms, Z-motion has the greatest friction of 7 to 8 N; in the MA 23, the greatest friction is in the X-motion).

Absence of backlash is very desirable for the elbow motion, especially when the lower arm is close to vertical, where telescopic arms have a master-to-slave backlash of 10 to 20 mm ($\frac{2}{5}$ to $\frac{4}{5}$ in.), and full-load deflections similar to that of of the MA 23. A unique feature is the variable-ratio force feedback, which prevents operator fatigue (as demonstrated on existing servo master slaves). Our fatigue test is based on stacking six 5-kg (11-lb) lead bricks into a two-row wall. Then, each brick is moved and the wall is rebuilt 0.5 m (20 in.) away. Each such reconstruction of the six-block wall constitutes one cycle. The cycle time drops at first with experience, then increases as fatigue sets in.

The fatigue limit is defined as the point at which the cycle time is longer than the starting cycle time. Tests on the MA 22 and MA 23 demonstrated that the fatigue limit was three times greater than that of the MA 11, as shown in Table III.

The Digital Multiplex Unit

Development started in 1973. The unit has been tested now on the MA 22 and the MA 23 without noticing any difference in comparison with the direct multiwire communication. The prototype is shown in Fig. 9 (3).

This system works in both directions [see Fig. 8 (10) and (11)] via a single coax tested up to 1 km ($\frac{2}{3}$ mile). Each unit is comprised of two cards for transmission and two cards for receiving, plus logic and power supplies. The master and slave units are strictly identical [see Fig. 3 (8) and (8')]. They are plugged directly into the standard servo-units.

At the present stage, this system has a capacity of two arms plus television pan and tilt drives. There are 16 to 24 analog channels provided for servo-signals, as well as 100 on-off channels. Analog data are sampled 600 times/sec. In case of any malfunction, both slave and master arms

TABLE I

The MA 23 Arm Specifications and Performances

	Master Arm	Standard Slave	HD Slave
Mass capacity, kg (lb)	5 (11)	10 (22)	20 (44)
Maximal force, N (lb)	50 (13)	120 (26)	240 (53)
Friction, N	1.5 to 3	2 to 6	6 to 12
oz	5 to 10	7 to 20	20 to 40
Deflection, mm/N	0.1 to 0.4	0.1 to 0.3	0.04 to 0.14
Full-load deflection, mm	5 to 20	10 to 30	8 to 30
in.	0.2 to 0.8	0.4 to 1.2	0.3 to 1.2
Inertia, kg	2 to 5	2 to 5	5 to 12
No-load accelerations		over 2 g	over 2 g
Payload acceleration —horizontal —against gravity		g to 0.8 g 0.13 g	g to 0.7 g 0.13 g
Velocity, m/sec		0.85 to 1	0.5 to 1
in./sec		34 to 40	20 to 40
Total weight, kg (lb)	90 (200)	90 (200)	180 (400)
Peak power needs, W	500	500	1000

Fig. 9. The MA 23 electronics: (1) master and slave servo-card for one motion; (2) plug-in power amplifier for one torque motor; (3) digital multiplex unit for transmission and reception; identical units at each end; (4) digital recording unit; (5) 3 M Company cassette.

TABLE II

The MA 23 System Performance Data in the Master-Slave Mode

	Master Friction	Full-Load Deflection	Reflected Inertia	Logarithmic Force Feedback Ratio	Temporary 20-min Load	Continuous Duty 60% Duty Cycle	Continuous Duty 100% Duty Cycle
Normal-duty 10 kg slave	2 to 7 N 7 to 23 oz	14 to 56 mm 0.3 to 2.2 in.	3 to 7.5 kg	1/1 at no-load 1/2 at full-load	12 kg 26 lb	10 kg 22 lb	7 kg 15 lb
Heavy-duty 20 kg slave	2 to 7 N 7 to 23 oz	10 to 50 mm 0.4 to 2 in.	3 to 7.5 kg	1/1 at no-load 1/4 at full-load	24 kg 53 lb	20 kg 44 lb	14 kg 31 lb

TABLE III

Lead Brick Wall—Fatigue Test

	Cycle Time (sec) Start	Cycle Time (sec) Best	Cycle Time (sec) Before Stop	Number of Cycles	Work Performed Total Load Moved 0.5 m in Specified Time (min)	Time to Move 1 ton 0.5 m (min)
MA 11 CRL handle	70	60	86	30	1 ton (2000 lb)—36	36
MA 11 PSM handle	50	45	58	49	1.5 tons (3000 lb)—39	26
MA 22 PSM handle MA 23 PSM handle	60	40	60	130	3.9 tons (8500 lb)—100	26

Note: This is executed with two hands.
The force feedback ratio is 1/3 in MA 22 and 1/3 for 11 lb with logarithmic MA 23.

remain at the last memorized transmitted position.

A safety word recognition circuit rejects all bad words, and the system still operates with 50% rejected words. It has been tested in a very high noise environment; i.e., Tungsten Inert Gas welding torch. This unit now will be tested and improved for operation of up to 9 km (6 miles) in single coax for the undersea program.[9]

The Digital Recording Unit

The second version of this unit appears in Fig. 9 (4); the first version appears in Fig. 3 (9). The unit uses a 3 M Company cassette (5) that records more than 20 min for one arm. Programs are recorded at 60 samples/sec on a single track by multiplexing digital position data from the master arm. This is recorded in a real operation, and playback reproduces the same dynamics with the same load. By using the computer-aided system, it will be possible to play back, with any load, a record taken with no load.

Tape recordings already are used in industrial robots; however, in the MA 23 recording from a master arm is applied for the first time. Many improvements are under development by our group to provide a self-adaptive reading in position and velocity.

We are conducting experiments with Metal Inert Gas welding by tape recording. This applies to remote maintenance[10] and to non-nuclear applications. There is now an extensive effort for development in this field.

Computer Control of MA 23 (Ref. 4)

Computer control of the MA 23 is under development at the University of Montpellier, where one standard-duty slave arm is installed [Fig. 3 (10)]. Coupling to a T-1600 minicomputer is operational—position and force feedback of the motions are put into the minicomputer—dynamic control via the complete modeling of the arm is to be tested by the end of 1976.

This allows either accommodation to environmental changes during reading a taped program, or control from non-manual input. By using a keyboard console, the operator can type orders in the arm coordinates or in cartesian space coordinates. Languages have been developed for this, and we are improving them. Because of the performance of the MA 23, computation appears simplified in comparison with open-loop servo arms.[11,12] This program soon will use proximity sensors to automatically pick an object tracked at a limited distance (1 ft).

Standard-Duty Slave-Arm Pressurized Booting Assembly

At the end of 1976, the first test operation will be carried out in the Phenix hot cell.[13] The MA 23 can replace any mechanical master slave and, in addition, can reach inaccessible places. A self-contained ventilated booting assembly has been developed [Fig. 3 (3) and Fig. 4]. The plastic booting is removed to show all the inner equipment.

The arm enters the upper maintenance cell. Hooked onto an attachment (8), it is lowered into the main cell through the transfer slab. A foot (9) allows placement on any horizontal surface. The Ateliers Chantiers de Bretagne (ACB) powered manipulator uses special fingers to grip it on the upper attachment (8) or on the two rear attachments (10). Because of the center of gravity overhang, the 100-kg (220-lb) MA 23 arm needs a 250-kg capacity (550-lb) heavy-duty manipulator.

A blower (11) supplies an air flow of 20 m^3/h (12 ft^3/min) into the booting to cool the manipulator motors. A glove box filter (12) is in the inlet, and four smaller exhaust filters (13) are close to the motors on the rear plate (14).

Fig. 10. The MA 23 test in programmed forging: (1) operator controlling the hammer; (2) hammer; (3) master handle used to remotely operate any cycle to be recorded; now master arm is off; (4) slave arm operating from a recorded program.

The booting of the body is fitted around this plate. Attachment (8) and foot (9) penetrate the plastic booting, while ring (15) protects against the moving counterweights and fits the arm booting for easy exchange. The booting outline is shown by line (16).

Hot-Cell Installation Principles

The first in-cell test in the Phenix hot cell uses the normal-duty arm "handled" by a heavy-duty powered manipulator. Installation in large facilities like waste vitrification, fusion facilities, etc. is based on either a telescopic bridge mounting, a wall-mounted boom, or a vehicle-mounted boom.

CONCLUSION

The MA 23 system now is ready for application. Remote maintenance and inspection is needed in reactors (in particular, remote welding), as well as in reprocessing and waste vitrification. New experimental facilities (fusion) also require remote operation on a large scale for dismantling and repair work (once again, welding is very important).

The modular concept of the MA 23 system uniquely provides for the possible introduction of advanced automation to nuclear remote handling.

APPENDIX—NON-NUCLEAR APPLICATIONS

Advanced Automation

The MA 23 is being used in drop forging experiments to develop an easily programmable robot. Figure 10 shows an experiment conducted in June 1976. The operator (1) is controlling the hammer (2). The master handle (3) remains still, while the slave arm (4) executes a recorded cycle.

A medical project utilizes a standard-duty arm, under computer control, to evaluate an advanced control system for use by quadriplegics. The first medical arm is in the project stage.

Undersea Remote Handling

Figure 5 shows a model of the ERIC II cable-controlled remote vehicle for 6000-m-deep (20,000-ft) submergence, being developed by the French navy. The MA 23 undersea model (1) will look like the heavy-duty arm [Fig. 3 (4)]. The entire arm, except for inside the motor block housings, which are filled with a special liquid under ambient pressure, works at ambient pressure in seawater. For long or delicate trips, the arms are retracted into the body of ERIC II. For two years, material testing has been conducted in both a corrosion and pressure environment. This year, construction with titanium castings and special bearings will start. The first tests are expected in 1978.

ACKNOWLEDGMENTS

This work is sponsored by the French Commissariat à l'Energie Atomique. The authors especially acknowledge the Delegation Centrale Sécurité, Inspection Diversification, Délégation Diversification.

We especially thank the Société La Calhène, the Université des Sciences et Techniques du Languedoc, the Laboratorie d'Automatique (LAM), the Centre d'Etudes et de Recherches de Techniques Sous-Marines (CERTS), and the Ecole Supérieure d'Ingénieurs en Electronique et Electrotechnique (ESIEE) for their cooperation.

There are many more outstanding participants in this project than there were authors. We especially thank Mm. Collombat, Germond, Gil, Grenier, de Montaignac, and all of the machine shop staff under M. Gre in the Section Equipement pour Milieux Hostiles, Mm. Brossard, Cazalis, Glachet in La Calhène, Mm. Dombre, Fournier, Khalil, Liegeois, Molinier in LAM, and twelve students in the ESIEE.

REFERENCES

1. J. VERTUT et al., "Variable Geometry Wheeled Teleoperator," *Proc. 20th Conf. Remote Syst. Technol.*, 303 (1972).

2. J. VERTUT, "Virgule—A Rescue Vehicle of a New Teleoperator Generation," 2nd Conf. Industrial Robots, Birmingham, Dec. 3-21, 1974.

3. J. VERTUT et al., "MA 22 Compact Bilateral Servo Master Slave Manipulator," *Proc. 20th Conf. Remote Syst. Technol.*, 296 (1972).

4. J. VERTUT et al., "Bilateral Servo Manipulator MA 23 in Direct Mode and Via Optimized Computer Control," 2nd Remotely Manned Syst. Conf. (1975); to to be published in *Mechanism and Machine Theory*.

5. J. VERTUT et al., "Through the Wall Master Slave Manipulator with Indexing for Backward and Forward Movement," *Proc. 12th Conf. Remote Syst. Technol.*, 67 (1964).

6. R. C. GOERTZ et al., "The ANL Model 3 Master Slave Electric Manipulator. Its Design and Use in a Cave," *Proc. 9th Conf. Hot Lab. Equip.*, 121 (1961).

7. R. C. GOERTZ et al., "ANL E4A Electric Master Slave Manipulator," *Proc. 14th Conf. Remote Syst. Technol.*, 115 (1966).

8. M. GALBIATI et al., "A Compact Flexible Servo System for Master Slave Electric Manipulator," *Proc. 12th Conf. Remote Syst. Technol.*, 73 (1964).

9. J. CHARLES et al., "ERIC II Deep Submergence Cable Controlled Teleoperator," 2nd Remotely Manned Syst. Conf. (1975); to be published in *Mechanism and Machine Theory*.

10. T. RAIMONDI, "Remote Handling in the Joint European Torus (JET) Fusion Experiment," *Proc. 24th Conf. Remote Syst. Technol.*, 188 (1976).

11. P. COIFFET et al., "Mechanical Design and Computer Configuration in the Computer Aided Manipulator Control Problem," 12th Annual Conf. Manual Control, University of Illinois, Urbana, Illinois (1976).

12. P. COIFFET et al., "Computer Aided Control of Force Reflecting Manipulators," VIIe IFac Symp.: Automatic Control in Space, Preprints Vol. 2, 724 (1976).

13. L. THEVENOT et al., "The Phenix Fast Breeder Reactor Fuel Dismantling Hot Cell," *Proc. 24th Conf. Remote Syst. Technol.*, 219 (1976).

Presented at the American Nuclear Society 25th Conference on Remote Systems Technology, 1977

SM-229–A NEW COMPACT SERVO MASTER-SLAVE MANIPULATOR

CARL R. FLATAU *TeleOperator Systems Corporation*
Bldg. 8, Flowerfield Industrial Park
Mill Pond Road, St. James, N.Y. 11780

KEYWORDS: *manipulators, remote handling, servosystem*

ABSTRACT

A new compact force-reflecting servo master-slave manipulator, the SM-229, has been developed to serve the requirements of a variety of new installations. Attention has been given not only to the performance of this new manipulator, but also to commercial production requirements so that its availability to various projects will be much improved. The first pair of SM-229 have been produced and satisfactorily tested for use in the Los Alamos Scientific Laboratory-Meson Physics Facility.

INTRODUCTION

The need for a compact and extremely dexterous manipulator with excellent mobility potential has been discussed previously and repeatedly.[1,2] Since then, progress in other fields has made this need more urgent.[3] In the past, with the exception of a few laboratory models of servo master-slave manipulators, these needs remained largely unfilled. We had to rely on either mechanical master-slave manipulators with their restrictive volume coverage or open loop manipulators with their low dexterity. So we were always faced with the dilemma of having either high dexterity with minimal mobility, or high mobility with minimal dexterity.

The SM-229 was conceived as the first member of a family of bilateral force-reflecting manipulators to break this dilemma by offering high-dexterity, high-mobility potential in one compact device that is both producible and maintainable. In addition, it was designed to produce all of this in one clean package, devoid of external mechanisms that, at best, are a liability in remote operations.

AREAS OF APPLICATION

Some of the fields in which this lack has been apparent include: particle accelerators, experimental fusion reactors, reactor fuel fabrication and fuel development, certain reactor maintenance tasks, and a number of new fields. These applications all have the following common characteristics, governing the choice of manipulator to be used:

1. The installations typically cover quite large volumes (areas) in which remote work must be done. The volumes are far in excess of those that can reasonably and economically be covered by mechanical master-slave manipulators.

2. The installations are usually very costly, requiring handling task time to be minimized. Since the tasks required often are also intricate, there is dual motivation for use of a high dexterity manipulator.

3. Despite the large operating volumes involved, little room remains for manipulators, since equipment fills much of the available spaces.

DESIGN CONSTRAINTS

The above characteristics lead to the following manipulator design constraints:

1. utmost available dexterity

2. maximum mobility potential over large volumes

3. very compact design for work in crowded quarters
4. better than commonly available articulation because of crowded quarters
5. clean external shape for ease of operation in crowded quarters
6. electrical actuation to prevent residual fluid leakage from hydraulic systems
7. relatively modest capacity (about that of a Model 8 manipulator)
8. good producibility and maintainability.

Examination of these constraints shows that most of them would probably not be a liability for any other application. A possible exception might be the need for increased capacity for some applications, with a possible loss of compactness.

DEGREES OF FREEDOM

The choice of kinematic arrangement was largely restricted by the decision to use conventional wrist and tong arrangements, thus improving producibility, reducing development risk, and presenting the user with familiar interfaces. Further analysis shows that this decision, and the restriction to seven degrees of freedom with articulated design, reduces the options for lateral movement to only two. One of these was used in the Brookhaven manipulator,[2] where lateral movement resulted from rotation about a vertical axis at the shoulder, while the second was used in most other electric master-slave manipulators,[4] where this movement results from rotation of the upper arm. In this case, the latter arrangement was chosen since it yielded a somewhat superior design arrangement. The arrangement is illustrated in Fig. 1.

SIZE AND RANGES OF MOTION

The compactness requirements dictated minimal upper and lower arm length. Since we wanted master and slave to use a maximum of common parts, the minimal usable lengths were strongly influenced by workability of the master. The upper arm is 457 mm (18 in.) long and the lower arm is 559 mm (22 in.) long. Even with these relatively short lengths, the volume coverage achieved without external mobility is ~3.7 m³ (130 ft³). This is because of exceptionally large articulation achieved in the manipulator, particularly the elbow motion whose range is 3.66 rad (210 deg). This large range of travel is possible because of the unique guide pulley arrangement and despite the fact that tapes and cables for wrist

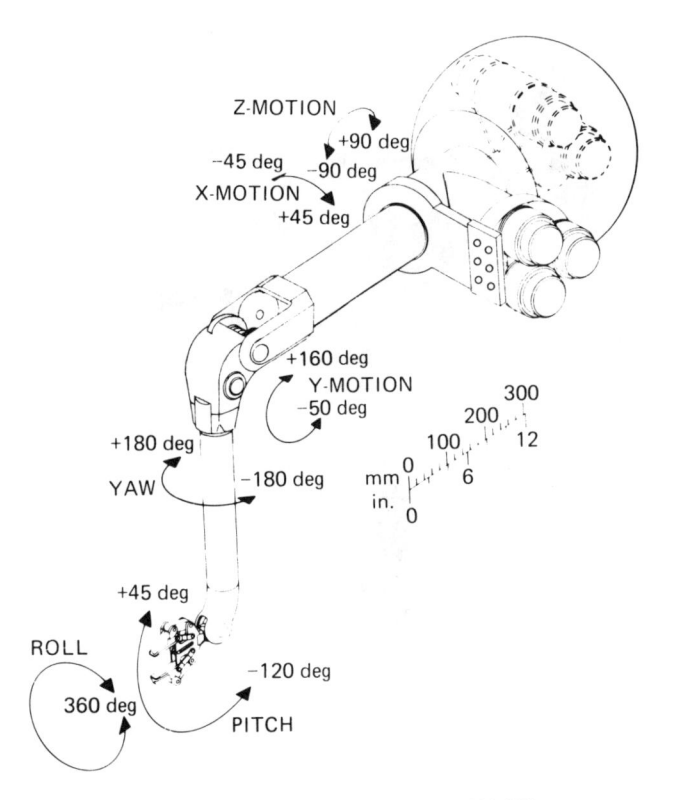

Fig. 1. Ranges of motion for the SM-229.

and tong actuation pass through the elbow. All the motion ranges are shown in Fig. 1. The weight of the slave arm is ~40 kg (90 lb), which is much less than the weight of any other electric master-slave manipulator.

DRIVE ARRANGEMENTS

Each of the seven degrees of freedom has a separate servosystem, except in two cases where two servosystems cooperatively drive two motions so that a servosystem drives one of the two inputs to a differential. Each servosystem consists of a motor on the master and slave, a position transducer on the master and slave, and gear reductions and drive connections on the master and slave. The motors are similar to those described in Ref. 5, but are refined somewhat further. The servosystem is a conventional position-position system refined in detail rather than by using a new overall scheme.

The seven drive systems are divided into two groups. One group of three has motors and gear reducers situated in the shoulder box driving the X, Y, and Z motions (Fig. 2). Note that all three motors are stationary. The other group of four drives are in the counterbalance assembly driving

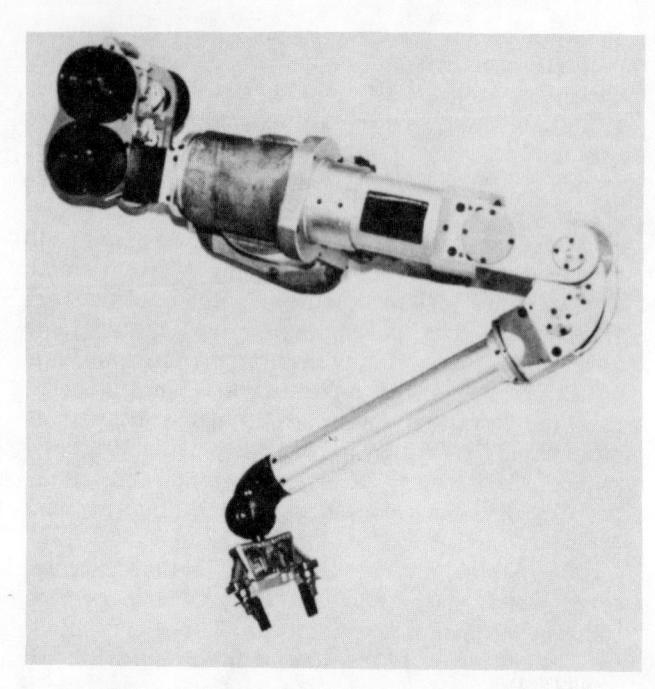

Fig. 2. Overall view of the SM-229 slave arm.

the tong, the two wrist motions, and the yaw (azimuth) motion.

The master and slave are essentially identical, with the exception of the tong or master handle, respectively, and the overall gear ratio, which is a factor of 2.2 lower in the master than in the slave. This ratio change is always done in the first stage, with center distances remaining the same in master and slave. This approach was chosen to maximize part similarity for ease of production.

THE COUNTERBALANCE ASSEMBLY

The X, Y, and Z motions are counterbalanced by an articulated counterbalance assembly that contains the four motors and gear reduction units driving tong, wrist, and yaw motions. This is not a new approach; it was used in MA-22. However, packaging is much refined, so neither excessive weight nor inertia are associated with this scheme.

Four motors and gear reductions in the counterbalance assembly furnish the mass required to

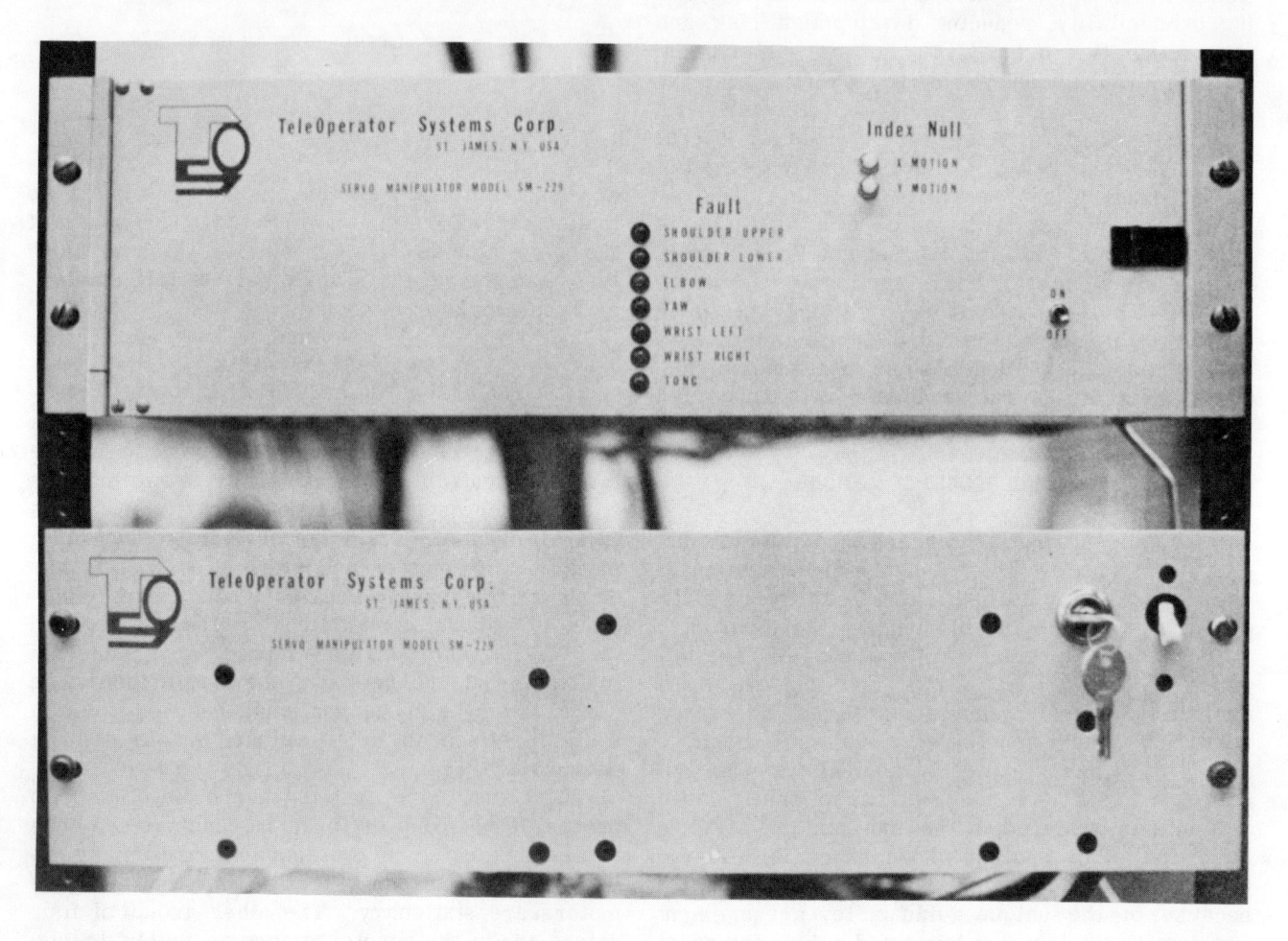

Fig. 3. Control panel for the SM-229.

exactly counterbalance the manipulator. This approach contributes substantially to the low weight achieved.

For the counterbalance to function properly, it is driven through the same angles as the forearm by a cable drive contained in the upper arm. This angular alignment between the forearm and counterbalance assembly is used to compensate the tape drives coming out of the counterbalance assembly into the forearm. This unique method allows unusually large travel of the forearm.

THE SHOULDER DRIVE

The three motors driving the X, Y, and Z motions in the shoulder form a stationary group in the shoulder casting. The two motors in front drive the X and Z motion through a shoulder differential. Motivation for using such a device is twofold: first, it allows these two motions to be driven cooperatively from two stationary motors, and second, this cooperative driving arrangement allows the overall gear ratio for these two motions to be substantially smaller than individual drives would allow, thus reducing both friction and inertial forces perceived by the operator. This is possible because the two motions in question are mutually perpendicular, and no conceivable force vector of magnitude within rated capacity can simultaneously require a maximal torque on these two motions. With the differential drive and high torque of the motors, the highest gear ratio associated with any one motion is as low as 38.4:1.

The shoulder differential is a bevel gear differential with the bevel sun gear situated coaxial to the Z motion, or shoulder pitch motion shaft. When these two bevel gears are rotated in the same directions, the planet bevel does not rotate on its own axis, but orbits in the Z motion direction and actuates that motion. However, with the sun bevel rotating in opposite directions, the planet will rotate on its own axis and actuate the X motion, or shoulder roll motion, via transfer sector gears.

The elbow, or Y, motion is actuated from a third motor in the shoulder box, driving the elbow through a shaft coaxial to the Z motion shaft via two bevel gear pairs and a shaft coaxial with the upper arm. The servosystem completes the scheme for the third member of a three-way differential. Thus, a clean arm devoid of external push rods, cables, or other undesirable protuberances has been achieved.

TABLE I

Specifications—Dynamic Characteristics of Model SM-229

	X	Y	Z	Yaw	Pitch	Roll	Terminal Device
Maximum moment arm, in.	25	25	45	6.25	5.5	4.5	---
Maximum available torque or force at slave	575 in.-lb	575 in.-lb	995 in.-lb	140 in.-lb	130 in.-lb	110 in.-lb	32 lb
Maximum perceived friction at master[a]	6.0 oz	6.0 oz	6.0 oz	60 in.-oz	55 in.-oz	55 in.-oz	---
Typical apparent no-load mass or inertia at accelerations 0.3 g[a]	5.5 lb	6.0 lb	6.5 lb	200 lb-in.	135 lb-in.	100 lb-in.	30 lb
Typical apparent no-load mass or inertia at high accelerations (3 g)[a]	3.5 lb	4.0 lb	4.5 lb	130 lb-in.	75 lb-in.	70 lb-in.	20 lb
Typical maximum no-load linear or angular velocity[a]	36 in./s	33 in./s	36 in./s	4.4 rad/s	8.7 rad/s	9.8 rad/s	10 in./s
Typical compliance of complete system rad/in.-lb[b]	6.5×10^{-5}	8.5×10^{-5}	5.5×10^{-5}	2.5×10^{-4}	2.7×10^{-4}	3.5×10^{-4}	---
Typical motion range[c]	±45 deg	+160 deg / −50 deg	±90 deg	360 deg	165 deg	360 deg	3.25 in.

[a] For x, y, and z motions perceived friction, velocity, and apparent mass are a function of the angle the lower arm makes with respect to the upper arm. The numbers given are for the maximum velocity position.
[b] At maximum load.
[c] Referenced from a vertical position to the forearm and horizontal position of the upper arm: (+) = up, (−) = down.

CONTROLS

The servosystem used on the SM-229 consists of conventional position-position loops for each of the degrees of freedom. For applications requiring extra high sensitivity, this system will be augmented with discrete force transducers.

The following startup and fail-safe procedures are incorporated. For the startup procedure, the slave units remain braked and without excitation until the master has drifted into synchronism in a special low gain mode. When synchronism is achieved, the "enable" condition exists for the operator to release the brakes. In this way, the unit can be stopped and braked in any condition and restarted without involuntary movement of the slave arm.

For fail-safe operation, proper synchronism is sensed continuously. If the allowed synchronization error is exceeded, a warning light for the motion in question is lit. If the error does not correct itself, all brakes are automatically applied. A fault light will also be activated to aid malfunction diagnosis.

The controls for master and slave are each housed in a 133-mm ($5\frac{1}{4}$-in.) × 432-mm (17-in.) module that is augmented by power supplies (Fig. 3). The upper drawer contains amplifiers and the lower drawer has auxiliary power supplies.

PERFORMANCE

The pertinent and customarily quoted performance parameters are listed in Table I.

CONCLUSIONS

The SM-229 is the first in a family of new, compact, electric master-slave manipulators designed to be relatively easily manufactured and clean in shape. Room is left for further development and expansion of the model line. We hope to report on this expansion and application experience in the future.

ACKNOWLEDGMENTS

We would like to acknowledge the contribution of Vincent Kovarik for doing the electronics, and the many designers and machinists who helped immensely under difficult conditions. We would also like to express our appreciation to the staff at Los Alamos Scientific Laboratory–Meson Physics Facility. Their cooperation was very helpful.

REFERENCES

1. C. R. FLATAU, "Development of Servo Manipulators for High-Energy Accelerator Requirements," *Proc. 13th Conf. Remote Syst. Technol.*, 29 (1965).

2. C. R. FLATAU, "Compact Servo Master-Slave Manipulator with Optimized Communication Links," *Proc. 17th Conf. Remote Syst. Technol.*, 154 (1970).

3. T. RAIMONDI, "Remote Handling in the Joint European Torus (JET) Fusion Experiment," *Proc. 24th Conf. Remote Syst. Technol.*, 188 (1976).

4. R. C. GOERTZ et al., "ANL Mark E4A Electric Master-Slave Manipulator," *Proc. 14th Conf. Remote Syst. Technol.*, 115 (1966).

5. C. R. FLATAU et al., "MA22—A Compact Bilateral Servo Master-Slave Manipulator," *Proc. 20th Conf. Remote Syst. Technol.*, 296 (1972).

Reprinted from the Proceedings of the 1984 National Topical Meeting
on Robotics and Remote Handling in Hostile Environments,
American Nuclear Society, pages 147-154

KEY WORDS: control, manipulator, remote maintenance, servomanipulator, television

The State-of-the-Art Model M-2 Maintenance System*

J. N. Herndon
Fuel Recycle Division
Oak Ridge National Laboratory
Oak Ridge, Tennessee
615-574-4302

D. G. Jelatis and C. E. Jennrich
Sargent Industries, Incorporated
Central Research Laboratories
Red Wing, Minnesota
612-388-3565

H. L. Martin and P. E. Satterlee, Jr.
Instrumentation and Controls Division
Oak Ridge National Laboratory
Oak Ridge, Tennessee
615-574-6336 and 615-574-5685

ABSTRACT The Model M-2 Maintenance System is part of an ongoing program within the Consolidated Fuel Reprocessing Program (CFRP) at Oak Ridge National Laboratory (ORNL) to improve remote manipulation technology for future nuclear fuel reprocessing and other remote applications. Techniques, equipment, and guidelines which can improve the efficiency of remote maintenance are being developed. The Model M-2 Maintenance System, installed in the Integrated Equipment Test (IET) Facility at ORNL, provides a complete, integrated remote maintenance system for the demonstration and development of remote maintenance techniques. The system comprises a pair of force-reflecting servomanipulator arms, television viewing, lighting, and auxiliary lifting capabilities, thereby allowing manlike maintenance operations to be executed remotely within the remote cell mockup area in the IET.

The Model M-2 Maintenance System incorporates an upgraded version of the proven Central Research Laboratories' Model M servomanipulator. Included are state-of-the-art brushless dc servomotors for improved performance, remotely removable wrist assemblies, geared azimuth drive, and a distributed microprocessor-based digital control system.

*Research sponsored by the Office of Spent Fuel Management and Reprocessing Systems, U.S. Department of Energy, under contract No. W-7405-eng-26 with Union Carbide Corporation.

By acceptance of this article, the publisher or recipient acknowledges the U.S. Government's right to retain a nonexclusive, royalty-free license in and to any copyright covering the article.

INTRODUCTION

A major goal of the CFRP at ORNL is to improve remote manipulation technology for future nuclear fuel reprocessing and other remote applications. Techniques, equipment, and guidelines that can improve the efficiency of remote maintenance are being developed. This goal is supported by the application of force-reflecting servomanipulators to provide manlike manipulative capabilities for performing maintenance tasks in the remote environment. The Model M-2 Maintenance System, installed in the remote cell mockup area of the Remote Operations and Maintenance Demonstration (ROMD)[1] Facility within the IET at ORNL, is an important demonstration tool for these remote concepts. The M-2 system was developed as a cooperative project between Central Research Laboratories (CRL) and ORNL. The CRL designed the mechanical systems and motor/amplifier components and performed the overall system fabrication; ORNL designed the distributed microprocessor-based control system and all system software.

SYSTEM DESCRIPTION

The Model M-2 Maintenance System consists of two major assemblies: (1) the slave package, which is the "in-cell" working portion of the system, and (2) the master station, where the human operator interfaces with the system to execute and monitor maintenance tasks. The slave package, shown in Fig. 1, incorporates a pair of CRL Model M-2 force-reflecting servomanipulator slave arms for performing in-cell dextrous manipulative tasks. Each slave arm has a 23-kg (50-lb) continuous capacity and a 46-kg (100-lb) time-limited (peak) capacity. An auxiliary hoist with a 230-kg (500-lb) capacity is

Fig. 1. Model M-2 slave package.

located between the two slave arms to perform the heavier lifts associated with reprocessing plant-type equipment maintenance. These features allow the servomanipulator arms to be used for "man-like" tasks such as positioning and operating tools, lifting light (less than 23 kg) loads, and guiding heavy loads lifted by the auxiliary hoist.

In order to provide operator viewing at the remote task site, one center and two overhead television cameras are provided. Each overhead camera is mounted on a pan-and-tilt mechanism located at the end of a positioning arm with extend/retract and rotate motions. The camera positioning flexibility provided by these four motorized degrees of freedom allows the operator to select orthogonal camera views, thereby optimizing viewing perspective and enhancing depth cues. Each overhead camera also has a motorized zoom lens. The center camera is fixed and equipped with a wide-angle lens to allow general viewing of the manipulator tongs at the task site.

A slave package azimuth drive provides 540° of rotation for orientation of the overall system at the task site. A convenient attach/detach capability enables the slave package to be mounted on different transporter systems.

The master station, shown in Fig. 2, incorporates a pair of CRL Model M-2 force-reflecting servomanipulator master arms for operator input. Each master arm has a 11-kg (25-lb) peak capacity. Switches on the master handles allow for slave tong lock, slave arm lock, master-to-slave "all joint" indexing, tool control, and a unique deadman function. These functions are described in the section on servomotors and controls.

The operator interfaces with the control system primarily through a CRT and touch-screen system. Operating mode selection, force ratio selection, camera/lighting control, and system status diagnostics are provided by this operator interface. This interface is described in depth in the servomotors and controls section.

MECHANICAL SYSTEM

The Model M-2 servomanipulator arms are based on a proven servomanipulator design that has been upgraded with state-of-the-art motor technology. The master and slave arm designs are based on the CRL Model M servomanipulator developed in the early 1960s at Argonne National Laboratory as the ANL Mark E4A.[2,3] The original Mark E4A kinematics, shown in Figs. 3 and 4, are for the most part retained. The slave and master arms each have the standard seven degrees of freedom (including grasp) arranged in the "over the wall" or "elbows up" configuration as shown in Figs. 1 and 2. The motions at the master and at the slave are in one-to-one correspondence.

Each master and each slave joint are driven by their own individual servomotor drive units. Motions in the X, Y, and Z directions on both master and slave are gear and linkage driven. Eccentric mounts allow adjustment of backlash and friction. The slave arm azimuth, elevation, twist, and tong motions are driven by a combination of the servomotor drive unit gearboxes and 1/16-in. stainless steel cables. All slave servomotor drive units are located behind the Z-motion axis.

Each servomotor drive unit incorporates a brushless dc servomotor, encoding sensors, brake, cooling fan, and enclosed gearbox. The servomotor drive units are easily removed and replaced as an assembly.

The slave cable-driven motion gearbox drums are located outside of the shoulder assembly (as shown in Fig. 5), which allows ease of maintenance for the cable drives. The cables are routed down the outside of the shoulder,

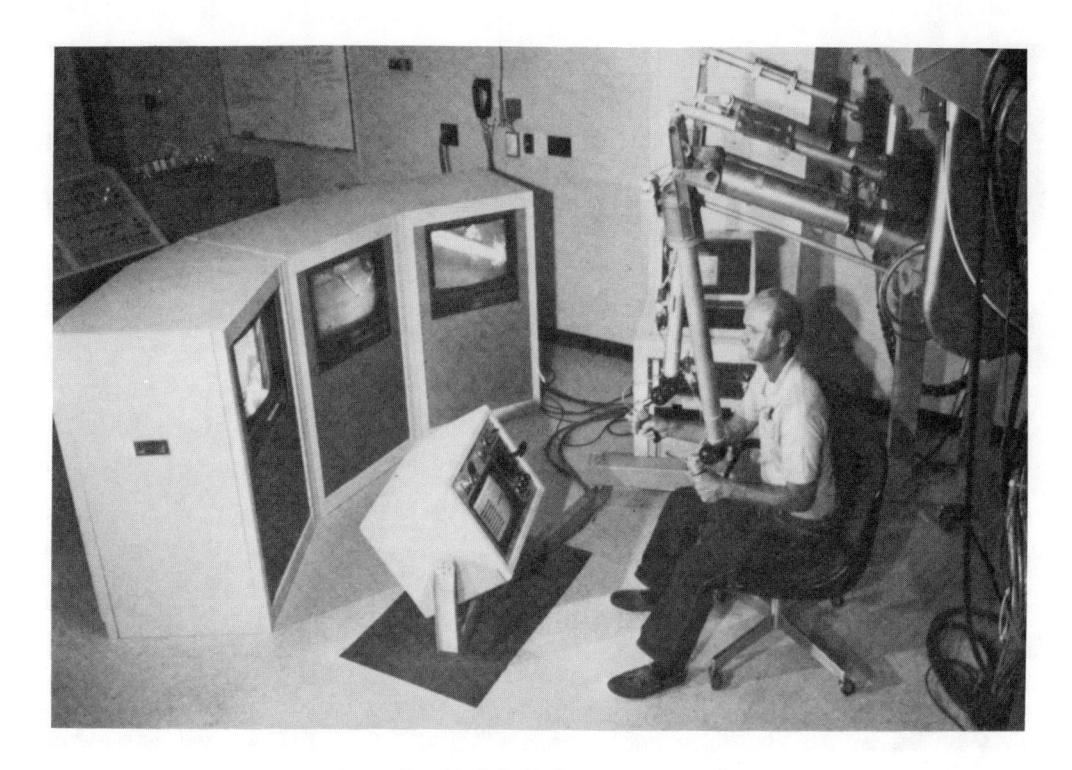

Fig. 2. Model M-2 master station.

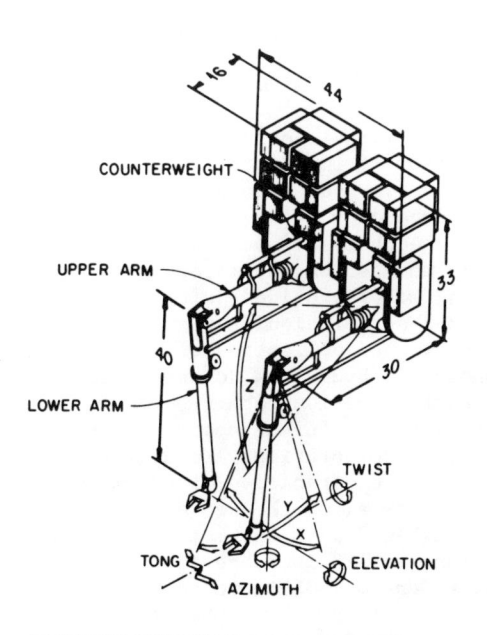

ARGONNE MARK E4A SLAVE MANIPULATORS

ARGONNE MARK E4A MASTER MANIPULATORS

Fig. 3. Argonne Mark E4A slave manipulators.

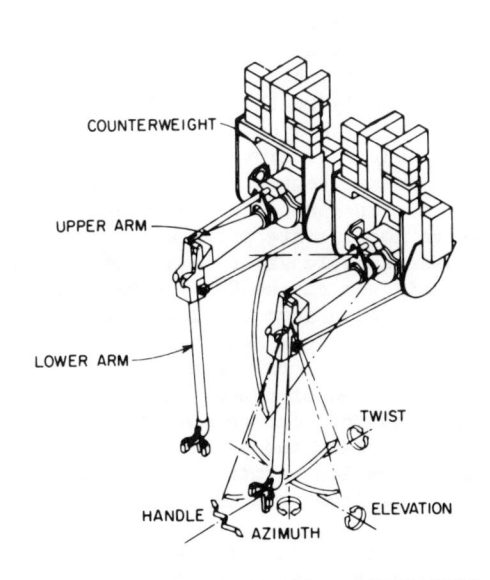

Fig. 4. Argonne Mark E4A master manipulators.

Fig. 5. Model M-2 slave servomotor drive units.

through the center of the Z-motion axis, and then down the center of the slave arm.

In order to upgrade to the peak 46-kg capacity, the slave arm azimuth drive has been modified. The former azimuth cable drum, which attached directly to the lower arm tube, has been replaced with a cable drum and pinion assembly mounted external to the lower arm tube driving a gear attached to the lower arm tube. The standard mechanical master/slave roll-pitch-roll wrist motions are provided. The heavy-duty, remotely detachable wrist joint and gear-driven tongs, available on the CRL system 50 manipulators, have been incorporated on the Model M-2 slave. This remotely detachable wrist joint assembly allows removal of the wrist joint and tong (the portion of the slave arm most vulnerable to mechanical damage and contamination) without disturbing the slave cable drives. The slave drive cables terminate in drums near the bottom of the lower arm tube.

The Model M-2 master arms, shown in Fig. 2, were originally designed with very light structures to minimize effective inertia.[2] This feature has been retained. The X, Y, and Z motions, as indicated earlier, use gear and linkage drives for force transmission. The master arm wrist azimuth, elevation, twist, and squeeze motion force transmission is accomplished using steel tapes. Tapes are used to reduce friction for better sensitivity.

SERVOMOTORS AND CONTROLS

The original design for the Model M servomanipulator[2,3] incorporated two-phase, ac servomotors. Four motors were used for each slave servomotor drive unit, and two motors were used for each master servomotor drive unit. A synchro mounted on one motor and a tachometer on a second motor provided joint encoding. The Model M-2 servomanipulator arms have been upgraded by the application of modern motor technology. Brushless dc servomotors with rare-earth samarium-cobalt permanent magnets, manufactured by Inland Motor Division, Kollmorgen Corporation, are utilized. The higher torque capacities of the master and slave motors allow the servomotor drive units to be packaged with one motor each. In addition, an 11% increase in continuous capacity and a 100% increase in peak capacity have been attained over the original design. The X, Y, and Z motion maximum velocities are more than double those of the earlier design. The velocity signal for the servomotor drive unit is provided by a tachometer and position is provided by a digital encoding technique.

Table 1. Performance and operating characteristics of Model M-2 servomanipulators

	X (shoulder roll)	Y (elbow pitch)	Z (shoulder pitch)	AZ (yaw)	Elevation (pitch)	Twist (roll)	Tong
Master maximum torque or force	111 N	111 N	111 N	17 N·m	15 N·m	12 N·m	111 N
Slave maximum continuous operating torque or force	222 N	222 N	222 N	56 N·m	47 N·m	40 N·m	445 N Squeeze
Peak slave torque or force	445 N	445 N	445 N	108 N·m	90 N·m	77 N·m	667 N Squeeze
Maximum no-load linear or angular velocity	>1.5 m/s	>1.5 m/s	>1.5 m/s	>344 deg/s	>400 deg/s	>344 deg/s	>1.0 m/s
Slave arm motion range	±45°	±45°	±45°; 255° total with indexing	±210°	+40°; −125°	±180°	Tong 0.08 m; handle 0.07 m

Table 1 summarizes of the dynamic performance characteristics of the Model M-2 servomanipulator arms.

The Model M-2 Maintenance System incorporates a distributed, microprocessor-based digital control system. This system represents the first successful implementation of an entirely digitally controlled servomanipulator and is a significant advance in servomanipulator technology. This control system is described in depth in refs. 4 and 5. Many benefits are derived from use of digital control techniques. First, hardwired analog control designs for the M-2 maintenance system would have resulted in a cable bundle requiring approximately 300 conductors between masters and slaves. Cable handling for a bundle of this size over a large-volume remote application such as a reprocessing plant would be very difficult at best. Using a digital serial link reduces the signal transmission requirements of the M-2 system to four coaxial cables for control and three coaxial cables for television signals.

Second, the digital control system is inherently flexibile by virtue of its software implementation. In addition to the standard master/slave mode, the M-2 servomanipulator has the capabilities for:

- selective robotic operations;
- multiple slave-to-master force ratios with individual joint tuning for each force ratio;
- unique "all-joint" index of master relative to slave, which allows a position offset between master and slave arms for reduced operator fatigue when operating near extremes in the slave coverage range;
- unique controlled low-power and low-velocity synchronization of master arm to slave prior to entering master/slave mode; and
- deadman switch on master grips which removes power from the master arms when the operator removes his hand.

Third, digital electronics designs are generally less susceptible to drift and thermal stability problems than are analog systems and so have reliable, enhanced performance. Finally, real-time diagnostics that report to the operator the occurrence and location of faults are incorporated in the Model M-2 Maintenance System. This allows the operator to modify his actions to prevent a full system shutdown and aids in the location of failures when they occur.

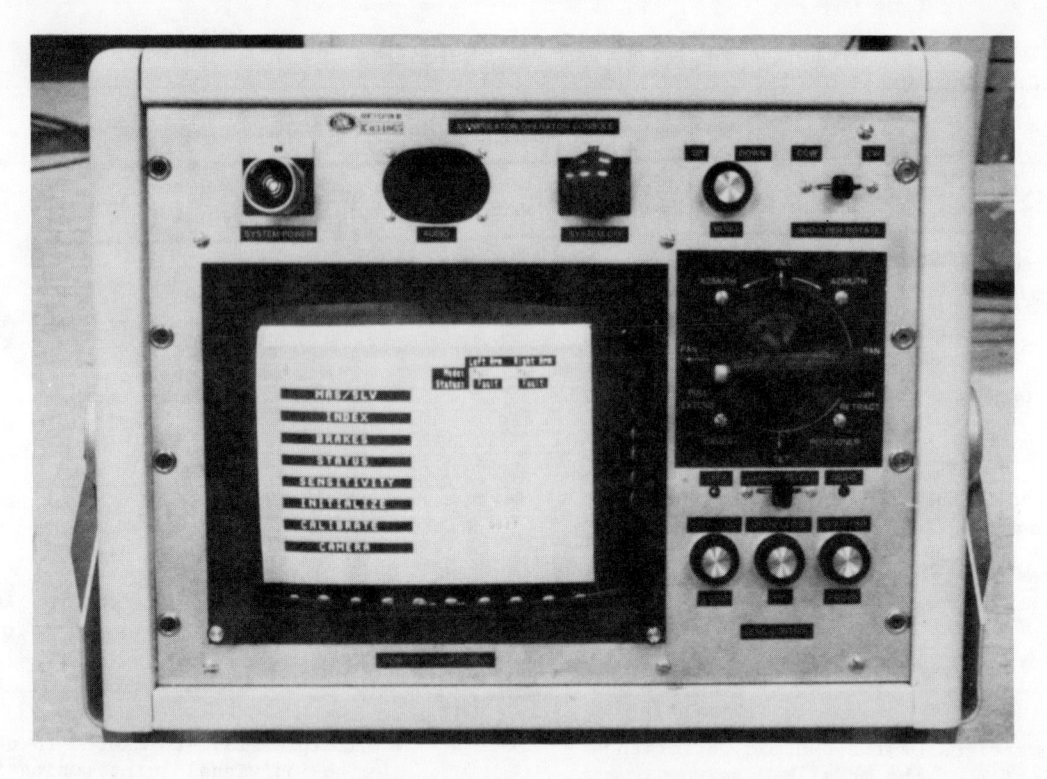

Fig. 6. Model M-2 operator console.

An additional benefit of the digital control system is the ability to implement a menu-driven, touch-screen/CRT operator interface. The operator console is shown in Fig. 6. All operating mode selections, force ratio selection, camera/lighting power, automated slave position calibration, and system status reporting are controlled through the touch-screen interface. The menu-driven system reduces operator training time and panel requirements. Experienced mechanical master/slave operators can learn to execute very complex maintenance tasks with little initial training on the M-2 system. Camera positioning is controlled with a four degrees-of-freedom joystick. Lens zoom, focus, and iris are controlled with rotary potentiometers.

IET INSTALLATION

The Model M-2 slave package is installed on an overhead three degrees-of-freedom positioning telescoping tube trolley in the ROMD remote cell mockup area within the IET facility, as shown in Fig. 7. This portion of the IET facility contains prototypical reprocessing equipment and systems designed to be operated and maintained remotely, and it is here that servomanipulator-based remote maintenance can be demonstrated and evaluated. Remote handling features of prototypical reprocessing equipment can be tested,

Fig. 7. ROMD transporter system.

and equipment mean-time-to-repair data for plant design applications can be acquired. The evaluation of these features and capabilities can be used to design future maintenance systems.

The Model M-2 master station is installed in the ROMD control room adjacent to, but separate from, the cell mockup area. The manipulator operator can view remote maintenance tasks on three color television monitors (Fig. 2). The operating team consists of two operators, as shown in Fig. 8. The primary operator is responsible for performing the dextrous maintenance tasks with the servomanipulators, and the secondary operator is responsible for controlling the telescoping tube transporter, overhead 10-ton cranes, and television monitor matrix switching. Either operator can control camera position and the auxiliary hoist.

OPERATING EXPERIENCE

The Model M-2 Maintenance System in the IET facility has performed very successfully. Servomanipulator-based remote maintenance has been demonstrated as practical for very complex remote operations. The remote maintenance capabilities and flexibility of servomanipulators have been demonstrated in successful maintenance campaigns on a solvent extraction centrifugal contactor module and on a mechanically complex automated remote sampler vehicle. Experienced mechanical master/slave operators can successfully execute these very complex tasks after only minimal training.

A technically revealing Model M-2 activity was the successful implementation of a finite-difference approximation of joint velocity from position data. This allowed elimination of the tachometer velocity signal from the control loops. The replacement of the noisy tachometer velocity signal with the estimated velocity signal permitted larger control loop gains, which in turn significantly increased the overall manipulator force sensitivity, and eliminated problems with thermal-related tachometer drift.

Future plans for the Model M-2 Maintenance System include (1) additional maintenance campaigns on prototypical reprocessing equipment; (2) software development and refinement including selective robotic slave operations and automatic camera tracking routines, to enhance operator performance; and (3) various evaluations of the operating characteristics and features of the servomanipulator and their effects on the human operator.

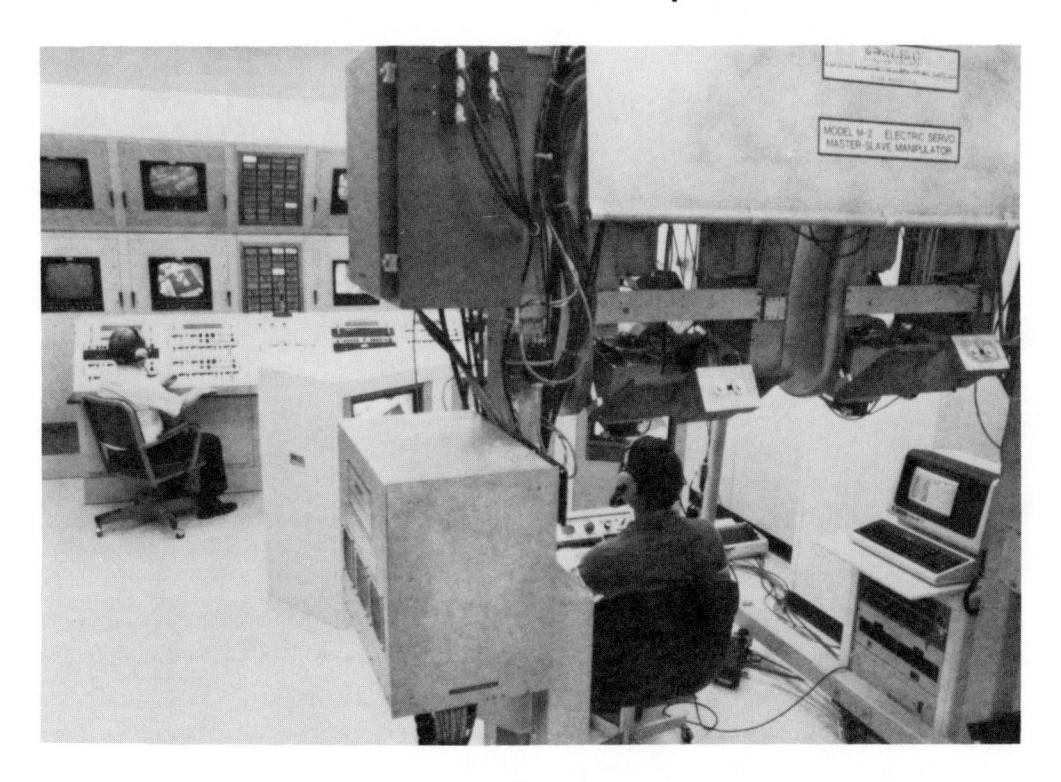

Fig. 8. ROMD control room.

REFERENCES

1. H. W. Harvey et al., "The Remote Operation and Maintenance Demonstration Facility at ORNL," Proceedings of the 26th Conference on Remote Systems Technology, American Nuclear Society, 1978.

2. Ray Goertz et al., "ANL Mark E4A Electric Master/Slave Manipulator," Proceedings of the 14th Conference on Remote Systems Technology, American Nuclear Society, 1966.

3. Demetrius G. Jelatis et al., "A TRIAC Output Control System for Two-Phase Servodriven Master/Slave Manipulator," Proceedings of the 25th Conference on Remote Systems Technology, American Nuclear Society, 1977.

4. H. L. Martin et al., "Distributed Digital Processing for Servomanipulator Control," Proceedings of the 30th Conference on Remote Systems Technology, American Nuclear Society, 1982.

5. P. E. Satterlee, Jr. et al., "Control Software Architecture and Operating Modes of the Model M-2 Maintenance System," Proceedings of the ANS Topical Meeting on Robotics and Remote Handling in Hostile Environments, April 1984.

Presented at the American Nuclear Society 26th Conference on Remote Systems Technology, 1978

MINIMAC—THE REMOTE-CONTROLLED MANIPULATOR WITH STEREO TELEVISION VIEWING AT THE SIN ACCELERATOR FACILITY

EYKE WAGNER and ALBIN JANETT
Schweizerisches Institut für Nuklearforschung (SIN)
CH-5234, Villigen, Switzerland

KEYWORDS: *remote handling, viewing, maintenance*

ABSTRACT

The Schweizerisches Institut für Nuklearforschung, an accelerator facility with high proton beam intensity, has acquired a special piece of equipment with which rescue operations can be carried out in highly irradiated areas in case of unforeseen events. The MINIMAC is a one-arm remote-controlled power manipulator without force feedback. It is operated under surveillance by a television system consisting of a main working camera with changeable stereoscopic or monoscopic (three-dimensional/two-dimensional) high-resolution imaging and some additional surveillance television cameras of standard quality. Experience in handling improvements in the adjustment of the stereo controls and an electronically controlled impact wrench are described.

INTRODUCTION

The Schweizerisches Institut für Nuklearforschung (SIN) is a national center for basic research in nuclear and particle physics and their applications.

The current accelerator system is capable of providing high-intensity proton currents of up to 100 µA and consists of an injector cyclotron for the preacceleration of protons to 70 MeV, and a ring accelerator—an isochronous cyclotron of novel concept—for final acceleration of protons to 600 MeV.

As a result of the high proton beam intensity, the activation of material in the area of main proton beam losses is extremely high and does not allow direct access of personnel even when the beam is turned off. Radiation levels on the order of 100 to 1000 R/h are measured 1 m (3.05 ft) from the source. These areas are surrounded by local shielding, and for all predictable maintenance in the high-beam loss regions, special tools and equipment are provided.

In case of unforeseen events that call for removing radioactive material, MINIMAC, the manipulator unit, can be used anywhere in the experimental hall where speed and precision have second priority compared with high versatility.

MINIMAC

The MINIMAC is a manipulator unit consisting of:

1. a one-arm power manipulator manufactured by Programmed and Remote Systems Corporation,[a] Model 3000, with a 3-m (9.15-ft) vertical movement of the four-tube telescope. This manipulator is driven by electric motors without force feedback. Only the grip force is controlled by an adjustable electric clutch. The load capacity is rated for 445 N with the manipulator elements in any position.

2. a rail frame carrying the manipulator bridge with the manipulator, stereo television camera (Fig. 1), additional tools, tool changing equipment, and two surveillance cameras. The size of the frame is 4 × 6 m (12.2 × 18.3 ft), corresponding to the length of the roof shielding beams of the proton beam cavity.

---

[a]Programmed and Remote Systems Corporation, St. Paul, Minnesota 55112.

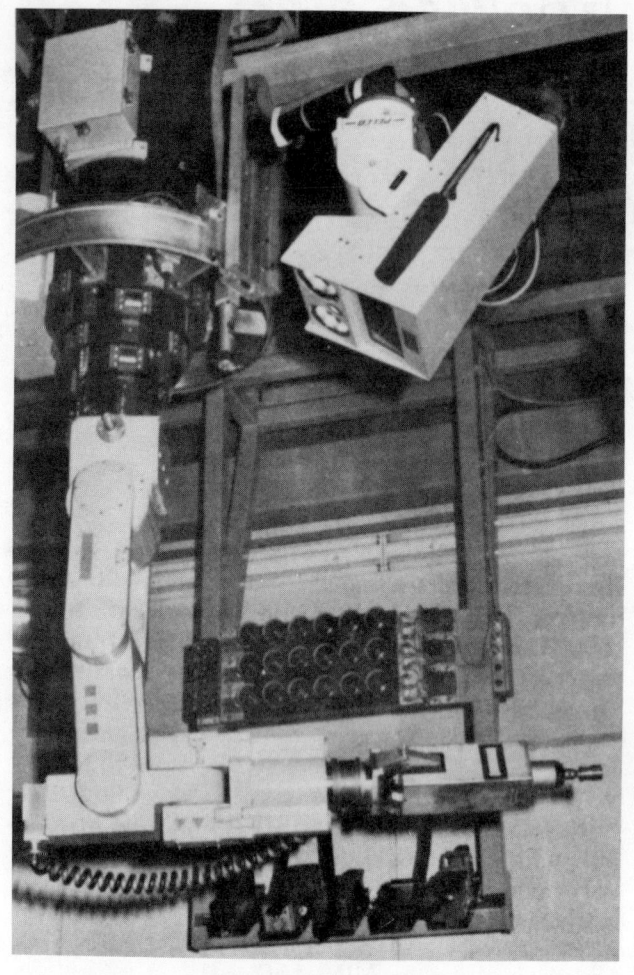

Fig. 1. The three-dimensional/two-dimensional television camera mounted on a remote-controlled carriage with movement along a circular rail, the manipulator hand with an impact wrench, and, in the background, some additional tools and changing equipment for sockets.

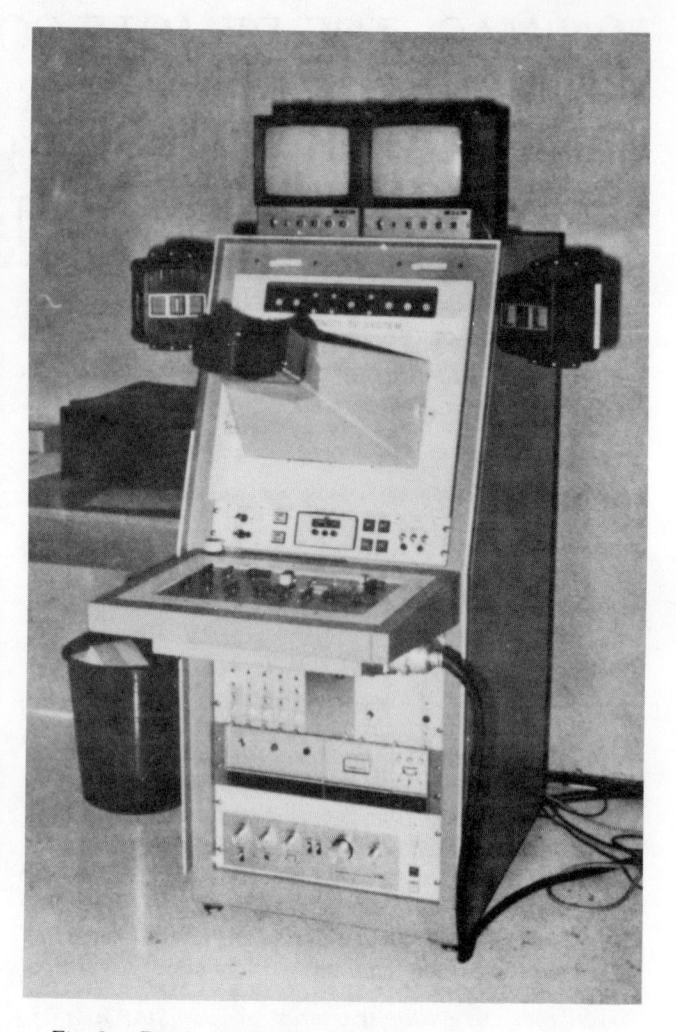

Fig. 2. Display and control console with stereo hood.

3. a control console containing the devices for three-dimensional/two-dimensional and surveillance cameras, the manipulator control, and the stereo loudspeakers (Fig. 2).

SPECIAL TOOLS

In addition to a set of commercially available electrical tools provided with adapters, an impact wrench control device with the following special features has been developed at SIN:

1. When tightening the bolts, the pulses can be counted and their frequency can be varied. One instrument can be calibrated to reproduce desired torque for bolts.
2. When removing bolts, the impulses stop automatically as soon as the bolt is loose. The bolt can then be removed with the manipulator hand so that it will not be lost because of the uncontrolled rotation of the impact wrench.

This is achieved by an SCR-controlled power feedback from a tachometer. The impulse frequency is controlled by the current in the feedback loop.

VIEWING SYSTEM

The viewing system consists of several surveillance television cameras and the three-dimensional/two dimensional "working" camera mounted on a carriage moving along a circular rail on top of the manipulator.

The stereo camera is a high-resolution black and white camera, Model 7120 from Cohu, California. It was modified for stereoscopic format by the Stereotronics Television Company. The corresponding zoom lens has a focal range of 10-to-1, and is television Model 10C from Pelco, California.

The stereo captor in front of the lens is from Stereotronics, Model 5010. It reproduces two separate images onto the lens and vidicon by using an optical block consisting of mirrors, prisms, and polarized filters. In this case, only ~40% of the screen surface is available for each picture. The convergence angle is remotely adjustable for distances from ~0.6 m (1.8 ft) to infinity with a stereo base of 12 cm (4.7 in.). For the two-dimensional mode, this optical block can be moved out of the light path to get a high-resolution image on the total screen surface.

The camera housing contains the television camera with zoom lens and stereo captor, as well as two tungsten halogen lamps. A one-directional microphone for stereo sound transmission is on each side of the enclosure.

The display and control console is equipped with a 43.2-cm (17-in.) monitor, Conrac Model RQB 17/RS. A stereo hood, which can be swung out of the way for direct two-dimensional viewing, is hinged on the side. It is positioned in the operator's main field of vision. A desk immediately below the monitor contains the manipulator controls. Stereo loudspeakers are fitted on each side of the console. Two television monitors, both of which can be connected to any of five surveillance cameras, are placed on top of the console.

PRACTICAL EXPERIENCE

Stereoscopic viewing of remote-controlled operations without force feedback is, in our view, essential. The risk of damage to neighboring components would be too large, especially when using mobile manipulators in unfamiliar areas.

The use of stereoscopic viewing appears to be indispensable, particularly for large movements of the manipulator hand, while the two-dimensional picture could be sufficient for local manipulations.

Since viewing stereoscopic pictures does not correspond to normal vision, it can lead to eye fatigue for repeated changes of viewing distance. By switching over to the two-dimensional mode, the eyes can relax and, in addition, a picture of higher resolution is obtained.

The time taken to switch between the two-dimensional and three-dimensional modes is 30 s for this installation. Experience shows that this time must be drastically shortened.

Any alteration of the zoom setting must be accompanied by a corresponding adjustment to the mirror system convergence angle even at constant viewing distance. The automatic coupling of the zoom and convergence mirror drives is therefore an important improvement.

Although the coupling ratio is slightly dependent on viewing distance, we find that over a limited range—in our applications between 1.5 and 5 m (4.58 and 15.25 ft)—a constant ratio can be used. This feature reduces operator's fatigue and speeds up operation.

A videotape recorder, which is continuously running for all operations, has proven useful for training purposes as well as for repeated playback of critical situations.

For viewing three-dimensional pictures, our operators almost always use the stereo hood since they can work with less eye strain due to the defined eye position. Furthermore, the added eye strain caused by looking away from the screen when wearing prism glasses is avoided.

The advantage of prism glasses, which allow several people to view simultaneously, can only be used to a limited extent since the optimal "stereo region" is fairly small.

In spite of the above-mentioned room for improvement, we consider the television system chosen in 1976 to be very suitable for our purpose and expect it to fulfill all our future requirements.

Proceedings of the 1984 National
Topical Meeting on ROBOTICS AND REMOTE
HANDLING IN HOSTILE ENVIRONMENTS

KEY WORDS: design equipment, remote
maintenance, servomanipulator

An Advanced Remotely Maintainable Force-Reflecting Servomanipulator Concept*

D. P. Kuban
Engineering Division
Oak Ridge National Laboratory
Oak Ridge, Tennessee 37831
615-574-6387

H. L. Martin
Instrumentation and Controls Division
Oak Ridge National Laboratory
Oak Ridge, Tennessee 37831
615-574-6336

ABSTRACT A remotely maintainable force-reflecting servomanipulator concept is being developed at the Oak Ridge National Laboratory as part of the Consolidated Fuel Reprocessing Program. This new manipulator addresses requirements of advanced nuclear fuel reprocessing with emphasis on force reflection, remote maintainability, reliability, radiation tolerance, and corrosion resistance. The advanced servomanipulator is uniquely subdivided into remotely replaceable modules which will permit in situ manipulator repair by spare module replacement. Manipulator modularization and increased reliability are accomplished through a force transmission system that uses gears and torque tubes. Digital control algorithms and mechanical precision are used to offset the increased backlash, friction, and inertia resulting from the gear drives. This results in the first remotely maintainable force-reflecting servomanipulator in the world.

I. INTRODUCTION

The Consolidated Fuel Reprocessing Program (CFRP) at the Oak Ridge National Laboratory (ORNL) is responsible for the development of advanced nuclear fuel reprocessing. Remote maintenance technology development that will increase future plant operational availability, reduce personnel radiation exposure, and reduce environmental impact is an integral part of the CFRP. The Remote Control Engineering and Special Remote Systems tasks are major activities which are working toward a new remote maintenance system[1,2] based on force-reflecting servomanipulators and television viewing as principal maintenance tools. It is believed that the increased dexterity of servomanipulators will significantly increase the remote maintenance system's work efficiency, as well as allow more difficult tasks to be performed.[3]

This paper presents a new *remotely maintainable force-reflecting* servomanipulator concept which is intended to bridge the gap between the complexity of current force-reflecting manipulators and the degree of reliability and maintainability deemed necessary for use in reprocessing environments.

II. MOTIVATION

Servomanipulators and television viewing can provide a large-volume remote handling capability which approaches that of mechanical master/slave manipulators (MSMs) and shielded-window viewing.[4,5] Increased dexterity will influence remote facility design and operation by increasing the number of admissible work tasks, increasing the time efficiency, and enhancing the maintenance system's ability to respond to unexpected problems. These influences should reduce process equipment mean time to repair (MTTR) and thus increase overall plant availability. Servomanipulators' kinematics and dexterity should also allow equipment designers to reduce the degree of special remote provisions they normally are required to include.

Servomanipulator systems can extend MSM capabilities to a large-volume facility since the interconnection is electrical rather than mechanical between the master and slave arms, which allows the slave systems to be mobile using a transporter. The value of MSM-like dexterity as an additional large-volume remote handling tool is clear. The underlying concern is manipulator system reliability, particularly for the slave system. The added benefits of

---

*Research sponsored by the Office of Spent Fuel Management and Reprocessing Systems, U.S. Department of Energy, under contract No. W-7405-eng-26 with Union Carbide Corporation.

By acceptance of this article, the publisher or recipient acknowledges the U.S. Government's right to retain a nonexclusive, royalty free license in and to any copyright covering the article

servomanipulator-based maintenance cannot be realized unless their reliability and maintainability are consistent with overall plant goals. Remote handling system failures increase plant downtime if they prevent process repairs to be completed. Consequently, process equipment MTTR is a function of both the remote handling system's functional capabilities and its inherent reliability. It is important for manipulators to operate *when* they are needed. In this regard, conventional servomanipulator slave designs have been inadequate. Current designs have used metal tape (or cable) and pulley drives to deliver motor torque to specific joint rotations on the arm.[6,7,8] This force transmission technique is an extrapolation of the tape linkages used in MSMs. Tape drives enhance teleoperation parameters by minimizing static friction and inertia, unfortunately at the expense of reliability and maintainability. Experience has shown that the tensioned tapes undergo fatigue failure after a few hundred operating hours. Tape replacement is a complex maintenance task which is almost impossible to accomplish remotely.[9] Replacement also results in significant personnel exposure. Overall system mean time between failure (MTBF) must be increased roughly an order of magnitude, and the slave manipulator itself must be remotely maintainable to reduce manipulator MTTR and minimize personnel exposure.

Recent advances in digital microelectronics and dc motors provide a foundation for more extensive use of gearing (rather than tapes) for motor torque transmission. Gear drives have been avoided in the past, particularly for the lower arm joints, because they undesirably increase friction, inertia, and backlash. These parameters are related to joint servo-control performance and affect force-reflection fidelity. Since force reflection is considered a critical performance factor in remote maintenance efficiency, microprocessor servocontrol was used to implement compensation algorithms which offset the undesirable effects. Since only the remote slave system must be gear-driven and modularized, the master controller can be designed to help reduce overall friction, inertia, and backlash also.

The motivation for developing this new servomanipulator concept is the belief that increased control electronic capabilities and precise mechanical design can be implemented to offset detrimental effects inherent in this modular design. To this end, a specific Remote Control Engineering task activity was established for the development of a new servomanipulator system, called the advanced servomanipulator.

III. THE ADVANCED SERVOMANIPULATOR

The development of an advanced servomanipulator (ASM) concept began with an initial in-house ORNL study and subcontracts with Teleoperator System Incorporated and Martin Marietta Corporation for conceptual design of the in-cell slave manipulator. The results of these design studies were used in formulating the directions of the ASM research. The initial concept was based on a dual-arm anthropomorphic package where each manipulator arm consists of 7 degrees of freedom. The seven-joint arm concept incorporated shoulder-mounted actuators with combinations of gears and nested-torque tubes to transmit torques to lower degrees of freedom. The concept was mechanically too complex to modularize, but later ORNL revisions rendered a feasible approach. This approach was particularly attractive in that the gear/torque-tube force transmission would facilitate modularization for remote maintainability.

A. System Description

The ASM slave is a two-arm, all-gear-driven servomanipulator. The single most obvious difference in this system from traditional servomanipulators is the anthropomorphic or "manlike" stance as shown in Fig. 1. Reprocessing plants generally consist of very large

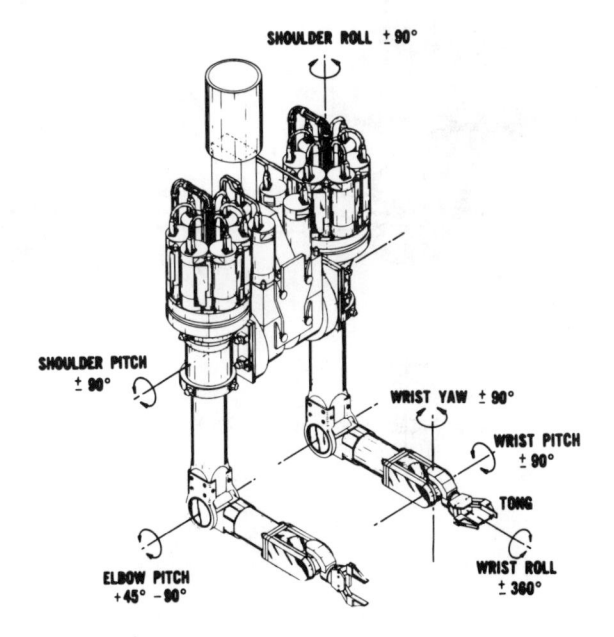

Fig. 1. Advanced servomanipulator

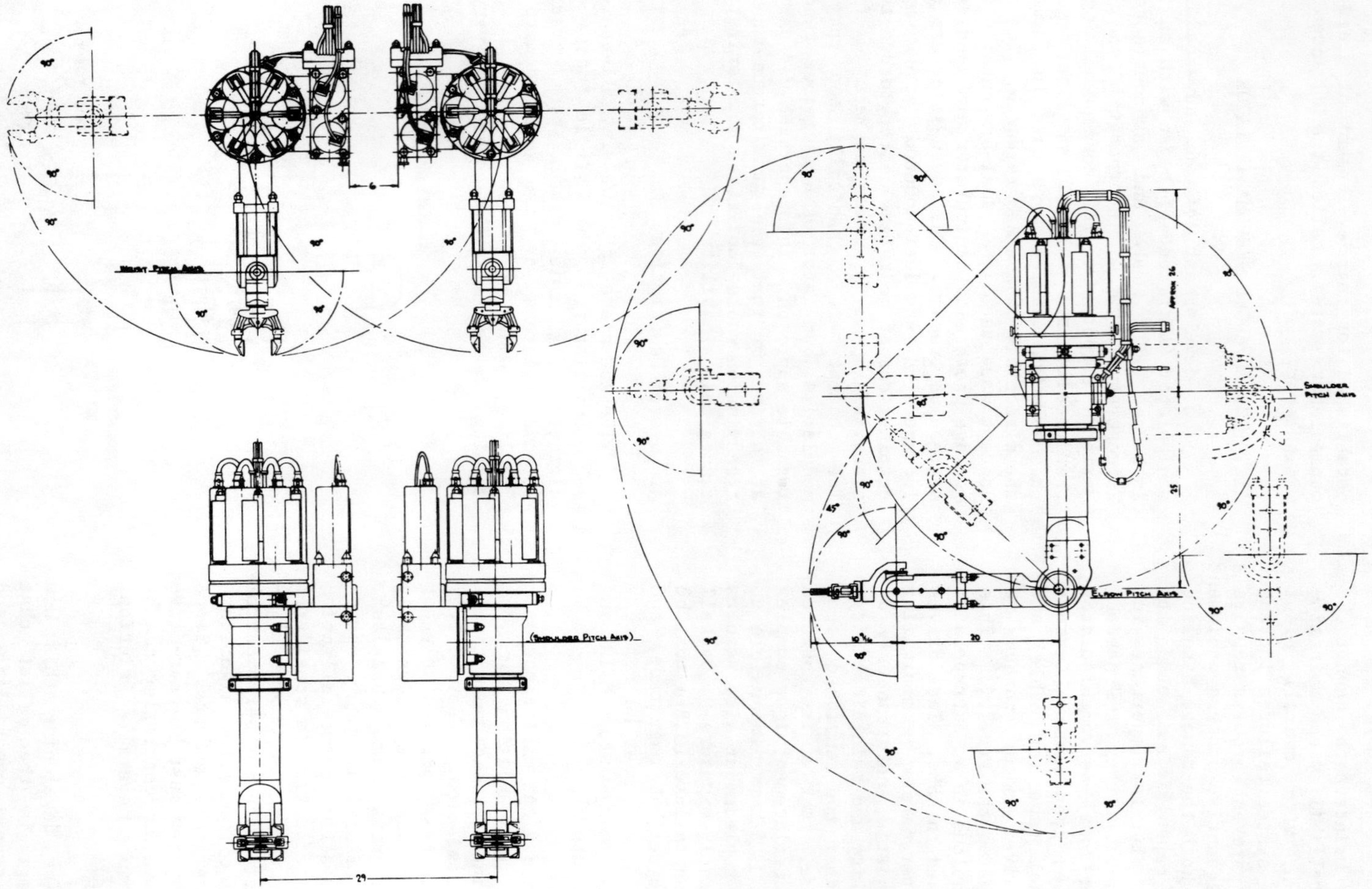

Fig. 2. Three-view motion range

chemical equipment such as tanks, heat exchangers and columns where horizontal manipulator access into the equipment is required.[1,10] The traditional "elbow-up" manipulator stance was derived from "over-the-wall" shielding penetration and was intended for tabletop operations. This kinematic configuration, so successfully applied to hot lab maintenance worldwide, is not amenable to horizontal reaching actions because of the obstruction created by the lower arm link.

The ASM kinematic configuration and ranges of motion are shown in Figs. 1 and 2. The upper 3 degrees of freedom (shoulder pitch, shoulder roll, and elbow pitch) provide wrist positioning in space. Three of the four wrist motions (wrist pitch, wrist yaw, and wrist roll) establish tong orientation. The fourth wrist motion, tong closure, supplies a grasping ability. This unique wrist design, with 4 degrees of freedom, provides very dexterous orientation of the tong by virtue of intersecting orthogonal axes and the pitch-yaw-roll kinematic sequence. Conventional MSM wrist kinematics (e.g., lower arm roll followed by wrist pitch and roll) could not be used in the elbows-down stance because the lower arm and end-effector axes of motion coincide at the normal operating position. When two joint axes align to form what is often called a singularity, they become redundant and difficult to control independently. The ASM lower arm does not rotate but has all the orientations occurring in the wrist body. In the ASM wrist, all the singularities occur at the extremities of the ranges of motion where they are of minimal concern.

The load capacity of the ASM slave arms is 23 kg, which is considered appropriate for reprocessing maintenance. The ASM design capacity presumes that it will be part of an integrated maintenance system including cranes, and is based on trade-off analysis of the packaging and envelope requirements of the manipulator and the reprocessing equipment needs. The ASM slave arms are designed with an end-effector maximum slew velocity of approximately 1.0 m/s and have sufficient dynamic response to follow an operator's normal range of input speed in real time.

Several other parameters associated with the ASM gear drives are critically important in terms of the ultimate performance of the manipulator. The disadvantages of the gear/torque tube force transmission system must be minimized for in the mechanical design and compensated for in the control development. These disadvantages include increased friction, increased backlash, increased inertia, and joint axis cross-coupling. Friction is of major concern because it directly affects the magnitude of the force-reflection threshold. The force-reflection threshold is that amount of backdriving force at the slave necessary to overcome static friction and to initiate master arm motion. Below this threshold the operator has no feedback from the remote environment. Therefore, minimizing the friction threshold is considered one of the most important performance-related design objectives. It is estimated that the ASM will have a force-reflection threshold on the order of 0.45 kg (2% capacity). For comparison, currently available servomanipulators have force-reflection thresholds ranging from 1 to 10% of capacity.

The sources of friction in the ASM are primarily bearings, gears, and motors. Friction magnitudes are dependent on bearing quality, gear quality, lubrication, and fabrication precision. Lubrication will be limited to light machine oil, and shaft seals are not used because they significantly increase friction. Gear friction is minimized with precision alignment, high-quality gears, and hardened gear materials. It has been estimated that the gear friction will be roughly an order of magnitude larger than the bearing friction. Drive motor friction must be minimized since its effects are amplified by the ratio and the efficiency of the associated gear train. A friction compensation algorithm will also be used to reduce the perceived static friction even further.

Backlash in gear-driven servo systems imposes undesirable nonlinear effects on the control system. This results in reduced positioning accuracy and can cause instability. Several methods can be used to reduce the magnitude and effects of backlash. Mounting the gear meshes with interference is often used in industrial robots to minimize backlash, but this significantly degrades force reflection. This approach is not possible for force-reflecting servomanipulators because the gear trains must be backdriveable. Other design solutions include selecting high-quality components that are manufactured to close tolerances, mounting feedback sensors at the drive motor shafts to close the control loop exclusive of backlash, distributing gear ratios such that the largest gear reductions occur near the output, and minimizing overall backlash by minimizing backlash in the master controller, which does not have to use gear drives. Each of these techniques was used to varying degrees in the design of the ASM.

A detailed parametric analysis was performed to determine the contribution of each mechanical component to the total system backlash. This study was the basis for the degree of precision chosen for each specific component. In general, the precision chosen is much better than standard components but is not the best that can be achieved.

Locating feedback sensors at the centralized motors circumvents most of the stability problems posed by backlash. This eliminates potential phase errors between motor position and actual joint position which could cause small amplitude jitter. The nonlinear load effects are still present, but the effect of gravity on the slave arm will reduce these effects in the majority of manipulator positions. Another advantage of motor-mounted sensors is the motor, brake, and sensors can be located in one package, enhancing the remote maintenance aspects.

A disadvantage caused by the centralized motors and magnified by the gear drives is joint axis cross coupling. The nature in which the gear drives transmit forces through manipulator arm joints results in the torques applied at the motor shafts becoming interrelated at the arm joints. Cross-coupling effects in the ASM are significantly larger (20 to 40% of the total torque) than in metal tape-driven servomanipulators. The implications of cross coupling are important and can result in more complex control requirements, especially for accurate force reflection to the operator. An analysis was performed to evaluate the impact of slave cross coupling on the master controller design and the overall control system. The most important conclusion obtained was that the master controller cross coupling must mimic the slave cross coupling to prevent the operator from being confused by sensing incorrect forces at the master controller. The analysis also showed that the cross coupling can be either additive or subtractive. The gear train designs are arranged so that cross coupling allows motor drives to assist one than another rather counteract each other.

Since gear/torque-tube force transmission systems have much larger inertia than conventional tape drives, special attention is being given to total weight minimization. Multiple control regimes will be implemented to reduce inertial effects and therefore operator fatigue. In addition, real-time servocontrol techniques will be used to eliminate mechanical counterbalancing weights and mechanisms. The ASM utilizes electronic counterbalancing by calculating the weight vectors of the various arm links in real-time based on joint position. The weight vector data are then transformed back to the motor-drive coordinate frame where they are treated as incremental additive torques necessary to offset the weight. The effective shoulder inertia was reduced 35% and the total arm weight was reduced 40% over a mechanically counterbalanced system. Electronic counterbalancing increases friction because of the additional gear train load. It also increases drive motor duty cycles, which leads to higher gear reduction and increased friction threshold.

Another important design trade-off occurs between gear ratios and motor size. Low gear reductions reduce perceived inertia but require large, heavy motors. High gear reductions with small motors are not only nonbackdriveable but result in large reflected inertia that is proportional to the gear ratio squared. The ASM design uses relatively large motors which operate at slow speeds through low gear reductions.

B. Components and Materials of Construction

The ASM will operate in an acidic environment requiring corrosion-resistant materials. All bearings will be 440 C stainless steel. Gears and shafts are fabricated from corrosion-resistant 17-4PH stainless steel, and various housings are passivated 6061-T6 aluminum.

Only spur and straight-bevel gears are used in the ASM to minimize friction and to achieve backdriveability. Bevel gears have slightly lower efficiency than spur gears but are necessary to transmit forces around joints. Helical, spiral bevel, and other gears which are known to be smoother and quieter are not used because they are less efficient. All gears conform to the American Gear Manufacturers Association (AGMA) standard and are either 16 or 12 diametral pitch. Gear stresses are limited to meet reliability and life requirements of the ASM. All gears are high precision (AGMA Quality Class 10C), based upon an overall manipulator precision and backlash analysis.

Precision antifriction ball bearings are used throughout the ASM, and they are almost exclusively the deep groove radial type. This low-friction bearing type was selected because of the vast number of sizes, variations, and suppliers available. The bearings are sized to provide a minimum 1000-h life at maximum load and speed with approximately 1% failures.

Since maximum load and maximum speed never occur at the same time, it is estimated that the ASM bearings will actually have an operating life of 10,000 h in normal use.

The ASM torque-tube drives were analyzed with respect to strength, deflection, and vibration natural frequency. Of these three, the deflection criterion was most severe, so the stresses are low and the stiffnesses (and natural frequencies) are high. The failure rate of these components should be very low.

Electric motors supply the drive torques for all the manipulator motions. The dc servomotors provide large torque capacity at low speed and are commonly used in precision servomechanisms. Two vastly different dc servomotor geometries were investigated: (1) the shell motor and (2) the rare-earth magnet, iron core motor. The geometries of these two motor types are shown schematically in Fig. 3. The key difference between the two geometries is the location of the motor windings where heat generation occurs. In the shell motor, the windings are at the exterior of the armature, where as in the iron core motor the windings are in the internal rotating armature. The motor parameters that are particularly important for force-reflecting servomanipulator applications are presented in Table 1. Of these parameters, the percent sensitivity is the most important factor because it determines the absolute friction threshold that the manipulator can ultimately attain.

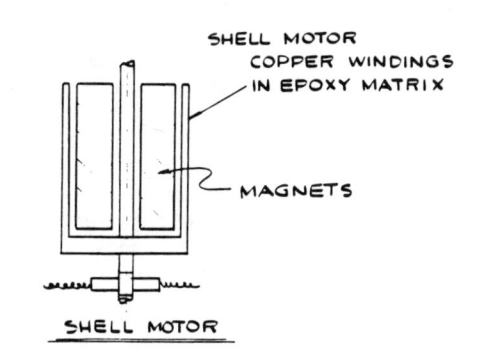

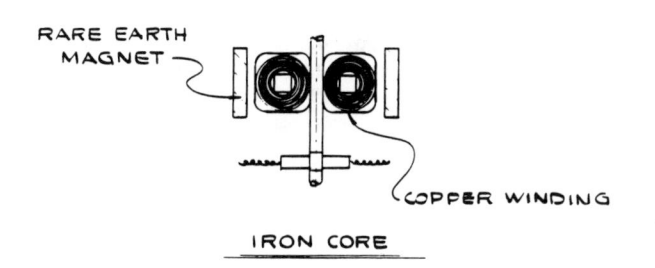

Fig. 3. Shell motor and iron core motor geometric comparison

Table 1. Important servomotor parameters

Parameter	Definition
1. Continuous torque capacity	Maximum torque that can be generated for a period of 2 h with an armature temperature $\leq 155°C$.
2. Friction torque	Shaft torque necessary to backdrive motor.
3. Percent sensitivity	(Friction torque/continuous torque capacity) x 100%.
4. Torque density	(Continuous torque capacity/motor weight) x 100%.

An extensive motor market survey with product testing and development was performed at ORNL. This has enabled several performance improvements to be made in the ASM motors. Comparative testing between the two motor types was used to make the final selection. The friction torque and heat dissipation characteristics of the shell motor proved to be superior. A geometry change placed all magnets in the center of the motor, allowing the largest coil lever arm for torque generation and the best conductive path for thermal dissipation. The percent sensitivity of the shell motor is only 60% of the iron core motor and is the primary reason that it was selected as the baseline motor.

All electrical leads are terminated in the motor modules, so cabling is required to provide power and signal wiring from the shoulder to the motors mounted on the upper arm. Since the cabling will have a finite life due to radiation embrittlement and fatigue, external cable routing was chosen to simplify remote replacement. To reduce the possible snagging, cables are routed at the rear of the manipulators and away from the work site. A free length of cable has been selected to allow shoulder pitch and roll joints to have motion ranges of $\pm 90°$. If additional motion is required, the cable jumper can be redesigned for the different motion changes. If a different stance (e.g., elbows-up) is desired for a special task, the cable jumper can be used after the arm stance is changed.

The electronic controls have been designed so that they can be located away from the arms and shielded from the radiation. More details about the ASM control system can be found in ref. 10.

C. Modularization and Remote Maintenance

The ASM is designed to minimize remote handling system MTTR by using a module replacement scheme. The ASM is constructed in modules each of weighing less than 23 kg so that they can be handled by another ASM. Modularity is possible because the gear/torque-tube force transmission concept allows separation of modules at gear or spline interfaces. The ASM concept is based on the philosophy that a manipulator failure can be isolated to a particular module malfunction. The failed module would then be replaced with a working spare in situ by another ASM. Through spare module replacement, the failed ASM system can be returned to operation quickly (relative to removing the entire arm to a decontamination/repair station). The failed module would then be transferred to a repair station for further evaluation. All the ASM modules are designed for remote replacement using fixtures and tooling necessary for synchronization of gear meshes and intermediate arm supports.

As shown in Fig. 4, the ASM consists of eight remotely replaceable module types. The largest subassembly in the ASM, the shoulder, is actually made up of two modules: the shoulder gear box module and the lower pitch sleeve module. The shoulder is subdivided so that the individual module weight is below the 23-kg ASM capacity. Also, it was felt that the failure rate of the pitch sleeve will be much lower than the more complicated gear box.

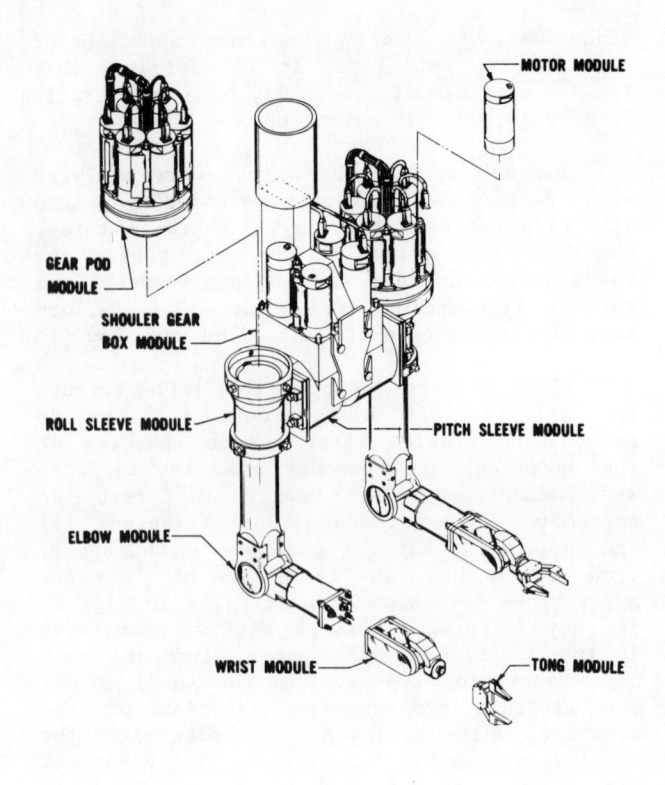

Fig. 4. Advanced servomanipulator modules

Since the pitch sleeve module supports the rest of the arm, it is designed to remain in place while the gear box is being replaced. The shoulder module provides the shoulder pitch motion for the arm. The shoulder pitch load is the largest and requires two of the standard ASM motor modules operating in tandem.

The roll sleeve module is attached to the shoulder. This module is basically a housing for the shoulder roll motion bearings and the internal torque tubes. These components have large design margins and should have a high reliability. The roll sleeve module can be removed from the shoulder pitch module with the elbow and gear pod modules attached. To replace the roll sleeve module, the elbow and gear pod modules must be removed first.

The gear pod module is removable from the roll sleeve and provides the appropriate gear reduction for the remaining 6 degrees of freedom. It also provides alignment for the remote coupling of the torque tubes and the motor modules.

The elbow module includes the upper arm, the elbow joint, and the lower arm. It is removable from the bottom of the roll sleeve through an attachment ring. It contains the torque tubes required to transmit power to the elbow motion and the four wrist motions and provides alignment for the remote coupling to the wrist module. The elbow module could be resized to provide different reach and capacity ranges for special applications.

Of all the modules in the ASM, the wrist module contains the largest number of gears, bearings, and other parts in a very small volume. It is also the most important module because it provides a degree of dexterity never before available in a manipulator. The ASM wrist has 4 degrees of freedom, whereas conventional MSM-type manipulators have only 3. This gives the operator a very human-like capability for performing his maintenance tasks. Since this module has the largest duty cycle and stress levels, it is expected to have the lowest reliability and therefore has been designed for remote replacement.

The tong module is based on a Sargent Industries, Central Research Laboratories Division, rotary drive two-fingered tong. This tong is modified to be remotely replaceable using a special changeout tool. This capability also allows direct attachment of special tooling to the wrist.

The motor module is identical for all joints. This simplifies spare parts inventory as well as remote replacement. The only disadvantage to using the same motor for all joints is that it increases the tong force reflection threshold. This is because the motor is larger than necessary; therefore, the friction is proportionately larger. The motor module includes all the necessary components, such as brake, position sensor, and tachometer required for joint control. The entire module will be replaced if any component fails. The motor module and its interface have been designed such that if significant commercial improvements are made in motors, sensors, or brakes, they can be incorporated by redesigning only the motor module.

D. Current Status

The initial design of the ASM slave manipulator has been completed. A 1-degree-of-freedom test stand has been built and extensively tested. Both the mechanical predictions and the performance of the control algorithms have been verified on this stand. The backlash and friction characteristics of this test stand have been as good or better than expected. Additional testing is continuing to optimize some remaining parameters. To gain further experience, two developmental units are being fabricated at ORNL. The ASM development arms will be used in preliminary testing with one unit acting as a replica master controller and the other as the corresponding slave. The objective of these tests will be to quantitatively evaluate basic gear-drive effects. Later, the two developmental slave arms will be combined as a dual-arm slave package and installed in an integrated demonstration of a complete advanced maintenance system including the mobile transporter, master controller, and operator station.[2]

IV. SUMMARY

It is believed that force-reflecting servomanipulators significantly improve the time efficiency and range of admissible tasks pertaining to remote maintenance and operations in advanced nuclear fuel reprocessing facilities. To be consistent with the basic objectives of the CFRP, servomanipulator reliability and remote maintainability must be improved. A new remotely maintainable servomanipulator, called the ASM, is being developed to meet these challenges. The ASM represents an entirely new concept that uses gears and torque tubes to transmit forces rather than conventional metal tapes. Improved control methods are used to compensate for gear and torque-tube shortcomings. The gear-drive approach allows the ASM to be subdivided into remotely replaceable modules. Remote modularization will allow failed manipulators to be repaired through (in some cases in situ) spare module replacement and thus decrease repair time. The mechanical design of the ASM slave manipulators has been completed, and fabrication of two developmental units is under way at ORNL. Experimental results are encouraging and fortify the expectation that the ASM can achieve the desired force-reflection performance while being more reliable and remotely maintainable.

REFERENCES

1. M. J. Feldman and J. R. White, "Remotex - A New Concept for Efficient Remote Operations and Maintenance in Nuclear Fuel Reprocessing," ANS National Topical Meeting on Fuel Cycles for the 80s, Gatlinburg, Tenn., October 1980.

2. W. R. Hamel and M. J. Feldman, "The Advancement of Remote Systems Technology: Past Perspectives and Future Plans," this volume.

3. J. N. Herndon and W. R. Hamel, "Analysis of the Options — Rationale for Servomanipulators Maintenance in Future Reprocessing Plants," this volume.

4. M. M. Clarke et al., "Human Factors in Remote Control Engineering Development Activities," Proc. 31st Conf. Remote Syst. Technol., vol. 1, Detroit, Summer 1983, (in press).

5. J. Vertut et al., "Contribution to Define a Dexterity Factor for Manipulators," 21st RSTD, 1973.

6. C. Flatau, "SM-229 - A New Compact Servo Master/Slave Manipulator," Proc. 25th Conf. Remote Syst. Technol., 1977.

7. J. Vertut et al., "The MA-23 Bilateral Servomanipulator System," Proc. 24th Conf. Remote Syst. Technol., 1976.

8. J. Herndon et al., "The State-of-the-Art Model M-2 Maintenance System," this volume.

9. J. Vertut et al., "Protection, Transfer, and Maintenance of the MA-23 Bilateral Servomanipulator," Proc. 26th Conf. Remote Syst. Technol., 1978.

10. H. L. Martin et al., "Control and Electronic Subsystems for the Advanced Servomanipulator," this volume.

CHAPTER 3
UNDERSEA APPLICATIONS

Reprinted from *Mechanical Engineering,* August 1982,
Copyright © The American Society of Mechanical Engineers

Robotics Undersea

Key concerns include position feedback, preprogrammed tool exchange and work operations, automatic television tracking, object and plane definition, and manipulator controls

Robert L. Wernli
Naval Ocean Systems Center
San Diego, Calif.

During the past two decades, the U.S. Navy has been a leader in the field of developing unmanned undersea vehicles and work systems. This technology base, developed under such programs as CURV III (Cable-Controlled Underwater Recovery Vehicle), RUWS (Remote Unmanned Work System) and WSP (Work Systems Package), has been transferred to industry. As a result, the use of similar remotely operated vehicles (ROVs) has been increasing in areas such as the North Sea where they are aiding divers in their work and completely replacing them in others. ROVs are being used by NASA to recover the solid rocket boosters of the space shuttle, and by the nuclear industry for the inspection of internal systems, a hazardous job previously performed by divers. However, the industrial acceptance of these systems is in its infancy. There is always a lag between development of a system

Fig. 1 SCORPI inspection vehicle.

and the confidence required by the user before becoming comfortable with it. Therefore, technological advances must be incorporated into systems at the earliest possible time to reduce the lag time of technology transfer and acceptance. In accordance with this philosophy, the Navy is continuing its quest to advance the state of the art in the area of ROVs and work systems.

One of the primary technological areas being addressed is the application of robotics to these systems. Robotics can have different meanings to many. However, it can generally be described as integrating the computer into a system, turning some or all control functions over to it, having it make key decisions, and reporting the results to the supervisory operator. The integration of the work system, the computer, and the supervisory operator is a critical area. Which one will control the operation? How much will it control it? How is control systematically passed from one to the other?

Classification of Systems

As in the use of the term robotics, the definition of a work vehicle or system may also be interpreted differently depending on the user. Therefore, the following classification of ROV capabilities is provided.

Inspection—All ROVs are capable of outside inspection of an object. However, this definition applies to vehicles capabale of inspection only, e.g., the Scorpi manufactured by Ametek, Straza Div. (Fig. 1). A vehicle with a capability of inside inspection, e.g., inside the cockpit of an aircraft lost at sea, would require a manipulator or arm that could hold a TV camera to be extended into the object, such as the RCV 150 by Hydro Products (Fig. 2).

Recovery—This definition covers any type of recovery operation, from simple grasping of a small object on the ocean floor, to recovery of a large object by means of a claw-type mechanism. When a recovery task becomes more complicated and involves, for example, placing slings around an object or rigging it with lines or snap hooks, it approaches but is not included in the category of work. If a system is not dedicated to making mechanical modifications to the object, it is classified as a recovery system.

Work—This category is defined as the performance of mechanical modifications to an object. Although these modifications may be for the purpose of recovery, they are still considered work operations, and a system with this capability woulld be considered a work system rather than a recovery system. An example is the Work Systems Package developed by the Naval Ocean Systems Center (NOSC) (Fig. 3).

By combining the manipulators required to perform the previous tasks, a means is then available to classify systems through the number of manipulators integral in the system (Fig. 4). For example, with an inspection system, no manipulators are required to actually see the object of interest, although it would be nice to have a single manipulator to retrieve a small object or perform some other minor task. For recovery tasks, at least a single manipulator is necessary, and a more desirable configuration would be one with two working arms. In other words, a system such as CURV III, with a single simple manipulator, could recover a torpedo, but it would be very hard for it to perform complete rigging operations. This would require a second arm for assistance, such as holding on to the point to be rigged with one arm while the other does the rigging. Unless the object is large (i.e., stable), and the vehicle is sitting on the ocean floor, no work can be done unless you have hold of the object to be worked on. This is dictated by tool positioning accuracies required to perform "real world" work. Holding on to the object is necessary to stabilize the work system with respect to the

Fig. 2 RCV-150 with manipulator.

Fig. 3 Work systems package mounted on the Pontoon implacement vehicle.

object. The argument for stabilizing the work system is quite evident in the industrial world, where robots are being used extensively for manufacturing or work processes. In industry, all systems are programmed around a defined set of locations, a "work cell." Within the work cell the location and orientation of all items—tools, machines, manipulators, parts, etc.—are known. This allows the manipulator or robot to efficiently perform its function with minimal assistance from the human supervisor. Ascertaining component positions is usually costly in the form of time, money, or both. Once this information is lost, some price must be paid to reacquire it. Often, the price is the intervention of the human operator into a task that could have been performed autonomously. Therefore, if given the choice of a system design to which computerized control methods are to be applied, a system that can retain the position information—i.e., establish a work cell—would be preferred. Unlike industry, undersea vehicles face a harsh, dynamic environment, and therefore require a multiarmed system that can create a work cell, i.e., a *work system*.

The controls required for a system to operate efficiently depend entirely on the type of tasks to be performed. Therefore, the system configuration, i.e., the number of manipulators, can be associated with a series of specific tasks. This association is provided in Fig. 5, in which the three generic types of systems (inspection, recovery, and work), the number of manipulators, and potential tasks are correlated. The "minimum manipulator requirement" is considered that which is definitely required to perform a given task, although through the use of brute force and ignorance some of the tasks can be completed with a simpler system. The "optimum manipulator requirement" indicates the number of manipulators that would be desired to efficiently perform the task. In general, a one-manipulator system would be considered an inspection system; two manipulators would classify it as a recovery system, while a three-manipulator system would be considered a work system. Therefore, to increase the work or task capability of a vehicle it is necessary for the manipulator suite to go through a metamorphosis as it changes from an inspection to a work system. With the definition of a work system provided, the application of the computer to it, and some potential benefits can be discussed.

Design Philosophy

As previously described, the heart of any work system is the manipulator suite. The question is which type of manipulator is required to perform underwater work tasks, especially if computer controls are being considered. The answer has two parts.

The first part concerns the sophistication of the system required to do the task. Most manipulators can fit into three classes based on their design and control system: Simple rate control (without position feedback); position controlled (with position feedback, e.g., terminus controlled or master-slave); and position-controlled with force-feedback. Studies have been performed by several researchers that relate the performance of these types of manipulators to task complexity. In general, as the manipulator becomes more anthropomorphic, the time to perform a task decreases and the complexity of the tasks that can be completed increases. Eventually, each type of manipulator reaches a practical threshold beyond which it can't realistically perform tasks (Fig. 6).

The second part of the answer addresses the usual tradeoff problems involving manipulator complexity, reliability, cost, efficiency, and capability. Figure 6 also shows the general relationships between these factors. Obviously, an efficient system capable of doing most tasks will increase in cost and complexity while decreasing in overall reliability.

Since the ultimate goal is to incorporate computer control, the selection process is narrowed. In order to use the computer, the position of all joints to be controlled must be known; therefore the manipulators without feedback are immediately eliminated. This constrains the designer to at least a medium level of complexity in manipulator design. Although the addition of position sensors on the manipulator may be a trivial task for most industrial applications, their addition on undersea manipulators becomes quite complex when they must be isolated from the water, pressure, and the usual hazards of working in unknown environments.

The requirement for force-feedback is, therefore, the only decision remaining. Although force-feedback manipulators have outstanding benefits, they also impose

System Type	Manipulators Required	
	Minimum	Optimum
Inspection	0	1
Recovery	1	2
Work	2	3

Fig. 4 System manipulator requirements.

Specific task classification	Optimum manipulator requirements	Generic task classification	Minimum manipulator requirement
Outside inspection	1	Inspection	0
Inside inspection	1	Inspection	1
Small object recovery by grasping	1	Recovery	1
Ordnance recovery	1	Recovery	1
Jetting	2	Recovery	1
Small object recovery by line attachment	2	Recovery	1
Large object recovery by claw attachment	2	Recovery	1
Cable clearance	2	Recovery	1
Explosive pad-eye attachment for recovery	2	Recovery	1
Object preparation and rigging for recovery	3	Recovery	2
Undersea structure inspection (with tools)	3	Recovery	2
Special purpose tooling use	3	Recovery	2
Offshore emergencies	3	Recovery	2
Component extraction from object	3	Work	3
Object installation and maintenance	3	Work	3
Large object recovery by multiple attachment	3	Work	3

Fig. 5 System configuration versus task capability.

fatigue on the operator, which is not desirable during long operations. In addition, this type of system will force a quantum leap in the levels of cost and complexity. Therefore three other factors should be considered. First, most tasks can be completed with a position-controlled manipulator without force-feedback, although at a subsequent cost in efficiency. Second, where force-feedback is absolutely necessary, it has been shown that it can be presented visually, allowing completion of these tasks, although with a time penalty. In this case, the master controller is not increased in complexity. Third, through the use of the computer, the manipulator design can be kept relatively simple while still providing force vector data. Thus, the typical force-feedback manipulator system is not

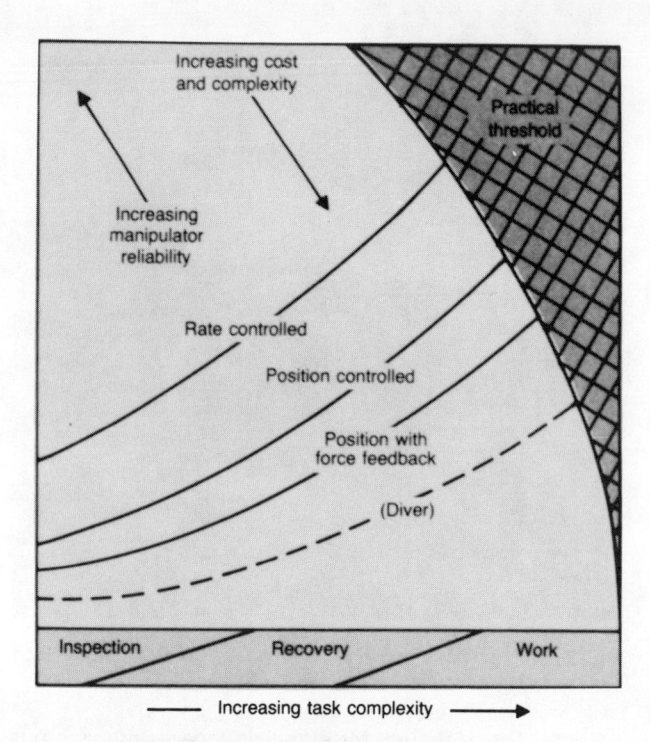

Fig. 6 Parametric manipulator relationships.

a basic requirement since most of its benefits can be acquired in other, more simplified ways.

Computer-Augmented Control System

After the design engineer addresses all the previous tradeoffs, he may often wonder how he is going to design a state-of-the-art system, or advance technology, when to do so will obviously condemn the system due to its high complexity, cost, etc. Fortunately, the addition of the computer to the system design reduces many of the previous problems while advancing systems technology and increasing system efficiency. For example, previous master-slave manipulators have used replica type controllers. These controllers have the same number of joints and degrees of freedom as the slaved manipulator, although they are of a smaller scale, and cannot be easily simplified. Through computer integration, mechanical complexity of the controller can now be placed into the software with the resulting simplification of the mechanical system. Through the use of algorithms for transforming coordinates, the controller can be reduced to a simple terminus type position-controller, as shown on the control console in Fig. 7. The computer controller may also allow simplification of the mechanical design of the slave arm. Often, mechanical complexity is increased to allow the manipulator to follow linear paths, or its dexterity is reduced to allow it to be more easily interfaced with the controller. Much of this can also be done with the computer by turning this requirement over to the control software itself, thus simplifying the manipulator.

With the computer-controlled work system justified, the engineer is left with a final decision: Which type of controller to use to best perform the high number of tasks required by the work system. Some systems, such as the Work Systems Package, were designed to perform many tasks completely without coming back to the surface to change tools. Design of such a system requires an in-depth look at the tool suite, being sure that you have the

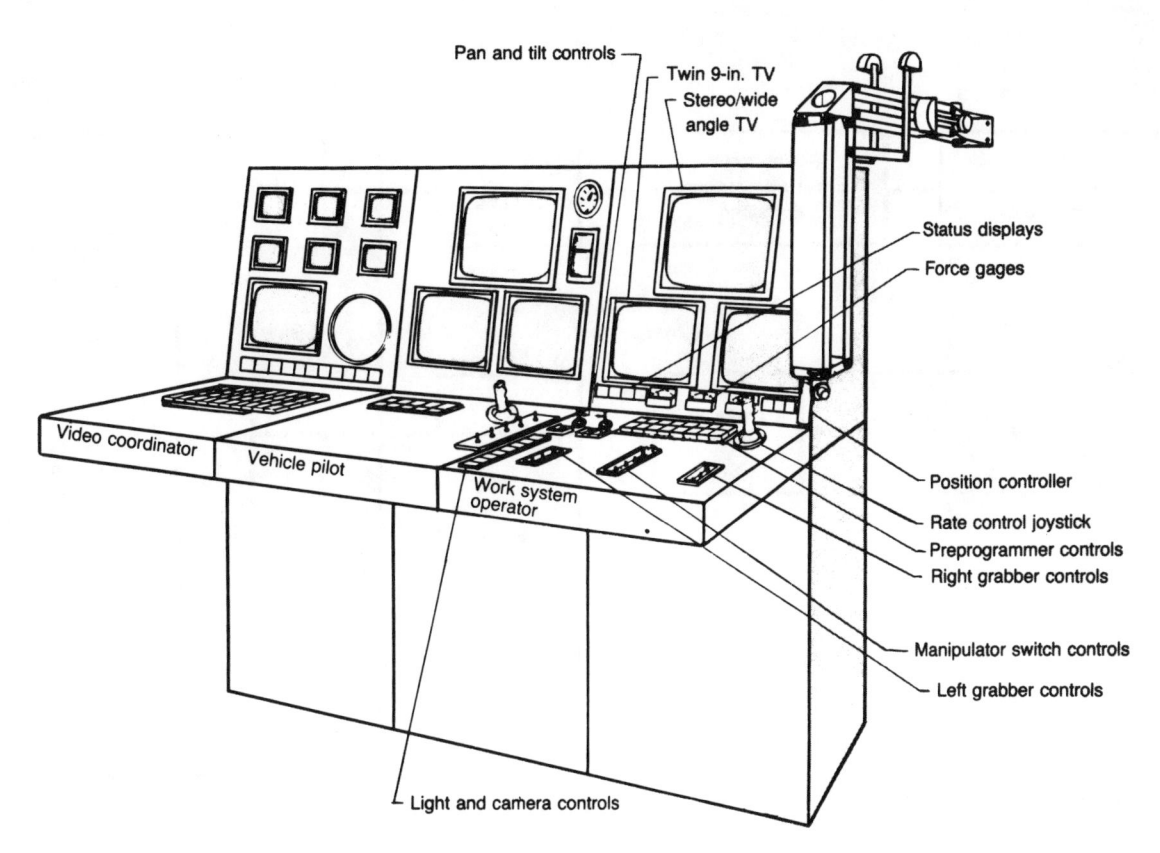

Fig. 7 Control console concept.

best tool for each job, and in some cases a few of the tools will back up others in the performance of a similar task. Based on the task at hand, the manipulator operator, much like the mechanic working on a car, will reach into his toolbox and obtain the tool *he* feels is best for the job in that situation. Since this choice is operator-dependent, it is often given a fair amount of design attention. This is the point where I feel that the design process often comes to a halt. The manipulator and tools have been chosen and are usually followed by a tradeoff to choose *the* best controller. What is overlooked is that the choice of the controller is more task- and operator-dependent than the tools themselves. The operator needs to be given a choice of which combination of input and output devices he desires to use for the task at hand. This becomes even more evident when the mission will be a long one, using more than one operator, since each may have a preference for which controller is the best. For example, with the choice of the computer as one of the controllers, the operator will obviously use it in any situation where it is considered beneficial. This will greatly reduce operator fatigue. Sitting at a control console for six hours operating a work system has a high fatigue level to begin with. Combine that with several hours of the operator using "body English" to move the manipulator on the TV monitor a "little bit closer" to the object, and you have one tired operator with an awfully stiff neck. The more a control system can reduce this fatigue level and provide the supervisory operator with an occasional break, the more efficient the overall operation will become.

Manipulator Control Modes. As discussed, the choice of the controller is dependent on the task at hand, the tool to be used, and the operator's preference. Therefore a multimode control console is desired that provides this versatility.

Switch Control (open loop). This control choice is included as a "fallback" option in the event of a position sensor or computer failure. Although a primitive method, this type of rate control has its advantages for the operator using certain types of tools in work situations. Often a slight movement by a single joint is all that is required, such as in drilling or cutting operations, and the operation would be hampered if other joints were moved. Therefore a position controller may not be desired in this situation, since a slight movement could cause other joints to become misaligned. It is also a less fatiguing form of control for the operator during long tasks. This type of rate control has been used successfully on the WSP.

Switch Control (closed loop). Through the addition of the computer to the control system (with position feedback), the capability exists to greatly improve the function of rate commands. By adding closed-loop rate control the computer can actually generate a sequence of position commands extending in the desired direction. The manipulator will be controlled through the calculated sequence of positions as if they were a preprogrammed sequence or one generated by a position controller.

A variation of this type of control is the capability of automatic tool advance. A sequence of positions can be generated to cause the hand (or tool) to advance in a desired direction. Advantages of operating in the closed-loop manner are:

• the switches, or joystick, can control true linear, orthogonal motions, with no cross-coupling of functions.

• there will be no creep in vane-actuated functions.

85

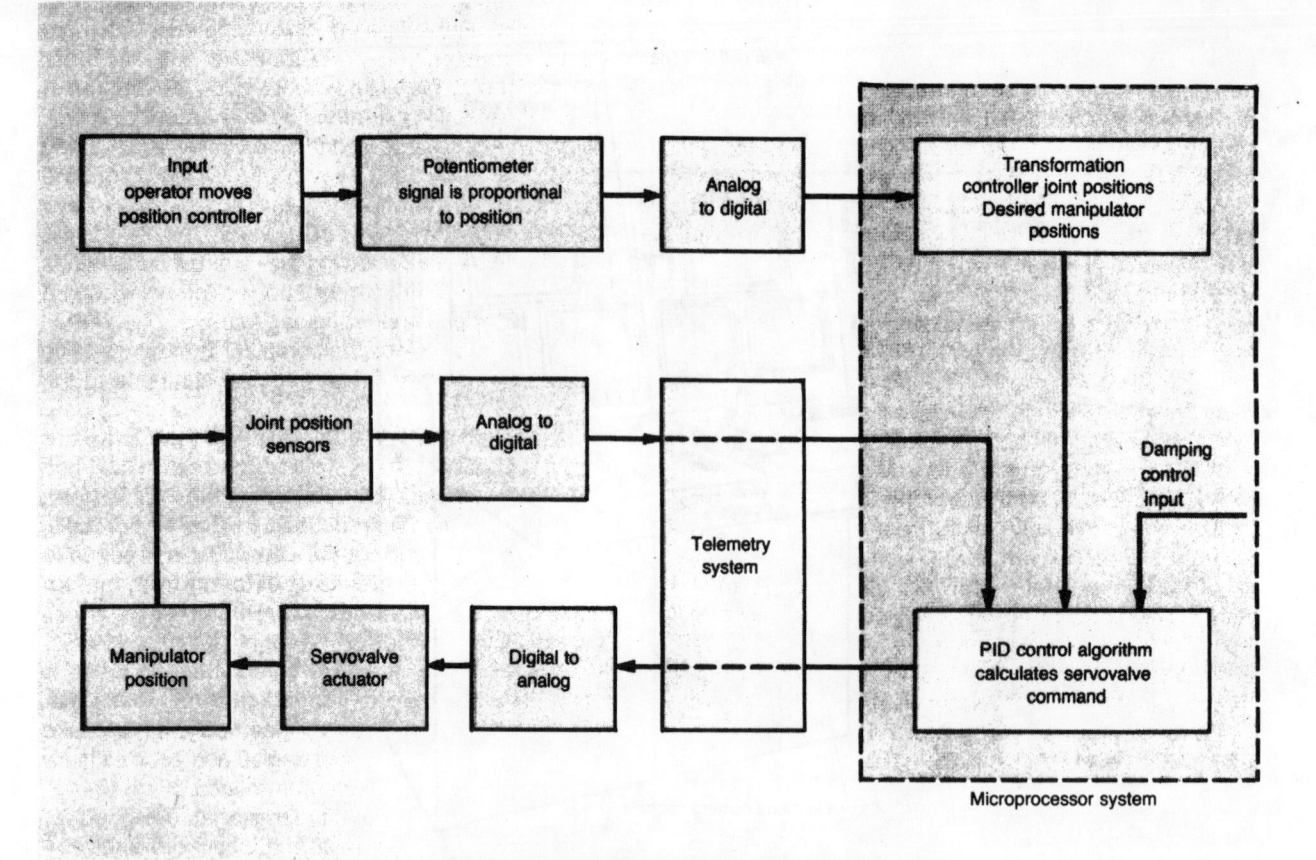

Fig. 8 Position-controlled manipulator flow chart.

- position and position commands can be monitored for sudden increases in error, as a simple collision warning.

Joystick Control. It would be desirable to have a joystick-type controller on the console to allow control similar to that with switches, but with the capability of controlling several functions at one time at a rate proportional to the displacement of the joystick. This would be beneficial in certain types of work where it is desired that the hand or tool follow a path in a straight line along a surface. This controller could operate the manipulator with respect to a set of coordinates located at the shoulder or wrist, depending on the task at hand. Once again, an excellent example of operator preference based on task requirements.

Position Control. This system has two key aspects: the method with which the computer controls the manipulator, and the method with which the operator controls the computer. It should be noted that the computer control of the manipulator actually applies to all closed-loop control methods.

The movement of the manipulator is caused by the computer's activation of servovalves on the arm, based on feedback from the position sensors. The servovalves will be driven at a rate proportional to the spacing of the commanded points to which it is to move and to the error between actual position and commanded position. This theoretically results in all functions arriving simultaneously at each desired position. However, the points to be transited will never actually be reached: the commanded points will continuously leap ahead as they are approached within some acceptable error (also a function of point spacing), like the electric rabbit at a dog race. Depending on the spacing of stored position commands, motions can be swift and smooth, or slow and precise. For this reason, the technique is called a "dog race."

Industrial robots typically use dc-torquemotors as actuators, so the signal is proportional to the actuator torque. In that case, as in using hydraulic *pressure* control servovalves, the signal should not be merely a function of the position error. The rate at which position error is changing, the error *derivative*, can be added to improve the dynamic response of the system. Thus, for example, if the error derivative is negative, implying the error is diminishing, the signal to the valve should be less than if the derivative is positive, indicating a growing error. A mechanical analogy to this control technique is a dashpot coupling the position controller and the manipulator arm. A correcting force proportional to the rate of change of error is transmitted to the manipulator via the dashpot.

Some compensation should also be made for steady-state position error caused by gravity, or "droop." Without such compensation, there would always have to be some error to create the actuator forces to resist weight. This error is often eliminated by adding a signal component proportional to the time integral of error, so that eventually even a small error creates a large enough restoring force to eliminate that error. When all three components—position error, integral, and derivative—are used, the system is called a PID controller.

A simpler control system can be developed if *flow* control servovalves are used. These produce essentially full system pressure at even very low signals; the *rate* of motion is controlled directly, not the torque. Therefore there should be no steady-state error, and integral compensation will not be necessary. Mathematically speaking, the integration occurs in the transfer function of the arm. Furthermore, since rate of motion is directly proportional to serovovalve signal, a simple error-proportional system should result in no overshoot. However, a differential term in the control algorithm would improve the response time of the manipulator. Tests or computer simulation would be required to determine the proper system gain, update rate, the effects of point spacing, and the potential advantage of differential compensation. An example is shown in Fig. 8.

In addition, during operation in any of the closed-loop modes, a collision by the manipulator can be sensed as a sudden increase in position error. The actual position of the manipulator will always be known and the computer could use this information to prevent collisions. However, this would require significant computer capability.

The method with which the operator controls the manipulator is also simplified through the integration of the computer. As previously discussed, earlier position controllers were more complex, usually a scaled-down version of the manipulator. They are often a strap-on or "harness" exoskeletal type controller, constraining to the operator and usually fatiguing. If force-reflecting, they are usually overly complex mechanically and also fatiguing. In addition, they do not lend themselves to the cramped quarters of a control van. Therefore a nonreplica type controller would be beneficial, since it would fit nicely over the control console, using the computer to perform the required coordinate transformations.

The manipulator geometry can't be changed, so that portion of the calculations is fixed; but by simplifying the controller, the other half of the calculations can be reduced. Designing the controller arm with a double parallelogram linkage causes the platform on which the pistol grip is attached to always remain horizontal. Therefore the axis of the potentiometer sensing yaw of the pistol grip is always vertical, and the pitch axis is always horizontal. This geometry simplifies the controller calculations to a few simple equations. Other minor changes to such a system could make it slightly more simple mechanically, or desirable from the human factors standpoint, but would cause the coordinate transformations to become much more complex. Thus, the controller shown in Fig. 7 is not optimized for mechanical simplicity, for computer design, nor for human factors; it is the result of a tradeoff among these three considerations.

As mentioned, the nonreplica controller can reduce the large swept volume required by other geometrically similar controllers. Since the motions of the nonreplica controller are converted to cartesian coordinates, its positions can simply be incremented by an arbitrary x, y, and z distance. If, in the course of a task, the controller handle reaches an awkward or uncomfortable position, the operator can electronically decouple it from the manipulator and reposition it to his liking. He merely holds down a button while he adjusts the controller, and all subsequent motion commands will be calculated relative to the position of the handle when the button is released. This process will be referred to as "ratcheting" because of the analogy to a socket drive tool. As a result of this ratcheting capability, the operator can remain comfortably seated, with elbow resting on the console, while controlling the manipulator hand near the ocean bottom, reaching to maximum elevation, or even working aft.

Programmed Control. Among the closed-loop options shown to be highly desirable is programmed control, in which the computer would be used to completely automate repetitious operations, such as tool exchange or other simple functions. This will not only reduce the amount of time to perform such operations by 50 percent or more, but will greatly reduce operation fatigue, once again giving the operator a break while the computer performs the more mundane operations. In addition, programs have been developed that will allow the operator to program the system at the work site, allowing the recall of subroutines to be repeated only a few times, such as following a path back to the work site and realigning itself.

System Controls. Once the computer is integrated into the system, other options now become available to the operator.

Automatic Pan and Tilt. The television cameras are mounted on actuator systems which drive them in the "pan and tilt" (P&T) motions. By automating the P&T system, the operating time can be reduced by up to 17 percent.

For automatic pan and tilt, the computer is required to perform a function that is entirely different from that for preprogrammed control. Instead of remembering a series of commands, it must perform a rapid and relatively sophisticated set of trigonometric calculations in order to generate those commands. Specifically, given the manipulator geometry and the P & T geometry, it must take the manipulator's measured joint angles and geometrically transform them into P & T joint commands.

There are high-level robotics languages designed for versatility in performing transformations from one arbitrarily defined device to another. The solution to this general problem requires very complex and time-consuming matrix algebra. However, for specific devices with fixed kinematic properties (e.g., the manipulator and P&T assemblies), the generalized transformations can be reduced to a simpler set of computations, so the specific kinematic equations for the system can be derived. By starting with the shoulder of the manipulator as the origin, and working outward segment by segment and joint by joint, it is relatively simple to find the cartesian coordinates of the hand. Then, as a sort of inverse problem, since the P&T location is known, the pan angle and the tilt angle necessary to aim the camera at the hand can be found.

Although Fortran would probably not be chosen as the final programming language, it was used to test the complexity and validity of such a solution. The program required about a dozen Fortran lines. It involved 10 trigonometric sines and cosines, and a couple of arc tangents. Both sine and cosine functions can be performed extremely rapidly and relatively accurately by using a single "look up" table in which the computer would have stored the values of these functions. This eliminates the slow process of calculating them using an approximating expansion. The arc tangent can be handled similarly, but requires several times as much computer memory for achieving the same accuracy.

Since the solution to this transformation is a collection of sums and products of trigonometric functions, the accuracy of the results will degrade with each such operation through built-up computational errors. Therefore, a microprocessor using a 16-bit word size—such as an Intel 8086—might better fit such needs rather than a standard 8-bit one. The program language most compatible with the Intel is PL/M. The location of the computer would optimally be in the topside van, rather than on the vehicle.

This places no special burden on telemetry bandwidth, since high-resolution position signals will be sent up the cable, rather than commands sent down. Low-resolution valve commands will also be sent down the cable, but this is a tiny fraction of total bandwidth. Locating the computer topside is in keeping with a general philosophy of having as much hardware as possible on the surface, rather than on the vehicle. It also provides the ultimate backup: In the unlikely event of a computer failure, a new card can be easily inserted. In the more likely event of a position sensor failure, operation can continue in the open-loop mode.

Object Definition and Visual Displays

A major problem with operating undersea vehicles is that they must often work in turbid water conditions, often caused by the vehicle itself as its thrusters disturb the bottom sediment. In a low current environment, this could be a potential disaster, since the water may not clear for some time. However, the computer can once again come to the rescue. Through the use of computer algorithms, the tip of the manipulator hand can be located, and this position data used to "paint a picture" of the object in the computer. As the manipulator touches its way through the scene, the contact points can be stored to produce a graphical representation of the object. Should the object be unfamiliar, but the task simple, the mission may be completed. However, should the object of interest be known in advance, the possibility of storing it graphically in the computer exists. Then, by defining some key points, the computer memory of the object can be fitted to the data on the screen, providing a complete "picture" of the object. At this point, the operator can work on the object or turn the work over to the computer through preprogrammed subroutines.

Control Console. Figure 7 illustrates the concept of a control console with three stations: video coordinator, vehicle pilot, and work system operator. The first two stations are drawn schematically, just to show their relationship to one another and to the work system controls. The video coordinator would be responsible for general monitoring, sonar operation, navigation, and selecting the video allocations needed by the pilot and operator. The pilot must fly the vehicle, guided by the coordinator, and position it with respect to the work in close cooperation with the work system operator. Only the right station, the work system controls, are addressed herein.

The primary controls for the manipulator and grabbers (stronger, less dextrous manipulators designed to aid in position keeping) are labeled in the lower right of Fig. 7. The various controls have been previously described. The position controller is a fully counterbalanced mechanical device which bears no resemblance to the manipulator, but from its joint position the computer will generate manipulator joint commands that cause the manipulator hand to follow the operator's hand. The rate control joystick will operate like an aircraft control stick; it will be spring-loaded to neutral, and displacement fore-aft, left-right, up-down will cause the computer to generate hand motions in the corresponding directions at rates proportional to stick displacement. The programmer controls will be used to store and command positions, position sequences, hand motions (such as tool advance), or whole tasks (such as tool replacement). The manipulator switch controls will be used, like the joystick, to control true orthogonal motions in a closed-loop, computer-generated position sequence or, in the event of computer or position sensor failure, the switch will command specific functions in an open-loop manner. The grabber controls would always operate in an open-loop manner. Manipulator and grabber switches are recessed to prevent inadvertent actuation.

The pan-and-tilt controls consist of two position-control joysticks. These are located on the upper left lap panel for ease of reach by the operator's left hand. Below these are switches for selection and control of cameras and lights, a function that can also be handled by verbal request of the video coordinator.

The primary displays are dual 9-in. (22.9 cm) TV monitors and a 15-in. (38 cm) monitor on which can be displayed the views from the various TV cameras or computer-generated graphic displays. Below the monitors are the usual system status displays and the manipulator analog force gages. The force feedback data and other system status information could also be displayed graphically on the video monitors.

Although there are multiple controllers on the console, it should not be considered to be complex with reduced reliability. Quite the contrary, since the computer controls all manipulator motion, the multiple controllers represent only redundant methods of directing the computer. The desired type of controller to be used is up to the operator, and should one fail, it results in only the loss of that input device with its resultant impact on operation efficiency. The control console is very versatile with built-in redundancy, an asset to any work system.

The Future

The control system discussed does not yet exist, although all aspects of it are within the state of the art and most have been developed or tested by various research organizations. Therefore the quest is to effectively integrate these subsystems into a work system designed to itself be integrated into an operational environment. Integration of such a system will not come easily, since the developers and the users must be aware of and design around the others' needs. This will result in the omnipresent lag time in transferring state-of-the-art technology to solving operational problems in the field. Yet there are more advances on the horizon. The control system of the future will turn more over to the computer. The operator will be able to verbally control certain functions while the computer provides the most critical display for that operation, be it visual, position, force, etc. The work load will be reduced to one where the fatigue factor has been eliminated, and the operator becomes a true "supervisory operator," with the computer handling the remainder. Eventually, with advanced vision systems, the work system may approach autonomy; however, it will hopefully fall just short of the capabilities of HAL, the rebel system of the movie *2001—A Space Odyssey*.

Acknowledgments

The author wishes to acknowledge fellow engineers Clay Davidson and Richard Uhrich for their contributions to the project on which this paper is based.

Based on a paper contributed by the ASME Computer Engineering Division

ered with permission from *Mechanism and Machine Theory*, Volume 12,
J. Charles, J. Vertut, "Cable Controlled Deep Submergence Teleoperator System,"
Copyright © 1977, Pergamon Press, Ltd.

Cable Controlled Deep Submergence Teleoperator System

J. Charles[†]

and

J. Vertut[‡]

Abstract
ERIC II, cable controlled deep submergence teleoperator system, is designed for remote observation, investigation and intervention from a surface ship, with a 6000 m depth capability. The system is in development at CERTSM in Toulon Navy Yard, France on contract of Ministere des Armees; its main parts comprise first the heavy ancillary subsystems; cable handling gear, main cable, tether, PAGODE recovery fish, data and power transmission and second the ERIC II teleoperator fish and its control module. Special attention was paid at man-machine interface problems in the early stage of development and the result is the current development of "telesymbiotic" oriented hardwares: head mobility with TV and microphones sensory feedback, force feedback dexterous arms on sponsorhip of CEA, Saclay, France, agility concept in the fish dynamic control with inertia feedback by kinesthetic motorized sticks also with CEA cooperation. First significative real world experience on underwater dexterous manipulative tasks was gained in late 1974 with great success. First experimentation of ERIC II is scheduled for early 1978.

1. Introduction

ERIC II, cable controlled deep submergence teleoperator system, is designed for remote observation, investigation and intervention from a surface ship, with 6000 m depth capability. The first proposals for the design were issued in 1971, when ERIC I, first underwater teleoperator of the French Navy, began its final stage of shop experimentation: French "Ministère de la Défense", with the financial contribution of CNEXO for a complementary deep submergence photographic system, gave its authorization for the design and construction of ERIC II system in August 1972. CERTSM, Centre d'Etudes et de Recherches Techniques Sous-Marines, in Toulon Navy Yard, is project director.

A deep submergence teleoperator poses heavy demand on cable handling, launch and recovery, power and data transmission and it is necessary to pass by these ancillary subsystems before any materialization of the teleoperator itself. When the teleoperator were finally on design, a cooperation began between CEA, French Atomic Energy Commission and CERTSM which has led to the merging of a common way to solve remotely manned problems on a tentative general scientific approach: application are on remote head, dexterity and agility.

2. General

ERIC II system is a part of the main project PARC–ERIC II in development at CERTSM in Toulon Navy Yard. The purposes of this project are multiple: it covers the updating of deep submergence technology gained by the French Navy with the early bathyscaphe FNRS III and the current one ARCHIMEDE, the generalization of the experience on cable controlled systems which began with ERIC I, the acquisition of a general capability to search, investigate and work on the sea floor whatever the location on the ocean up to 6000 m depth. ERIC II system is the part which fulfills the needs to investigate and work on a point of the seafloor with a remotely manned fish.

[†]DCAN Toulon–CERTSM, 83800 Toulon Naval, France.
[‡]C.E.A., Saclay, France.

Figure 1. ERIC II system.

Architecture of ERIC II system (Fig. 1) includes from the fish up:
ERIC II, teleoperator fish which is the key element.
The buoyant tether which provides the local path for energy and data transmissions.
PAGODE, "fish-house" for launch and recovery of ERIC II, which is the dead weight for the main cable and the storage means for the tether.
The main cable which links PAGODE and the surface ship.
The cable handling gears aboard the ship.
The shelters which provide all the ancillary power and data transmission needs and the man-machine interfaces.
Overall characteristics of ERIC II system are summarized in Table 1.

3. Main Ancillary Subsystems

To have an overview of the ERIC II system, it is useful to review in more details the main subsystems which are necessary parts and form the more expensive items of the project.

3.1 *Main cable*

The main cable has two functions, first power and data transmissions, second load sustaining; the cable has a purely coaxial structure with electrical member inside and load member external.

Table 1. ERIC II system characteristics

ERIC II	Depth, nominal	6000 m
	Depth, proof	7000 m
	Range, horizontal	300 m
	Speed	1 m/s
	Power, max	100 kW
	Weight, in air	4500 kg
Main cable	Diameter	32–35 mm
	Length	8000 m
	Dead weight lower extremity	4000 kg
	Load, nominal	10,000 *daN*
	Load, rupture	19,000 *daN*
Support ship	Deck space	300 m^2
	Deck load, total	130 tons
	Power, total, ERIC II System	700 kW
	Displacement, min	1200 tons
	Length, min	70 m
	Width, min	11 m
	Working speed	0–2 knots

Electrical member has a central core with one copper divided strand, a polythene dielectric extruded on the core, a braided outer conducter with aluminium wires and an external jacket in hypalon rubber.

Load member is a double helical, armor with strands either in ultrahigh strength stainless steel or in KEVLAR (trademark Dupont de Nemours).

General characteristics of the main cable are summarized in Table 2.

3.2 *Cable handling gear*

The cable handling and fish recovery gear are the heaviest parts of the system.

In operation, the gantry is cantilevered astern of the ship and the cable passes to and fro on the main pulley which is free in roll along its upper horizontal tangent to compensate for the roll of the ship. In launch and recovery, the gantry is pivoted by hydraulic rams and the fish is maintained between the legs (Fig. 2).

The linear traction machine exerts the load on the cable with minimum strain: each wheel has an hydraulic motor, and all the motors are supplied with the same hydraulic pressure to undergo the same torque by pair of wheels and thereafter the same force on the cable.

Each pair rotates with a slightly different speed to allow for the progressive longitudinal stretching of the cable along the machine. The hydraulic supply is servoed by a main variable displacement pump to follow the heave and pitch motion of the ship, so that the vertical motion of the cable in water is suppressed.

The slack compensator is moving to and fro with the motion of the cable and maintains the necessary constant force for reeling on the storage drum.

The whole storage drum is laterally moved by a servo system to assume correct spooling of the stored cable; the drum is stopped when the fish is in operation.

General characteristics of the cable handling gear are summarized in Table 3.

Table 2. Main cable

Electrical	Characteristic impedance	56 Ω
	Outer and inner resistance	2 Ω/Km
	Attenuation 12 MHz	8 dB/Km
	Nominal working voltage	6000 V, a.c.
	Corona	7000 V
Strength	Overall diameter, steel	32 mm
	Overall diameter, KEVLAR	35 mm
	Nominal flexure radius	1.6 m
	Flexure life, 10,000 *daN* load	25,000 min

Figure 2. Cable handling system.

Table 3. Cable handling gear

Gantry	Pulley diameter	3.2 m
	Distance between legs	4.5 m
	Height below pulley, vertical	3.7 m
	Height, total, deployed	8 m
	Load capacity	12,000 daN
Traction Machine	Dimensions	12 × 1.5 × 1.7 m
	Force	4000–10,000 daN
	Speed, servoed	±2.5 m/s max
	Speed, cable paying and retrieving	±1 m/s
	Power, hydraulic	400 kW
Slack Compensator	Dimensions	10 × 5 × 3.4 m
	Tension force	300 daN
	Cable length compensation	±4 m
Storage reel	Dimensions	4.4 × 3.5 × 4.3 m
	Speed	±1 m/s
	Storage capacity, 38 mm cable dia.	8000 m
	Weight, total installation	100 tons

3.3 *Pagode*

PAGODE, from the name of a seashell, is suspended at the cable lower extremity; essentially, it is an open box with the tether reel on the top. PAGODE forms a relay between the heavy main cable, and the mobile and agile fish ERIC II, and the necessary dead weight to maintain the main cable down. PAGODE solves also the long deceiving task to launch and recover a neutrally buoyant fish: ERIC II is firmly enclosed in its heavy and strong "fish-house" which acts as a lift between the bottom and the surface; the assembly is progressively coupled with the motion of the ship as it comes near the surface, by the gentle suppression of the servo action of the cable traction machine on the lowest part of the main cable, Fig. 3 and Table 4— depict scale model and characteristics of PAGODE.

Table 4. PAGODE characteristics

Dimensions	6.35 × 3.7 × 2.6 m
Reel capacity	300 m, ϕ 45 mm tether
Weight	4.0 tons in air

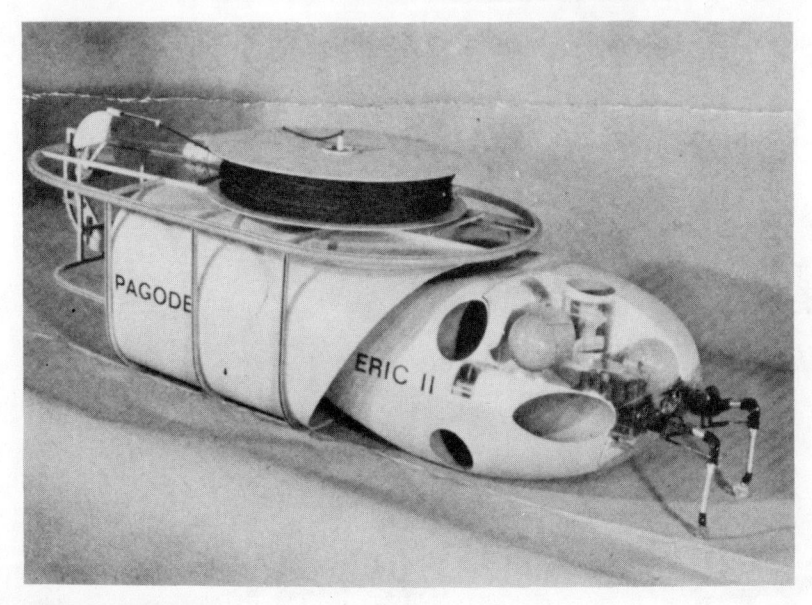

Figure 3. PAGODE.

3.4 Tether

The tether is a buoyant and flexible version of the main cable with the same coaxial arrangement and a lesser structural strength. It is covered by a layer of flexible buoyancy material.

3.5 Transmissions

Power and data transmission capacity is the basic need which dimensionnalizes all the preceeding ancillary subsystems and therefore pertains close scrutiny. Experience gained with ERIC I insufficient propulsive power has given the basis for a 100 kW ERIC II power capacity; the need, for electronic purpose, of a 10% voltage variation between no-load and full-load conditions has led to 6000 V a.c. working voltage; the continuous pressure for the lightiest electrical machinery aboard the fish has made a 400 Hz line frequency mandatory. On the other hand, generous data capacity projection and provision for equipments which have their intrinsic cadence (sonar, underwater navigation) has led to a composite data transmission structure, with analog specific channels and general purpose digital channels. Transmission, a part for power in a up-down direction and TV in a down-up, is bidirectional and managed by the solely two conductors coaxial cable.

Table 5 gives transmission frequency distribution (digital transmission recurrence is 100 Hz).

4. ERIC II

ERIC II *E*ngin de *R*econnaissance et d'*I*ntervention à *C*able, is the logical extrapolation in capability and depth of ERIC I fish. In the early stage of development of ERIC II, a teleoperator minded approach was chosen, and followed up to the current detailed final design stage.

4.1 Fish description

The general requirement of ERIC II is to mimic as close as possible the overall capabilities of a human diver and to do so by remotely manned means. The general aspect of ERIC II (Fig. 4) is a flat ellipsoïd with two tails and a main propeller between them.

The main propeller is gimballed with 2 degrees-of-freedom relative to the fish: the first axis is lateral between the tails, the second axis is orthogonal and consists of a circular track inside the propeller equatorial ring. The tether extremity is firmly attached along the axis of main propeller which, by its patented gimballing arrangement always follows the direction of the tension exerted by the tether. Inside the 2 pivots of the tails, stress transducers measure forces and torques between fish and main propeller, and control orientation and thrust of the propeller to null these forces and torques: the fish itself behaves as if it were a free body without first order tether reactions.

The fish propulsion is provided by 6 ducted propellers inside the hull: 2 longitudinal on each side for foreafter and yaw motions, 2 lateral in the middle vertical plane for lateral and roll motions, and 2 vertical in the front and stern positions for vertical and pitch motions. The 6 propellers have identical characteristics, with variable orientation blades and thrust transducers for servo purpose. The mobility of the fish is monitored on several ways: on a large area basis by bottom acoustic transponders and panoramic sonar, on a local basis by direction and vertical platform and doppler sonar, and indirectly by forward TV camera and lateral sonar.

Table 5. Transmission frequency distribution

		Frequency range
Power	100 kW 6000 V Monophase	400 Hz
Up-down direction	20 analog channels, 10 kHz bandpass	1.3–3.35 MHz
	64 digital channels, 10 bits	
Down-up direction	20 analog channels, 10 kHz bandpass	0.35–1 MHz
	64 digital channels, 10 bits	4.35–5.1 MHz
T.V., down-up	1 analog channel	6.6–11.6 MHz

Figure 4. ERIC II scale model.

Apart from its teleoperator aspect, ERIC II can manage many general purpose oceanographical instrumentation, and *in situ* bottom investigation systems. It can seat firmly on the bottom with a variable buoyancy ballast, or navigate in a truly automatic manner to follow a predeterminate path above the sea bottom. Table 6 summarizes characteristics of ERIC II.

4.2 *"Telesymbiotics"*

"Telesymbiotics", from "tele" *distant* and "symbolic" *living with and from*, would be the word concept covering the science, methodology and technique of using a remote general mechanical organism, teleoperator in a broader sense, in symbiotic relation with a distant man who feels and reacts as if he was in place of the mechanism but without the hostile fact or environment which prohibits his presence. In a deep submergence system, the anaerobic high pressure ambiance is the hostile environment and ERIC II tries to achieve symbiotic relations with its operator in head sensory, arm dexterity, and body agility aspects. The front face of ERIC is devoted at this telesymbiotic approach of the man–machine relations problem. This

Table 6. ERIC II characteristics

Dimensions	$5 \times 3 \times 1.8$ m
Main propeller thrust	1000 *daN* max
Ducted propeller thrust, each	100 *daN* max
Variable buoyancy, vertical	80 *daN* max
Speed, longitudinal	2 m/s max
Buoyancy, nominal	30 *daN*
Electronic modules containers each	80 l., spherical
Primary instrumentation	Forward TV camera Panoramic sonar Acoustical navigation Direction and vertical platform
Optional instrumentation, first items	Doppler sonar Lateral sonar Sea bottom measurements Sea water measurements
Total power	100 kW, 400 Hz a.c.
Weight, in air	4–5 tons

approach is time consuming, expensive and highly minded, for telesymbiotics does not suffer mediocrity because the man in the loop prefers an efficient control lever to an highly sophisticated mechanism which does not match perfectly with him.

5. Remote Head Concept Application

Head is the primary support of the more essential sensory transducers for perception of the external world, i.e. eyes and ears. In analysis of current tasks in target detection and identification, and in manipulation, head mobility relative to the body plays a fundamental and necessary part. Head motion allows visual scanning of a scene, binaural detection of the direction of a sound, depth appreciation by variations of perspective aspects and relations, obstructions overriding in close tasks. "Telesymbiotics" leads therefore to mimic body relative head motion.

5.1 *Head mobility*

Head mobility relative to human body has essentially 3 degrees-of-freedom in rotation, but with a coupling on the 3 translational degrees. Therefore, CERTSM has in development a 6 degrees-of-freedom platform (Fig. 5) which will be on the front of ERIC II and will bear one miniature TV camera and two microphones. This platform has motions range which approximates closely those of the shoulders and neck combination, and dynamics similar to those of the neck.

The man-machine interface is made by a 6 degrees-of-freedom transducer which measures head movements relative to the fixed space, e.g. the seat. This transducer has compensating springs to hold up its weight and sensory stimulators ones, so that the operator head suffers no permanent loading.

5.2 *Sensory feedbacks*

If, in a "telesymbiotic" perspective, head mobility stands for man control in the remote head concept, sensory feedback are carried out by transducers on the 6 degrees-of-freedom platform of ERIC II, i.e. TV camera and twin microphones, and by stimulators on the operator helmet, i.e. minature TV displays with optical transfer and earphones. Figure 6 shows the 6 degrees-of-freedom head mobility transducer; together with a mockup of the miniature binocular TV display: helmet has integral earphones.

Future development of remote head concept must lead to stereo TV and distributed microphones and earphones. Lack of sufficient bandpass on the main cable transmission system, could make, foveal system mandatory: study on an oculometer integral with optical

Figure 5. Remote head platform mockup.

Figure 6. Operator head control and stimulators.

TV display is on way with CEA–LETI at Grenoble, France. Nevertheless remote eye concept is yet to be made clear and mechanical problems are to be set and solved: hierarchical mobility with, in order, hull, remote head and remote eyes are looking very intricate.

6. Teleoperator Arms

Electromechanical arms were the first systems to truly behave in a telesymbiotic way. Some very successfull models of bilateral servoarms were built by atomic energy users and are in growing use in hot cells. Difficulties in underwater transposition and with underwater minded people have made use of bilateral servoarms a long waiting possibility.

Following demonstration of advanced bilateral servoarms by CEA, Saclay, France, a cooperative effort began in late 1973 on underwater application between CEA and CERTSM. Phase one of the development culminates in a public underwater demonstration of dexterity work in an aquarium at OCEANEXPO, Bordeaux, France, October 1974 (Fig. 7).

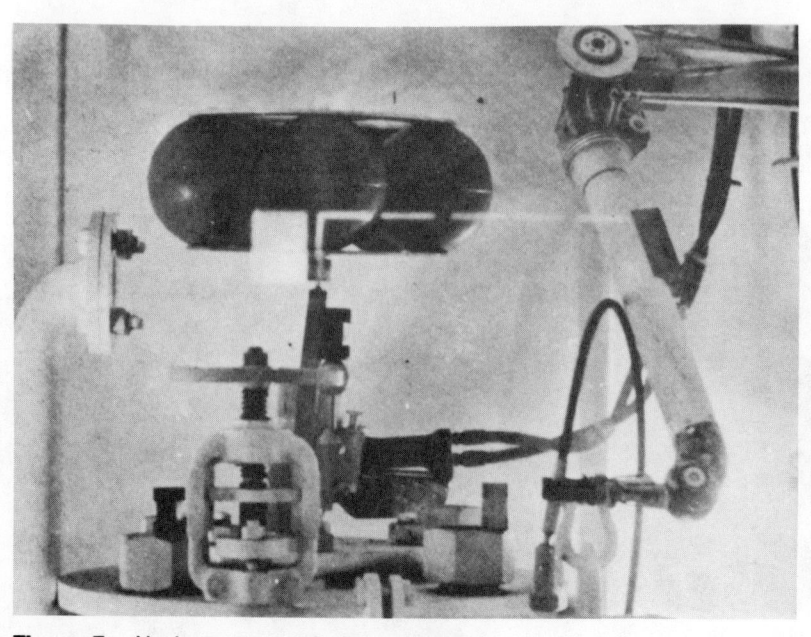

Figure 7. Underwater work demonstration.

6.1 Dexterity

CEA team has made the scientific demonstration of the superiority of bilateral servoarms in manipulative work. In the course of this development, as more and more knowledge accumulated on the specific needs in remote manipulation, dexterity was the central concept which came in view: dexterity is thought as the quality to move with exquisite fineness and accuracy and more with high acceleration and great speed, the prehensive element of an arm. Quantification of dexterity is expressed in terms of accuracy in relative positions of the master and slave hands, deflection ratios, maximum play and motion "graininess"; on the other hand, acceleration, speed and reach are key factors in dynamic evaluations. A final measurement of bilateral quality is the nominal mirror threshold mobility force: it is the force which, exerted on the extremity of the master or the slave, gives no movement of the distant extremity; this force is expressed in term of the distant extremity and therefore affected by the master–slave force ratio.

As an example, some projected characteristics of the bilateral MA 23 arm for ERIC II are summarized in Table 7.

On series of standard tasks with growing difficulty and intricacy, CEA showed that, with 1 for hand manipulation speed, speed of a dexterous bilateral arm is 5, speed of a servo non bilateral arm is 25, and speed of a non servo arm is 100; some tasks are merely impossible with a no bilateral arm.

6.2 Tools

Current approach for effective working with bilateral dexterous arms is to use external tools, powered or not: situation is the same than in normal life. The fine and precise capability of the arm allows, for example, to use impact wrench with hexagonal socket to screw or unscrew bolted assemblies: fitting of the socket and the nut poses no problem.

Bilateral dexterous arms suffer from a lack of a terminal device with equivalent sophistication. Soft palpation, or following and knowing geometrical features of a surface, is the dexterity limit. With a bilateral device which will fit and mimic hand-fingers capability, soft touch, or microgeometrical knowing, will be added and new developments on "fingering" aptitude will appear.

7. Mobility Problem, Agility

One of the more frustrating problem on mobile teleoperator is the poor performance of man as a remote controller of mobility.

7.1 Pilot disorientation

When the pilot is inside the vehicle a complex sensory evaluation of external world and of its relative motion takes place. Vision feedback is the first important one: vision feedback takes is total efficiency when sufficient field of view present; sufficient, that is to say forward visibility range many times superior to the stopping ability of the vehicle, lateral visibility range two or three times its close evolution ability, afterward visibility sufficient to know again the past track. Cognitive vision in vehicle piloting calls for memorization of the preceding and future anticipated situation, latent surveillance of sudden disturbance in all the field where stopping or

Table 7. Bilateral MA 23 characteristics

Relative static positions accuracies	Translation	±3 mm
	Rotation	±2°
	Deflection full load	50 mm
Dynamics no load	Acceleration	10 m s^{-2}
	Speed	2 m s^{-1}
	Reach	1.2 m
Load capacity	Master	6 *daN*
	Slave	20 *daN*
Threshold mobility force	Slave, by master	0.50 *daN*
	Master, by slave	0.20 *daN*

evolving ability are doubtful. Aural feedback is a natural aid and complement to vision feedback, especially in latent surveillance. Last but not the least, kinesthetic and postural feedbacks play their inconscious part: internal ear sensory system gives cues about linear and angular acceleration, and kinesthetic sensation between thighs and seat, between steering devices and arms, gives cues about dynamics and forces on the vehicle.

For a great part, all these sensory feedbacks lack for the pilot of a mobile teleoperator, or more generally of a remotely piloted vehicle. Pilot disorientation comes from the feeling to be external to the vehicle, to have no sense of local acceleration or attitude, and from very restricted field of view in underwater applications.

7.2 Agility

Efficiency of a remotely manned mobile system with sophisticated telesymbiotic devices is still doubtfull if its global motion is slow, coarse or difficult. In the development course of ERIC II system, agility as a fundamental quality of mobile teleoperators came in view: agility has to handle with angular and linear acceleration, attitude stability, external forces reaction sensitivity, path disturbances swallowing.

Agility characteristics of ERIC II are summarized in Table 8.

7.3 Moving display, inertia kinesthetic feedback

Telesymbiotic approach of mobility is harder to manage than remote head or arms. A good way would be to have the pilot in a 6 degrees-of-freedom cabin, like training flight simulators. Lack of place, funds and former habits has precluded this solution; in place, a semi-external control and command system was chosen.

ERIC II pilot deals with several specific visual or kinesthetic feedbacks and hand controls. The first visual feedback is a TV screen on the control console, which map the front field of view of ERIC II; the TV camera is fixed relative to the hull and can be zoomed. The second visual feedback is by the remote head system: pilot can enlarge his field of view by head motion, and can obtain some feeling to be inside the vehicle by the coherent relative motion of image and vehicle; it is a pseudo-internal sensory situation. The third visual feedback is a computer generated display which depicts in real time a moving drawing of ERIC II against a moving symbolic bottom: representation of ERIC II is moved in close relation with high frequency components of the 6 degrees-of-freedom of the vehicle; symbolic bottom is moved in relation with the low frequency components. Two symbolic obstacles or targets can be superimposed on the symbolic bottom at the disposition of sonar operator. The underlying idea with moving display is to give pilot an external point of view of ERIC II local motion: it is thought that pilot feels to be on a smooth constant path behind and above a roving and sinuating ERIC II.

Furthermore, hand controls impress kinesthetic feedback to the hands. These bilateral control sticks are generalization of bilateral arms and consist of six 1 degrees-of-freedom sticks and two 3 degrees-of-freedom sticks. The control sticks are designed by CEA, Saclay on an action parallel to the bilateral servoarm developments. The 1 degree-of-freedom sticks are simple powered sticks in a twin arrangement; one of the 3 degrees-of-freedom sticks has only angular freedom, the other combines angular and translationnal freedom. Controls and feedbacks are interfaced with ERIC II via a computer: several mode will be tried, e.g. direct propeller thrust control, angular and translationnal position, speed or acceleration vectors control, and some combination between these dynamic components. Stick motion feedbacks

Table 8. ERIC II agility

Linear acceleration	Longitudinal	0.5 m s^{-2}
	Transversal	0.5 m s^{-2}
	Vertical	0.3 m s^{-2}
Angular acceleration	Yaw	1.0 rd s^{-2}
	Pitch	0.6 rd s^{-2}
	Roll	0.4 rd s^{-2}
Speed	Linear	2 m s^{-1}
	Angular	1 rd s^{-1}

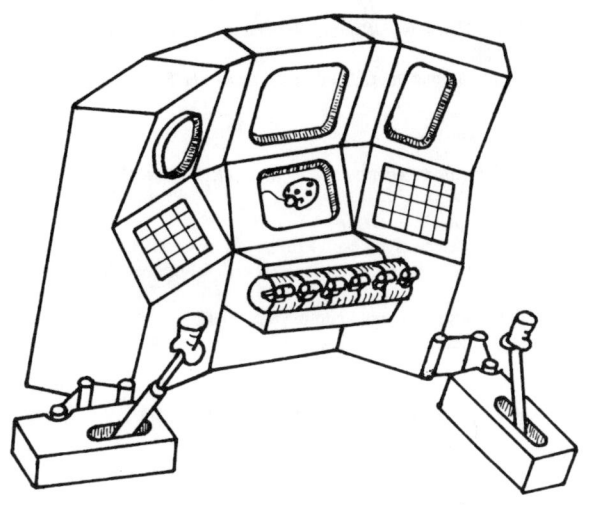

Figure 8. Pilot control console proposal.

will also be a representation of some dynamic characteristics of ERIC II with provision for kinesthetic feelings of hull inertia and external forces impressed on it. A tentative "self organized control" will be tried on the same way.

Remotely piloting a 6 degrees-of-freedom vehicle is not a simple matter and agility concept adds complexity furthermore. It is hoped that, with remote head, moving display and powered sticks hardwares, together with sufficient computer software, some advance will be gained on this most difficult field.

8. Conclusion

ERIC II, cable controlled deep submergence teleoperator, is quite ambitious on a technical point of view. His design and development has followed some constant and basic thought about remotely manned systems as telesymbiotic extension of human operator.

The ideas presented here are not new as the selected bibliography at the end testifies but systematic gathering under a common heading is less frequent: "telesymbiotics" can be a name for the new science of remotely manned systems. ERIC II will not have, in its early 1977 state for first underwater trials all the facilities presented, but basic structure and characteristics can manage future enhancement at the completion of each subsystems.

Acknowledgements—Approval of the whole project by the French Ministere de la Defense has allowed its development and construction. Thanks to financial and personal support of CEA Directors Board and Ergonomy Section of DRME, telesymbiotic system will be used with great enhancement of ERIC II capabilities.

References

Vision–Television
1. L. L. Sutro and J. B. Lerman, *Robot Vision*. Charles Stark Draper Laboratory, MIT, Report R-635, N 73 27936 (1973).
2. D. Jameson, L. M. Hurvich, H. Autrum, R. Jung and W. R. Loewenstein, *Visual Psychophysics* in *Handbook of Sensory Physiology*. Vol. VII/4, Springer–Verlag, Berlin (1972).
3. H. L. Task, *An Evaluation of the Honeywell 7A Helmet Mounted Display in Comparison with a Panel Display: Target Detection Performance*. Aerospace Medical Research Laboratories, Wright-Patterson Air Force Base, Ohio, AD 775 993 (1974).

Tact
1. J. C. Bliss, J. W. Hill and B. M. Wilber, *Tactile Perception Studies Related to Teleoperator Systems*. Stanford Research Institute, NASA CR 1775, N 71 22571 (1971).
2. G. I. Kinoshita, S. Aida and M. Mori, Pattern recognition by an artificial tactile sense. 2nd *Int. Joint Conf. on Artificial Intelligence*, pp. 376–384. The British Computer Society, London (1971).
3. Rabischong *et al. Biomecanique de la main*. Unité de recherches biomécaniques de l'INSERM, DRME contrat no. 72, 668 (1973).

Teleoperator–Kinesthetics
1. E. G. Johnsen and W. R. Corliss, *Teleoperators and Human Augmentation*. NASA, NASA-SP 5047 (1967).
2. J. Vertut *et al.* Contribution to analysis manipulator morphology coverage and dexterity. 1st *CISM-IFTOMM Symposium on Theory and Practice of Robots and Manipulators*. udine, **1**, pp. 277–300. Springer–Verlag, Berlin (1974).
3. R. S. Mosher, *Applying Force Feedback Servomechanism Technology to Mobility Problems*. Robotics Inc., AD 769 952 (1973).

4. R. Goertz, Manipulator systems development at ANL. *Proc. 12th Conf. Remote Systems Technology.* pp. 117–136. American Nuclear Society, San Francisco (1964).
5. H. Kleinwachter, The anthropomorphous machine Syntelman in atomic energy and space research. *Proc. Seminar 2nd Int. Symp. Ind. Robots.* pp. 101–110. Chicago, IITRI (1972).

Piloting–Simulation
1. J. H. Herzog, *Proprioceptive Cues and Their Influence on Operator Performance in Manual Control.* University of Michigan, NASA CR-1248 (1969).
2. *AIAA Visual and Motion Simulation Technology Conference.* Cape Canaveral, AIAA (1970).
3. J. L. Meiry, Space and deep submergence vehicles: integrated systems synthesis. *J. of Hydron.* **3**, 88–94 (1969).
4. S. L. Johnson and S. N. Roscoe, *What Moves, the Airplane or the World.* Aviation research laboratory Institute of Aviation, University of Illinois, Savoy Ill. AD 713 179 (1970).
5. R. L. Barron and R. A. Gagnon, Application of self-organizing control to remote piloting of vehicle. *Proc. 1st Nat. Conf. Remotely Manned Systems* (Edited by E. Heer). California Institute of Technology (1973).

General
1. J. Charles and J. Vertut, *La télésymbiotique.* 2eme Colloque international sur l'exploitation des deons, Vol. 8, pp. Bx 156. 1–26, Bordeaux (1974).

Reprinted from *Marine Technology*, January 1983

Experience with an Unmanned Vehicle-Based Recovery System

Robert L. Wernli[1]

The Naval Ocean Systems Center (NOSC) has been in the forefront of undersea vehicle and manipulator development since the early 1960's. Through extensive at-sea and laboratory test programs, methods have been developed to optimize these remote systems. NOSC's technological background is presented here with particular emphasis on the optimization of undersea manipulator and work systems. Methods of increasing system efficiency while keeping complexity to a minimum are also presented.

Introduction

THE ADVANCEMENT of today's technology often results in the exposure of man to hazardous environments. In his quest for protection, he has made great strides in the field of remote systems technology. Today, with the conquest of new frontiers, remote systems technology is playing a greater and greater role. Sophisticated manipulator systems are being built to work in the nuclear environment and for space exploration and development. The well-defined, mathematically structured realm of space is an ideal location for the application of this technology. An environment not so ideal, however, is that of the deep ocean. Mother Nature has not made man's conquest of the oceans an easy task. Corrosion, extreme pressures, unpredictable sea states, and severe ocean currents combine to provide an unstructured and hostile environment. Because of this, remote-system technology is playing a greater role in ocean exploration and development.

The debate of whether man is required at the worksite in a submersible is still on-going. But, in fact, almost all aspects of man's capabilities, except his ego, can be duplicated sufficiently to perform adequate underwater manipulation and work [1].[2] The increase in the offshore oil industry has resulted in remotely controlled vehicles and work systems replacing divers and manned submersibles in performing many underwater tasks. In the future, as more equipment is designed to be maintained or inspected by remote systems, their use and efficiency will increase. Although the diver will not be totally replaced in the near future, his time in the water can be greatly reduced by the proper integration and use of remote-systems technology. Ultimately, completely autonomous systems will begin doing tasks formerly requiring the "human touch."

Background

One of the pioneers in the application of remote systems technology to the ocean has been the U.S. Navy. Since the early 1960's, the Naval Ocean Systems Center (NOSC) has been in the forefront of undersea vehicle and manipulator development. The basic approach has been to keep the system simple and reliable and to keep the operator topside in a safe, comfortable, controlled environment. Through the application of this design approach, a range of vehicles and work systems has been developed [1–3]. These systems, which are discussed in the following paragraphs, have been operational proof of the Navy's design philosophies.

Snoopy. The *Snoopy* vehicles are small, lightweight, portable

[1] Ocean Engineering Department, Naval Ocean Systems Center, San Diego, California.
[2] Numbers in brackets designate References at end of paper.
Presented at the September 1981 meeting of the San Diego Section of THE SOCIETY OF NAVAL ARCHITECTS AND MARINE ENGINEERS.

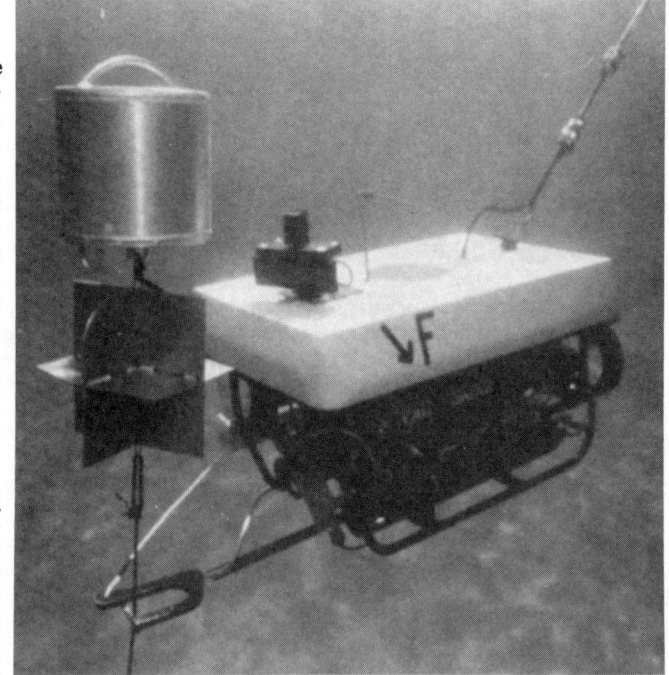

Fig. 1 The NAVFAC *Snoopy* attaches recovery line to target

submersibles, primarily intended to provide a remotely controlled underwater observation platform. As the first in the series, *Hydraulic Snoopy* is basically a small flying television camera capable of operation to 61 m (200 ft). It carries a small grabber for simple recovery tasks. A more advanced vehicle, the *Electric Snoopy*, was developed with the capability to operate to 457 m (1500 ft). It is 1.07 m (42 in.) long, 0.76 m (30 in.) wide, weighs 68 kg (200 lb) in air, and carries a line reel and grabber for recovery tasks. More recently, the NAVFAC *Snoopy* (Fig. 1) has been developed for use by the Naval Facilities Engineering Command during ocean construction work. It is similar to *Electric Snoopy* with the addition of a small scanning sonar system. During recent years, it has assisted in the recovery of three other tethered vehicles that were either lost or entangled on the ocean floor.

SCAT. The Submersible Cable-Actuated Teleoperator (*SCAT*) was initially designed to evaluate underwater head-coupled stereo television. A three-dimensional television display was installed in a helmet to which the motions of the television cameras on the bow of the vehicle were slaved. In this way, the vehicle operator was given the sensation of actually being in the *SCAT*. In addition, a simple, two-function claw was incorporated

101

Fig. 2 The *SCAT* being launched prior to underwater television inspections

Fig. 4 The *MNV* being launched during at-sea mine neutralization tests

to provide a recovery capability. The *SCAT* is currently being reconfigured as a light-duty inspectional work vehicle capable of operating to 610-m (200 ft) depths (Fig. 2).

CURV. The Cable Controlled Underwater Recovery Vehicle (*CURV*) was originally developed for recovery of ordnance items in 1965. The *CURV I* was outfitted with a simple claw built to recover MK-46 test torpedoes at depths below 457 m (1500 ft). The *CURV I* is well known for its assistance in recovering the nuclear bomb which was lost off Palomares, Spain in 1966, as a result of the collision of two U.S. Strategic Air Command aircraft. The *CURV I* vehicle has been replaced by the *CURV II*, with a depth capability of 762 m (2500 ft), and the *CURV III* (Fig. 3), with a depth capability of 3050 m (10 000 ft). The manipulators on these systems have a replaceable hand that easily allows replacement by cable cutter, snare, toggle bar, hook, or other hands of various sizes and shapes. This adaptability more than proved itself when the *CURV III* was flown to Cork, Ireland in 1973, where it assisted in the rescue of the *Pisces III*, the manned submersible that was stuck at a depth of 457 m (1500 ft). A makeshift toggle was used to attach the lift line and ultimately raised the submersible safely, recovering the two men below. The simple design of the *CURV* claw has provided over a decade of reliable, low-maintenance operation.

Fig. 3 The *CURV III* with ordnance recovery claw installed

Fig. 5 The *NP* used by NASA in the recovery of space shuttle rocket boosters

Fig. 6 The *RUWS*, which is capable of operating to 6100 m (20 000 ft)

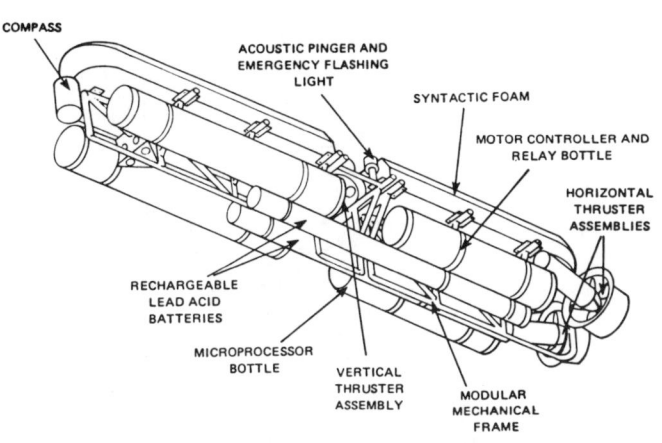

Fig. 8 *Freeswimmer* schematic

MNV. The Mine Neutralization Vehicle (*MNV*) was developed to classify and neutralize sea mines while being deployed from a minesweeper (Fig. 4). Location and classification is performed through the use of a high-resolution scanning sonar and an underwater television system.

NP. The Nozzle Plug (*NP*) vehicle was developed for the National Aeronautics and Space Administration (NASA) to assist in recovery of the solid rocket boosters (SRB) of the space shuttle program. This 4.27-m-high (14 ft) system, shown in Fig. 5, has a capability to fly into, seal, and dewater the partially submerged SRB, thus raising it to a position that will allow towing to a recovery site.

RUWS. The Remote Unmanned Work System (*RUWS*) is a 6100-m (20 000 ft) tethered vehicle system (Fig. 6). The *RUWS* work suit includes two manipulative devices (Fig. 7). A simple, heavy-duty, four-function arm called the *RUWS* grabber is used primarily for position-keeping or object recovery, while a seven-function bilateral master-slave manipulator provides a dexterous working arm.

To control the manipulator, the operator holds a pistol-grip controller and moves it to the position and orientation in space corresponding to that which he wishes the manipulator hand to assume. The *RUWS* vehicle carries several tools that can be acquired by the manipulator to do simple tasks, such as underwater cable cutting.

Freeswimmer. The NOSC/USGS (U.S. Geological Survey) *Freeswimmer* (EAVE West) is an unmanned untethered underwater submersible designed as a testbed platform for advanced pipeline and structures inspection and Navy search and recovery technology. The vehicle is designed to operate in a two-computer supervisory controlled configuration to provide demonstrations of advanced technology in both teleoperator and autonomous modes of operation. The vehicle itself (Fig. 8) is 2.7 m (9 ft) long, T-shaped, with open-frame configuration mounted to a series of syntactic foam blocks for buoyancy. The T-shaped frame was used to minimize total weight of the frame and was made in three sections to allow lengthening of the vehicle to accommodate 25 lb (11.25 kg) of additional payload per foot of extension. The long narrow configuration was chosen to allow for minimum drag in the water. Propulsion is provided by three thrusters, giving the vehicle three degrees of freedom in the water (two canted horizontal thrusters and one vertical thruster). The operating console is an Intecolor 8051 color graphics display terminal and its associated minifloppy disk drive, keyboard, and 24K of user memory. The entire vehicle system is currently being used to demonstrate advances in the technological areas of controls and displays, fiber optics communication links, supervisory controlled manipulators, and automatic pipe-following techniques [4, 5].

WSP. The Work Systems Package (*WSP*) is a work system comprising three manipulators, two television cameras, and 15 interchangeable tools along with the required support equipment (Fig. 9). It is adaptable to six different undersea vehicles. The system is capable of underwater tool exchange and can complete complex work operations without returning to the surface. For example, the simulated flight recorder recovery performed while operating with the *CURV III* used seven different tools and was completed in less than 2½ hr. (Fig. 10). The *WSP*, which is designed to operate to 6100 m (20 000 ft), is one of the most successful remote work systems ever developed for research and development. Considerable advances in remote work systems technology have been acquired due to the extensive amount of research performed with the *WSP*. Therefore, it is discussed later in more detail.

Manipulators. A simple, highly reliable, switch-controlled manipulator known as the linkage arm also has been developed by NOSC (Fig. 11). It is constructed through the use of a double parallelogram tubular linkage. This provides an arm with a high

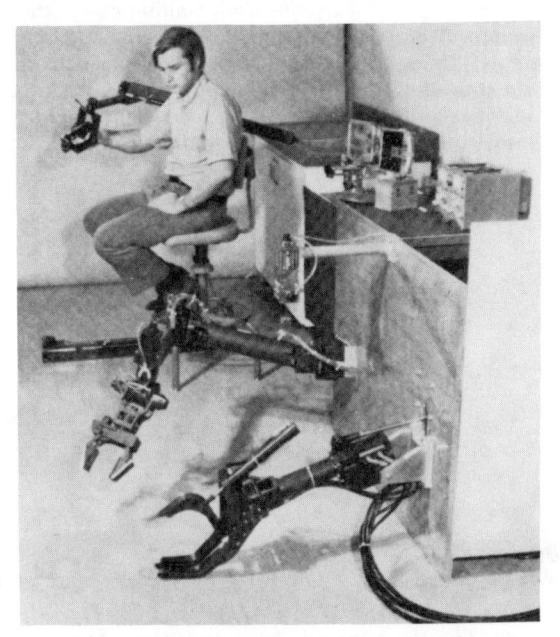

Fig. 7 The *RUWS* manipulator suit during laboratory testing

103

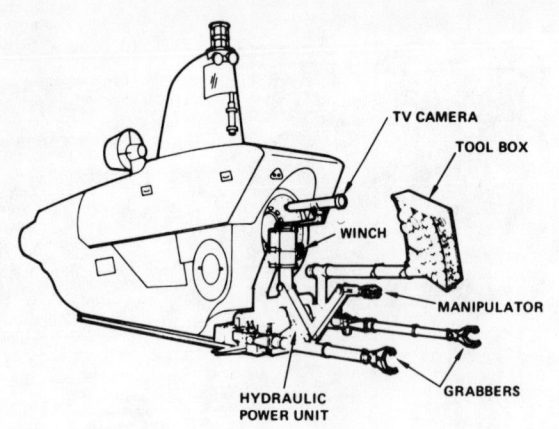

Fig. 9 The *WSP* as it would appear mounted to the *Alvin* manned submersible

strength-to-weight ratio, capable of lifting 23 kg (50 lb), while weighing only 34 kg (75 lb).

An improved linkage manipulator, the Nuclear Emergency Vehicle (*NEV*) manipulator was built for the former Nuclear Rocket Test Station, a joint U.S. Atomic Energy Commission/National Aeronautics and Space Administration (NASA) facility near Las Vegas, Nev. The *NEV* manipulator was designed for service on the nuclear emergency vehicle, for use in air only.

A summary of the manipulators developed by NOSC and their capabilities is presented in Table 1.

Work system design philosophy

Many areas of design must be taken into account when developing systems for remote work in the ocean. Since most of these are common to remote systems, that is, structure, propulsion, electronics, etc., they are not addressed at this time. More importantly, however, the design of the system that will actually perform the remote handling of work operations is discussed. This work system must be capable of the following:

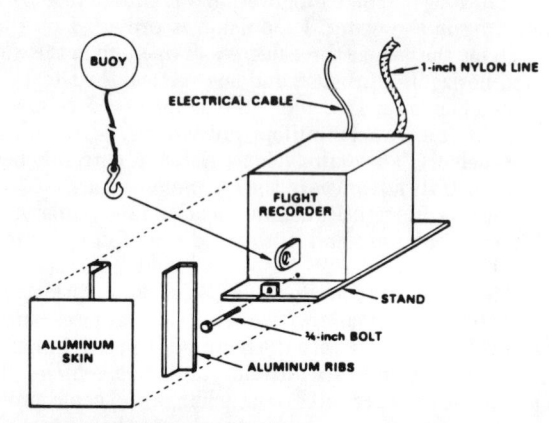

SEQUENCE OF OPERATION
1. EXTRACT THE DRILL MOTOR AND A 1-INCH DRILL BIT
2. DRILL ACCESS HOLES IN THE ALUMINUM COVER TO ALLOW SPREADER INSERTION
3. EXTRACT THE SPREADER, INSERT INTO THE ALUMINUM SKIN AND OPEN THE SKIN TO ALLOW INSERTION OF THE JACK
4. REPOSITION THE VEHICLE TO ALLOW USE OF THE JACK
5. EXTRACT THE JACK, INSERT, AND SPREAD APART THE ALUMINUM RIBS ALLOWING REMOVAL OF THE "FLIGHT RECORDER"
6. EXTRACT THE IMPACT WRENCH AND SOCKET AND REMOVE THE ¾-INCH BOLT FROM THE "FLIGHT RECORDER"
7. ATTACH A BUOY-LINE TO THE "FLIGHT RECORDER" AND REMOVE IT FROM THE TEST FIXTURE USING THE MANIPULATOR
8. EXTRACT THE CABLE-CUTTER AND CUT THE ELECTRICAL CABLE ATTACHED TO THE "FLIGHT RECORDER"
9. EXTRACT THE SYNTHETIC LINE-CUTTER AND CUT THE 1-INCH NYLON LINE ATTACHED TO THE "FLIGHT RECORDER" RELEASING IT TO FLOAT TO THE SURFACE

Fig. 10 Simulated "flight recorder" recovery scenario

1. Attach to and maintain work system orientation at the work site.
2. Provide the manipulation required to operate tools to perform the remote tasks.
3. Provide an adequate viewing system to allow efficient and safe completion of the operations.

The system must have this capability not only on the bottom, but also during midwater operations.

Previous submersibles usually had no more than two manipulator arms: one to hold the vehicle in position, and the other to perform work operations. This configuration caused the system to be pushed away due to the reaction forces of the work manipulator, usually resulting in tool breakage or intolerable completion times of required tasks. To alleviate this problem, the *WSP* was designed using three manipulators: two manipulators to act as grabbers or restraining arms, while the third and more dexterous manipulator was used for performing tool exchanges and work tasks.

Grabbers. The design of grabbers can be held relatively simple. Their primary function is to hold the work system in place, so they do not need additional elements such as extensive angular movements in every joint. The main problem with designing grabbers to act as restraining arms for a system is that not enough attention is paid to what is really being restrained. The grabbers must be designed for enough strength to hold the entire vehicle in place in the maximum expected cross current. The drag forces imposed on the vehicle by the cross current can be quite substantial and can easily damage the grabbers. When the work task is completed, it is also desirable to have a control which will open and retract both grabbers at the same time, thus eliminating the possibility of one grabber being damaged or caught when bearing the entire vehicle load while the other grabber is being retracted.

When designing grabbers, the type of objects to be worked on must be taken into consideration. Not all objects lend themselves to easy attachment of the work system. When working on the bottom or around objects with several appendages, grabbers with conventional-type claws can be used easily. However, if the object to be worked on is large with a smooth exterior, other techniques must be used. One such technique that is being developed is the use of suction pads for attachment to smooth surfaces. These devices lend themselves quite well to deep-ocean applications, where extreme ambient pressures combined with a simple suction pad can provide adequate attachment forces.

Manipulators. The dexterous work manipulator is the heart of the system. It must be capable of exchanging and operating tools and performing the required work operations with accuracy and in the time allotted. Although manipulators come in various forms and levels of complexity, from very lightweight, open-framed rate-controlled manipulators to more complex, master-

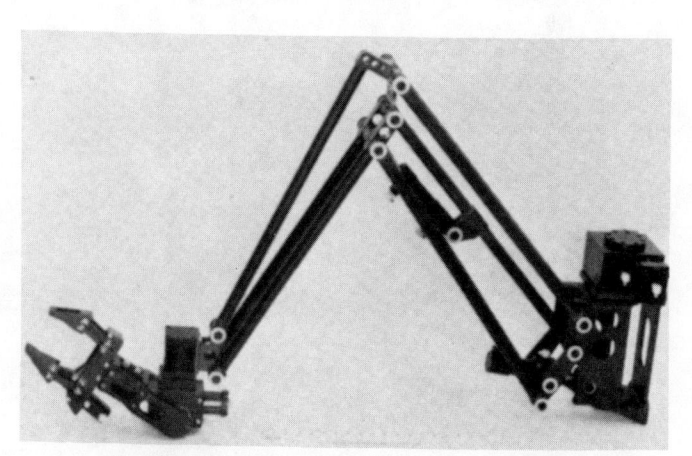

Fig. 11 The seven-function "linkage" manipulator

Table 1 Design characteristics of NOSC manipulators

Manipulators	Number of Functions	Weight in Air, kg (lb)	Lift Capacity, kg (lb)	Maximum Reach, cm (in.)	Operating Depth, m (ft)
SCAT claw	2	9 (20)	23 (50)	91 (36)	610 (2 000)
CURV I claw	3	45 (100)	182 (400)	127 (50)	610 (2 000)
CURV II claw	4	45 (100)	182 (400)	127 (50)	762 (2 500)
CURV III claw	4	45 (100)	182 (400)	127 (50)	3050 (10 000)
Linkage manipulator	7	34 (75)	23 (50)	140 (55)	2135 (7 000)
NEV manipulator	7	45 (100)	23 (50)	140 (55)	0
RUWS manipulator	7	27 (60)	20 (45)	127 (50)	6100 (20 000)
RUWS grabber	4	33 (73)	91 (200)	61 (24)	6100 (20 000)
WSP manipulator[a]	7	227 (500)	45 (100)	183 (72)	6100 (20 000)
WSP grabbers	6	113 (250)	113 (250)	274 (108)	6100 (20 000)

[a] Manufactured by PaR System Corp.

slave-type manipulators with proportional control and force feedback, the complexity of the manipulator must be tailored to the types of tasks to be performed. Most tasks involving the use of tools can be adequately performed with a simple, rate-controlled manipulator. For example, the manipulator on the WSP is a seven-function, rate-controlled, hydraulically actuated manipulator. Other tasks requiring large excursions of the manipulator and random motions such as rigging or valve turning may be more efficiently performed through the use of master-slave-type manipulators. However, the following should be kept in mind.

1. A master-slave-type system occupies much more space in the control room and can impose considerable restraints if operated in the pressure sphere of a manned submersible.

2. When performing tool operations such as drilling or tapping, which require holding the manipulator in a predesignated position for an extended period of time, the master-slave harness can become very fatiguing.

3. A more dexterous or master-slave-type manipulator generally results in a more expensive, complicated, less-reliable system, although it may do the job faster and more accurately.

Because of its importance to the work tasks, the manipulator is usually the first item considered for modification. In fact, this may not be the place to start designing a more efficient system. Recent studies have shown that when performing work at sea with tools, the manipulator is used only 30 percent of the time, while the operator spends 37 percent of his time in decision-making, 11 percent of the time in operating television cameras, and the remaining 22 percent of the time operating tools (Table 2) [6]. Therefore, other areas such as reducing operator decision time, eliminating the need for repositioning cameras, or increasing tool efficiency can have a large effect on the efficiency of the entire system. Although a more dexterous, faster-operating manipulator may aid in reducing operator decisions, the primary effect will be across only 30 percent of the total task time, that is, that time which is spent actually operating the manipulator. Thus a manipulator system that is twice as fast will not necessarily cut the total operational scenario time in half.

However, almost any method of increasing the efficiency of the overall system and thus reducing time and power consumption required by the work system is of great significance, especially when working with manned submersibles. For example, the WSP runs on 60-Vdc batteries, either its own or those of a manned submersible. Since manned submersibles have limited dive times, the impact of the task or mission to be performed on the battery supply of the vehicle is quite important, especially when considering the amount of time and power required to dive to 6100-m (20 000 ft) depths.

New technologies also are lending themselves to the performance of remote manipulation tasks. For example, through the use of minicomputers programmed to control manipulators, the amount of time to perform repetitive tasks can be considerably reduced. This can be of great benefit when undertaking such repeated tasks as tool exchanges performed by the manipulator. Results of the tests performed on the WSP using microprocessor control are presented in Table 3. The benefit to the operator can be seen easily. Routines have been developed in which the operator can push a button and a microprocessor can store the entire movement of the manipulator for future use. This can be of great benefit in complex path-following or in performing tasks

Table 2 WSP operational time distribution (percent)

Operation without tools (%)	Operator Decision	Manipulator Operation	Camera Pan-and-Tilt Operation	Tool Operation	Light Operation[a]
Average operation time	50	33	17	...	100
Low-speed pump idle time[b]	50	...	...	...	...
Low-speed pump duty time[c]	...	33	17	...	...
Total power consumption	32	27	14	...	27
Operation With Tools (%)					
Average operation time	37	30	11	22	100
Low-speed pump idle time	37	...	...	(22)[e]	...
Low-speed pump duty time	...	30	11	...	...
High-speed pump duty time[d]	...	...	...	22 (10)	...
Total power consumption	17	18	6	26	23

[a] Lighting = 0.75 kW
[b] Low-speed pump idle = 1.55 kW
[c] Low-speed pump duty = 2.00 kW
[d] High-speed pump duty = 3.97 kW (on-off only)
[e] It is assumed the manipulator is not being moved during tool activation.

Table 3 Comparison of WSP task times (minutes) under direct operator control and computer control

Task	Operators Inexp.	Exp.	Pro-grammer	Reduction, % Inexp.	Exp.
Acquire tool	5.18	2.12	0.90	82	57
Replace tool	3.24	1.42	1.31	59	8
Acquire bit	3.02	1.23	1.00	33	17
Replace bit	3.56	1.30	0.74	79	43

not known prior to the dive. Such a routine thus allows efficient integration of subroutine storage with actual operations. When considering programmed assistance, the designer must assure that the required programming time does not exceed the time in which the operator could manually perform the task, especially with tasks that are not too repetitive.

With the addition of position sensors to the manipulator, the minicomputer can then be expanded to include control of the viewing systems. It would be a simple task to instruct the camera pan-and-tilt units to automatically follow the manipulator hand position. Table 2 indicates that savings of up to 17 percent can be achieved by eliminating the manual control of the camera systems. This would have the additional benefit of allowing the operator to concentrate on the task at hand without having to stop operations to move or adjust the television cameras. These are but a few of the areas that lend themselves to computer control. Eventually, it is conceivable that preprogrammed submersibles with object locating and recognition routines will be entering the field of undersea work.

Recent at-sea testing

During Fiscal Year 1979, techniques for remote work and recovery operations in the deep ocean were evaluated. As a result, an early concept of a recovery system was established. However, the answers to several tradeoff questions were required to complete the concept. Additional input from practical at-sea tests of the concept would be required. It was decided that two systems currently exist which, if mated together, would closely meet the requirements for work capability, size, and thrust capability of that concept: the Work Systems Package (WSP) previously described, and the Pontoon Implacement Vehicle (PIV) (Fig. 12).

Fig. 12 Work System Package/Pontoon Implacement Vehicle

The Pontoon Implacement Vehicle was chosen as the mounting platform for the WSP. The PIV was developed as a part of the Large Object Salvage System (LOSS) at the Naval Coastal Systems Center, Panama City, Fla. The PIV is a cable-controlled, highly maneuverable vehicle with a high thrust capability and a 900-kg (2000 lb) variable ballast system.

The WSP and PIV were mated together and transferred to San Clemente Island (SCI) for testing. The purpose of the test was to investigate or develop applicable recovery techniques to be used in conjunction with a remotely controlled vehicle/work system. Results from the testing will be used in the formulation of a technology base which will provide the Navy with the capability to develop future systems to perform deep-ocean recovery operations [7].

An operational depth of 65 to 95 ft (20 to 29 m) was chosen to maximize documentation by divers. Tradeoff studies were conducted during FY-79 to determine the most appropriate method of rigging and lifting objects from the ocean bottom for the WSP/PIV. For testing purposes, objects were chosen to reflect the general characteristics of classes of objects which might require recovery. Studies were conducted to determine the most effective scenarios for attachment, rigging, and recovery of these objects using an unmanned tethered vehicle. It was assumed that the object had been located, marked, and photographed and that the recovery team knew, as accurately as possible, the condition of the object to aid in the choice of attachments. Several different techniques were utilized for the recovery exercises, depending upon the generic class of the target.

During the 30 days of testing at SCI, 14 dives were made with the vehicle which accumulated a total of 58 hr of in-water time. The operating experience gained with the vehicle/work system and the substantial amount of photographic documentation acquired have greatly enhanced the success of this series.

The basic approach to these tests was from an engineering standpoint. Given a recovery task, an engineering approach could be made to the task which would result in the development of simple and reliable techniques to ensure a successful recovery through the use of remote systems. Based on this approach, the following objects were successfully rigged for recovery and lifted to the ocean surface using those recovery techniques:

  a. Slinging and lift of an F4 aircraft.
  b. Claw attachment to and recovery of a jet engine.
  c. Rigging and recovery of a large steel object.

In addition, techniques were developed which successfully demonstrated the system's capability to perform the following:

  a. Rigging of objects (installation of lift lines, snaphooks, etc.).
  b. Performance of midwater maneuvering, docking, rigging, and recovery operations.
  c. Successful installation of lift slings on an intact aircraft.
  d. Object recovery using the vehicle variable ballast and thrust as the lift force.
  e. Remote implacement and deployment of a lift module which can be controlled by the work system or a microprocessor to generate a 4500-kg (10 000 lb) lift force.
  f. Object recovery using the lift module while under diver control.
  g. Installation of "toggle bolt" lift points through heavy steel plate.

The knowledge gained from these operations will enhance man's quest in extending his presence throughout the oceans via remote systems and is already being incorporated into the conceptual design of an advanced tethered vehicle/work system.

Conclusion

Design of a more efficient manipulator or work system does not necessarily mean a more complex or expensive system.

Through the use of simple, reliable systems with highly trained operators, great strides can be taken toward system efficiency. And, with the addition of today's computer technology, the system can approach automation, requiring only a supervisory operator and eventually only a programmer. Application of this technology to system design, combined with a "real world" engineering approach to the problem, can result in a system highly advanced in its capability. The ocean is one of the few frontiers remaining to man, and its conquest will be through the use of remote systems—systems that are as simple and rugged as the ocean itself.

References

1 Talkington, Howard, "Manned and Remotely Operated Submersible Systems: A Comparison," Technical Paper 511, Naval Undersea Center, 1976.

2 Uhrich, Richard W., "Manipulator Development at the Naval Undersea Center," Technical Paper 553, Naval Undersea Center, 1977.

3 "Ocean Technology: A Digest," Technical Document 149, Naval Ocean Systems Center, San Diego, Calif., 1978.

4 Bosse, Peter and Heckman, Paul, "The Development of an Underwater Manipulator for Use on a Free Swimming Unmanned Submersible," Technical Report 652, Naval Ocean Systems Center, San Diego, Calif., Oct. 1980.

5 Heckman, Paul, "Free-Swimming Submersible Testbed (EAVE WEST)," Technical Report 622, Naval Ocean Systems Center, San Diego, Calif., Sept. 1980.

6 Wernli, Robert, "Development of a Design Baseline for Remotely Controlled Underwater Work Systems," IEEE Oceans '78, Naval Ocean Systems Center, San Diego, Calif., 1978.

7 Estabrook, Norman B., Wernli, Robert L., and Hoffman, Robert T., "Recovery Operations Utilizing a Remotely Controlled Vehicle/Work System," MTS-sponsored Oceans '80, Naval Ocean Systems Center, San Diego, Calif., 1980.

Development of "Underwater Robot" Cleaner for Marine Live Growth in Power Station

KYOSUKE EDAHIRO

Takasago Machinery Works
Mitsubishi Heavy Industries, Ltd.
676 2-1-1 Shinhama, Arai-cho,
Takasago City, Hyogo Pref., Japan

1. Background of the Development

   At power generating stations with cooling system using sea water, fouling of water channel, pipelines and in-line equipments by marine live growth has long been major maintenance problem to cope with.
   Marine lives such as mussel, barnacle, oyster and bryozoa are distributed very commonly at the shore line in temperate to sub-tropical zone, and the incoming ova and larvae attach and grow-up on the system wall surface, especially on intake side of the system.
   The growth not only affects to system loss or performance, further jeopardizes the safety of the plant operation causing heat exchanger tube clogging and failure.
   To suppress and remove the growth, injection of chlorine at intake, use of anti-fouling coating, provision of screen system and debris filter have been widely applied.
   However, the trend is to minimize the chemical release for ecological reasons and neither screen nor filter is effective against the growth within the system. Thus the clean-up of the system during the plant shutdown is becoming serious burden at the stations not applying the chemical suppression, due to the enormous amount of marine growth to dispose and to the foul working environment.
   Under above situation, the effective substitute measure of suppression and removal has been sought in several approach (Table-1).
   But, up to date, practical method applicable on commercial basis has not yet been established.

2. Development of "Underwater Robot"

   To develop practical marine growth control system to be applied on the thermal power station cooling water channels, Tokyo Electric Power Co., Inc. (TEPCO) and Mitsubishi Heavy Industries, Ltd. (MHI) proceeded a joint R&D program in 1981 - 1982.

   As the basic principle, mechanical removal method was chosen in view of practicability and expected stable performance at actual service condition.
   The target of the functional specification was:

   1) Objective Marine Growth

      . Species: Mussel, barnacle, oyster, starfish and etc.
      . Surface applied: Open or closed channel/culvert
      . Max. removable growth: Up to 10cm in thickness

   2) Operating Conditions and Requirements

      . Applicable to servicing channel, velocity up to 2m/s.
      . Fully remote and safe operation (no underwater operator)
      . Easy handling/transportation for efficient cleaning along long channel
      . No major modification of channel required
      . Self sustained power supply
      . With wide operating flexibility and system expandability

   Prototype system was designed so as to be operated at long term runs on commercial basis.

The schedule of the program including continuing field pre-commercial test runs is shown on Fig.1.

3. System Description

The "Robot" system comprises following functional units.

1) "Robot" cleaner unit

2) Hydraulic/electric power unit

3) Hose and cable with reel unit

4) Control equipments

System concept and interconnections are shown on Fig.2

3.1 "Robot" Cleaner Unit (Fig.3)

A submergible, hydraulically powered, remotely maneuvered, power brushing-scraping mobile.

The unit is maneuvered as clinged to either of side wall, ceiling or bottom of the channel by reaction force generated by two axial pumps and four independently driven wheels permitting versatile motion in water.

Four power driven brushing wheels are arranged to enable to sweep wider area and also to clean hunched area of the channel.

Several types of brush and scraper are available to best suit for each species of marine growth and duty of the job (→ 4.2).

The body is made of FRP with integral buoyancy tanks to cancel the weight under submerged condition.

Color ITV camera, lighting and position sensors are provided for remote control from the control panel above ground.

> Weight: 520 kg (zero in sea water)
> Dimensions: 1780(L) x 1270(W) x 800(H) mm
> Body material: FRP
> Cleaning capability: 1000 m^2/h at medium duty
>   Max. 2100 m^2/h
> Mobile speed and brush wheel speed: adjustable

3.2 Hydraulic/Electric Power Unit

A packaged, high/low dual pressure hydraulic pump and generator unit driven by a diesel engine.

All the equipments are integrally assembled on a common bed, and are designed for self-standing operation.

The generator supplies all the necessary electric power to the control circuits, ITV, lighting and other electrical equipment.

External electric source is applicable instead of the diesel engine.

> Diesel engine: 65 ps/1500 rpm
> Hydraulic pumps: HP 160 kg/cm^2, 80 ℓ/min.
>   LP 30 kg/cm^2, 20 ℓ/min.
> AC generator: 3 kW

3.3 Hose and Cable with Reel Unit

Hoses for hydraulic oil, control cable and core wire are bound in a single sheath and wound on a power driven reel to enable smooth feeding and rewinding.
The hose assembly has float rings to compensate its weight to allow free motion of the robot in the water.

Maximum length: 100 meters

3.4 Control Equipment

3.4.1 Control Unit

A control console with maneuvering switches, control joystick, ITV monitor, posture monitor indicators and power unit remote control devices, is installed on a van type vehicle as a mobile operating station.
The van is also used as transporter for crew and accessories.
A strapheld portable control box is also available.

3.4.2 Submarine Sensors

a) Color ITV camera and lighting
b) Touch sensors (Limit switches)
c) Transducer for ultrasonic posture indicator

4. Operating Experiences and Test Results

4.1 Test Operation History

The test operation of the prototype system was conducted with series of sequence as shown on Fig.1

Through each series, operating know-how was gained, and minor problems arose were remedied by improvement modifications.

4.2 Major Items of Test and Their Results

4.2.1 Maneuverability in Water

Estimated adhesion force by axial pump thrust was max. 270 kg to permit the operation in channel water velocity of max. 2.2 m/s.
It was confirmed that the robot can safely be maneuvered in channel of water velocity up to approx. 2 m/s.

4.2.2 Test of Rotary Brush/Scraper

Fig.4 shows the typical type of rotary brush/scraper. Because the selection of the best type suitable for different growth level and species determines the cleaning capability and efficiency, 20 test brushes and scrapers of different type and configuration were prepared and tested on different conditions and combinations.

Test results are summarized as below:

	Species and Growth Level	Type	Material	Cleaning Ability (Line Speed)
1	Fully grown up barnacle and oyster, thick. 5 ~ 8 cm	Heavy duty loop type	Tool steel blade	420 m²/h (Approx. 5 m/min.)
2	Grown up mussel with smaller barnacle and oyster thick. 5 ~ 8 cm	Medium duty loop type or heavy duty brush type	Tool steel blade Hardened steel wire	1170 m²/h (Approx. 15 m/min.)
3	Larvae of mussel, barnacle and oyster thick. ~ 2 mm	Light duty Wire brush	Steel wire or Polypropyrene	2100 m²/h (Approx. 25 m/min.)

Fig.5a shows the surface of side wall of covered channel before cleaning. Main species is mussel, and ciona intestinals and starfishes are also seen.

Fig.5b shows the same place after cleaning.

4.2.3 Interval Test

The preferable mode of the robot application will be cleaning during the larvae phase of growth with appropriate interval.

This mode is more efficient because there is no need to dispose the big amount of removed growths and it permits higher cleaning speed.

To determine the appropriate operating interval the test was carried out during 1982 - 1983.

It could be concluded that cleaning once a month will be sufficient even in peak growth season.

5. System Expansion and Improvement

5.1 Debris Disposal System

Removed marine growths disposal system comprising debris suction pump with submergible (robot controlled) sweeping suction head, hose, debris water separator was tested and the commercial system will be available on request.

5.2 Future Improvement

At present, improvement of maneuverability at blind operation in turbid water is being proceeded.

Utilizing the obtained experience, the system will soon extend to robot cleaner system suitable for coated pipelines.

6. Afterwords

The Technology gained through the development could also be applied to similar underwater work area, and MHI will continue the effort for it.

MHI wishes to acknowledge to Tokyo Electric Power Co., Inc. for the permission of reporting the results from the joint R&D program between TEPCO and MHI.

The Underwater Robot Underworking

Fig.1 Development Schedule

Item \ Year	1980	1981	1982	1983
1. Ecological & Site Survey	▨▨			
2. Basic Design		▨▨		
3. Detailed Design		▨		
4. Manufacturing		▨		
5. Functional Test		▨▨		
6. Trial Runs at Power Stations			▨▨▨	
7. Interval Test				▨▨▨

TABLE-1 POSSIBLE ANTI-FOULING PROCESSES FOR POWER STATION SEA WATER SYSTEM

PRINCIPLE	PROCESS	DESCRIPTION	PROBLEMS	REMARKS
1. Chemical Inhibition	1 Chlorine/Hypochlorite Dosing	Most widely applied with stable effect, concentraion 2 ~ (10) ppm	To be reduced in ecological view (in the future)	
	2 Anti-Fouling Paint Coating	Paint containing cuprous oxide and/or organic tin is usually applied.	To be reduced in ecological view (Re-coating at several years interval required.)	
	3 Copper Alloy Lining	Used for ship hull anti-fouling, but not common for channel application	To be reduced in ecological view	
	4 Ozonation	Injection of ozon solution. Reported to be effective for slime fouling.	Availability of large capacity ozonizer.	
2. Physical Inhibition	1 Sonic/Ultrasonic Excitation	Reported to be effective for specific spectrum and phase of growth.	Low efficiency, high investment/ operating cost.	
	2 Ultraviolet Irradiation	- Ditto -	- Ditto -	
	3 Electro-Magnetic Field Treatment	Impression of strong magnetic field. Mechanism of inhibition is not clear.	Questionable in large scale application.	Magfree (R) (MHI) (Hydro Dynamics Corp. U.S.A.)
	4 Hot Water Recirculation	Heating over 45°C, holding more than several hours are required.	High operating cost. Disposal of hot water.	
	5 Increase of Velocity	Effective for pipeline anti-fouling. Velocity over ~ 3.5 m/s.	Increase of horsepower. Not applicable to channel	
3. Mechanical Removal	1 High Speed Water Jet	Jet nozzle sweep/traverse be required.	Low efficiency in case of under-water jet.	
	2 Rotary Brush/Scraper	Primitive method, but most stable effect is expected.	Life and effect of remover. Applicability to coated surface.	
	3 Shock Wave Excition	Underwater electrical discharge or explosion.	Not practical at large scale Operating safety.	Joint Research between TEPCO-MHI
	4 Use of Pig or Baloon	Actually applied for condenser tube cleaning.	Limited application to actual channel/pipeline system.	

Fig.2 System Concept and Interconnections

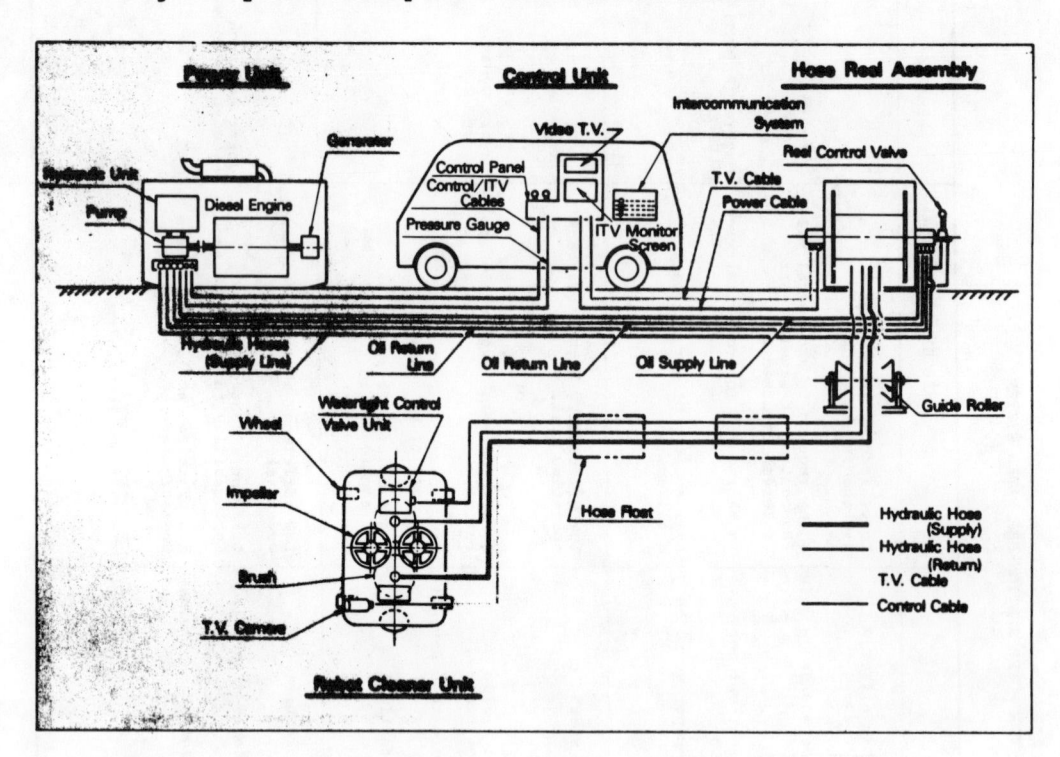

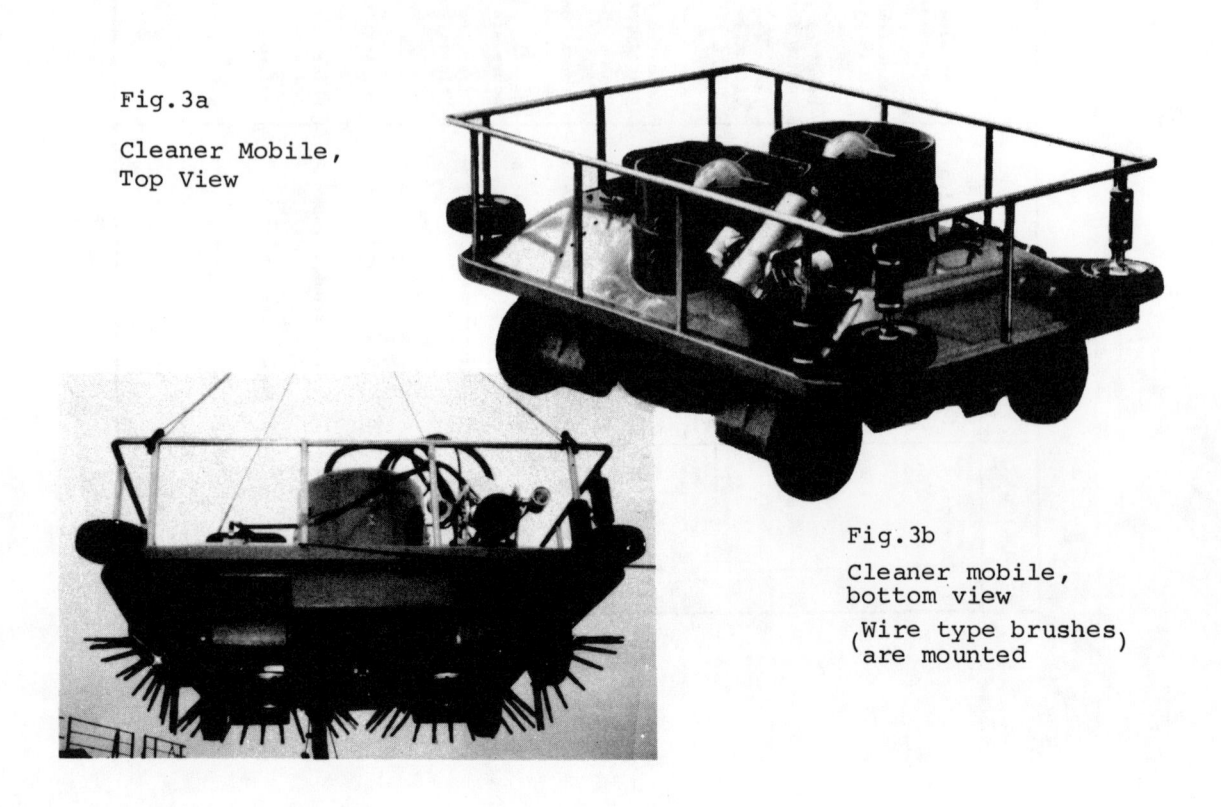

Fig.3a

Cleaner Mobile, Top View

Fig.3b

Cleaner mobile, bottom view

(Wire type brushes are mounted)

Fig.4 Type of rotary brush/scraper

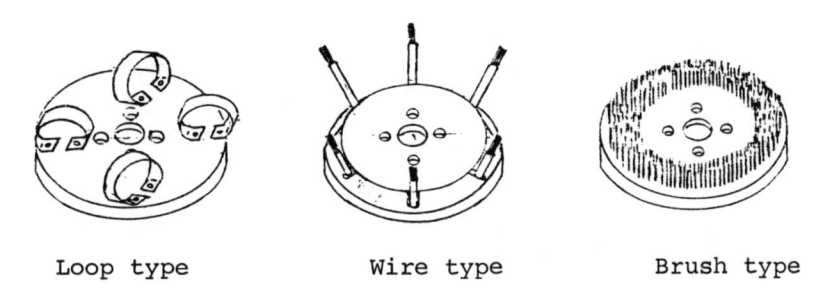

Loop type   Wire type   Brush type

Fig.5 Surface before and after cleaning

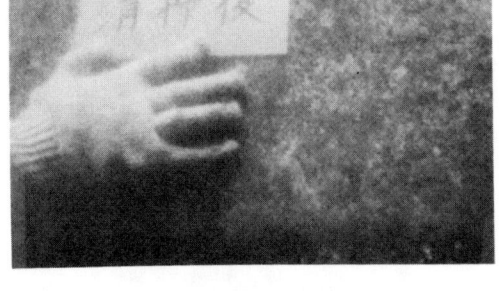

a. Before Cleaning   b. After Cleaning

115

CHAPTER 4

SPACE APPLICATIONS

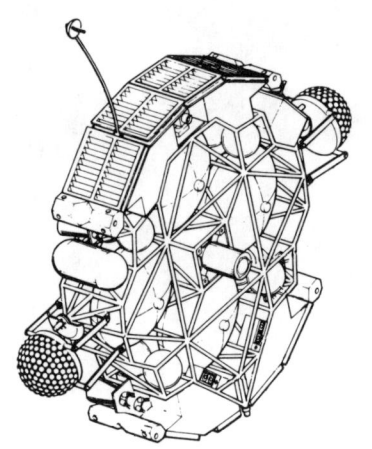

SYNERGY IN SPACE —
MAN-ROBOT COOPERATION

As space operations become space industry, robots will evolve into autonomous instruments, capable of propelling themselves to a repair site; carrying tools, spare parts, and sensors; diagnosing and repairing faults; performing structural activities; and communicating with human supervisors. These "telepresence" tools — containing high-quality sensory feedback mechanisms — will take us a giant step toward the thinking machine

SAM WALTERS
Associate Editor

On October 19, 1983, at NASA's 25th anniversary celebration in Washington, D.C., President Reagan forecast a national space strategy for the next 25 years and beyond. "We're planning an entire road," he said, "a high road, if you will, to the permanent occupation of space."

Described in the President's message were immense spacecraft assembled in orbit from subassemblies or individual parts brought up by shuttle orbiters, extremely large antennae, power generation equipment, large spacecraft for earth departure expeditions, and manufacturing plants, including materials processing facilities.

A SPACE FLOTILLA

It is not yet certain what the first space station will actually look like, since no space station design currently exists. So far, NASA has settled only on the concept of a "flotilla" (Figure 1), with a manned base in the center providing a human habitat with room for six to eight astronauts, a utility core, modular laboratory, and orbital service station. This base will have at least two unmanned platforms floating around it, bearing sensitive instruments such as telescopes and automated modules for zero-gravity materials production.

Within such a concept the shuttle is the basic assembly and logistics vehicle. In addition to carrying the various station elements to orbit, it will help assemble the space station and periodically return to the manned element, bringing supplies, fresh crews, and equipment. Transportation between the manned element and the free-flying platforms will be by a co-orbiting, unmanned space tug called an orbital maneuvering vehicle (OMV), a high-energy, self-propelling stage for transporting payloads to higher earth orbits or into the solar system.

A few examples of candidate configurations for a 1992 space station are shown in Figure 2. Certain common architectural features can be noted. The laboratory modules, habitat, servicing bases, and other elements are all modules similar to the European Spacelab whose size is determined by the dimensions of the orbiter payload bay.

The flotilla concept is based upon two radical approaches to spacecraft design. One is that of evolving tech-

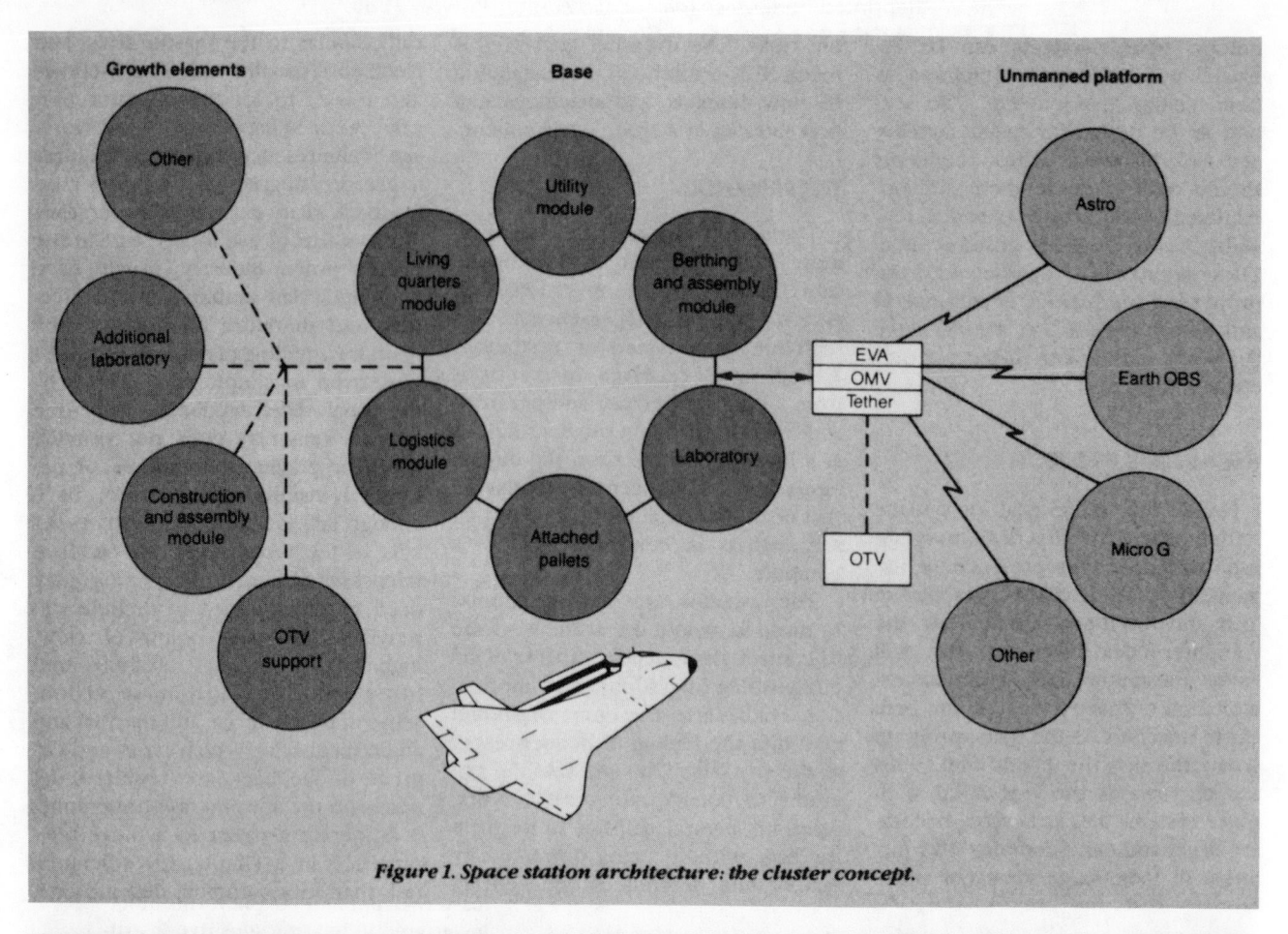

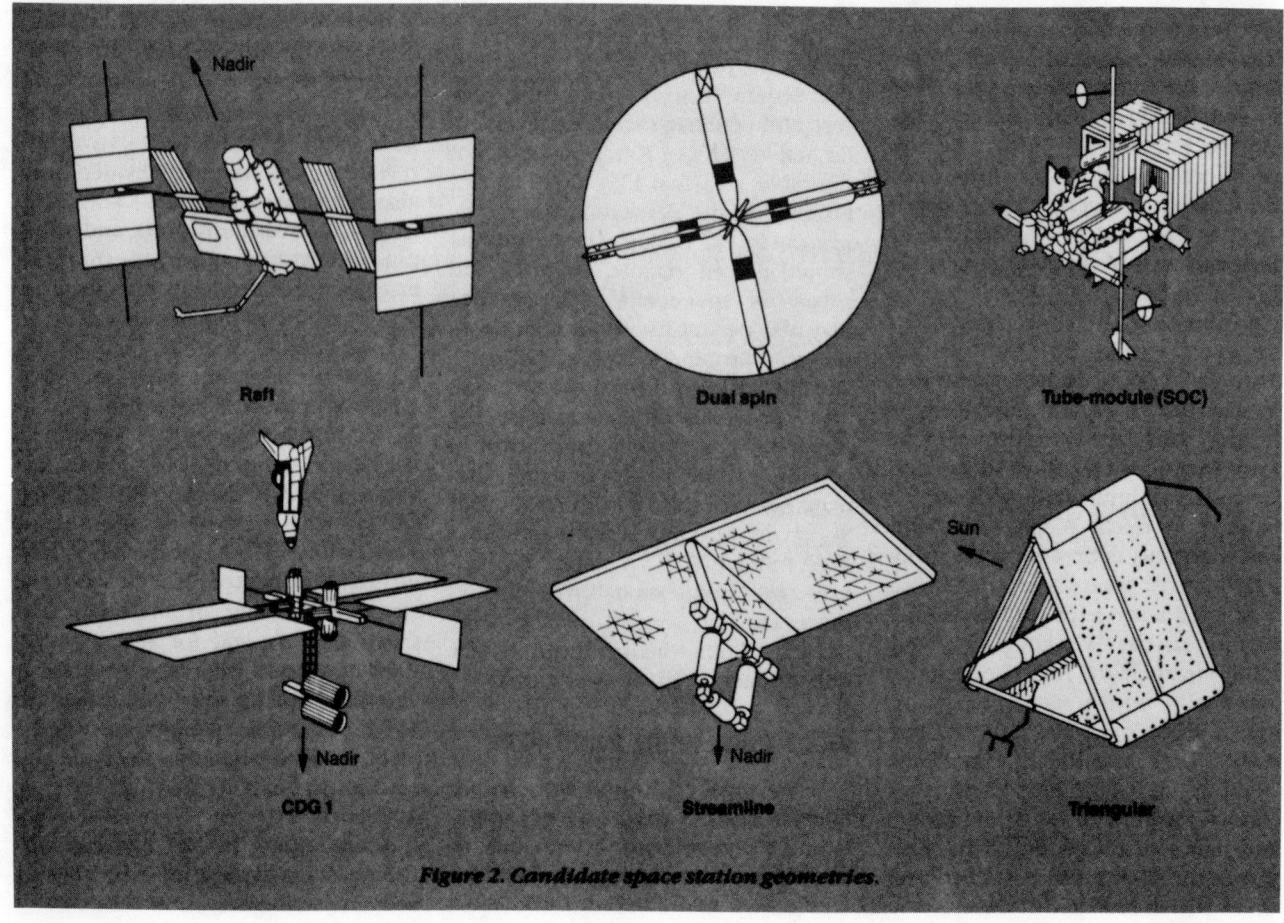

Figure 1. Space station architecture: the cluster concept.

Figure 2. Candidate space station geometries.

nology, where systems can be replaced with advanced equipment as new technologies emerge. The second is the design for maintainability and refurbishment. These concepts should reduce development and operational costs, greatly enhance reliability, and obviate obsolescence. These approaches, together with the proper use and balance of automation and human capabilities, should make the space station able to serve a variety of user needs.

THE ARAMIS PROJECT

NASA's long-range goal for a space station program is, first of all, to establish the permanent presence of humans in space. It is this presence of man and machines—particularly the computer-linked machines—that will enable the routine and continuous exploitation of space for maximum benefit to mankind. As the space program grows through the 1980s and 1990s and on towards the year 2000, with space systems and structures becoming larger and more complex, two factors will loom large: cost (or more properly cost effectiveness) and safety. Other factors will be non-interference requirements (permitting spacecraft operation compatible with operation of extremely delicate zero-gravity materials-processing equipment), and the contingency factor. All this spells out the need for automation, robotics, and machine intelligence systems that can be applied in increasing levels of sophistication to future space tasks.

Automating all space activities, however, would be prohibitively costly as well as being undesirable. Human flexibility and ingenuity in dealing with partial information or novel situations can never be entirely replaced by automated systems, no matter how much is spent on them. Less ambitious efforts are under way. Over the past several years, NASA has conducted a series of studies at MIT, Martin Marietta, Vought Corp., and other members of their contracting community, on the sensitive parameters of space industrialization. These studies have identified automation, robotics, and machine intelligence systems (ARAMIS) as an important contributor to the productivity of orbital factories, and as a potential asset in supporting humans in space-assembly tasks. The eventual goal of this research is a thorough understanding of how humans and machines can best interact in a space environment.

TELEPRESENCE

The word "telepresence" means remote presence, just as "teleoperation" means remote operation. One way to think of telepresence is as a high-fidelity teleoperator system. A teleoperator receives instructions from a human operator, and performs some action based on the instructions at a location remote from the human operator. The concept is similar to that of an industrial robot, except that a human is in control instead of a computer.

The operator uses motions similar to those he would use at the worksite to control manipulators capable of accomplishing operations. The information available to the operator should give him the feeling of being present at the worksite. This permits the operator to concentrate on the work, using his natural abilities to perform the task, without being distracted by unnecessary differences between actually being present and using a remote system.

The purpose of a telepresence system is to perform space operations that require human intelligence, control, and dexterity when extravehicular activity (EVA) is not possible, not desirable, or when EVA alone cannot accomplish the desired mission. Telepresence or teleoperated systems should permit remote assembly and repair of spacecraft. Telepresence may also permit the solution of unanticipated problems. Skylab, Apollo 13, and the successful repair of the Solar Max spacecraft all demonstrate the importance of human capabilities in solving problems. Fortunately, humans were on board both Apollo and Skylab to perform repairs and Solar Max was within EVA range, but failures will unquestionably occur on spacecraft that are out of EVA range or time limits—hence the need for telepresence in space operations.

BEING THERE WHILE BEING HERE

Telepresence includes the components of current master-slave manipulators: a control station with one or two master arms; a remote worksite with one or two slave arms, geometrically similar to the master arms; and feedback (usually video, sometimes also force) to let the operator perceive what is happening at the worksite. Telepresence, however, requires a greater degree of dexterity and feedback than current teleoperators. The systems in use today (e.g., in the nuclear power industry) usually have two-finger claw grabbers as end-effectors, and therefore do not give the operator a feeling of natural manipulation, even in simple tasks. Similarly, the usual video feedback (from one or two cameras) does not provide depth or parallax perception, or peripheral vision; some do not have enough resolution to show sharp details in the work scene. To achieve telepresence, current systems may need to be upgraded to include stereovision, movable points of view, high-resolution zones of focus and low-resolution peripheral vision, sense of touch, force, and thermal and audio feedbacks. Which types and degrees of feedback are required depends on the specific task to be done; it is therefore easier to achieve telepresence in a simple, low-tolerance task than in a complex, delicate one. The basic criterion is that the interaction between operator and worksite must give the operator the impression of "being there."

1995 AND BEYOND

NASA plans can be divided into near-term (through 1995) and long-term. The near-term goals and plans can be further divided into three areas: spacecraft servicing, structural assembly, and contingency events.

Spacecraft servicing is the most important area for near-term telepresence application. NASA is firmly committed to servicing such spacecraft as the space telescope (ST), the advanced X-ray astronomy facility (AXAF), and the long duration exposure facility (LDEF). Servicing is also virtually mandatory for large-scale space processing of materials, for space stations, and for space operations in general. Such large-scale projects may not be fully developed by 1995, but the technology must be in place prior to beginning full-scale operations in order to provide servicing as needed.

According to the MIT study group, the servicing system must be able to accomplish the following basic tasks:

- Operate mechanical connections
- Operate electrical connections
- Operate latching devices
- Grasp objects
- Position objects
- Operate cutting devices
- Operate welding devices
- Grapple docking fixtures or handholds
- Observe spacecraft/component

As listed, these tasks are general in nature, and in application each could be either very simple or very complex. They are intended to be an inventory of basic mechanical operations that can be combined to handle near-term spacecraft servicing, structural assembly, and contingency events.

THE ORBITAL MANEUVERING VEHICLE

After a decade of study, Martin Marietta, Vought Corp., and other members of NASA's contractor community are at the threshold of designing and building a free-flying, remotely controlled tool called the orbital maneuvering vehicle. With special adapters it will be able to carry out enough of the necessary servicing activities for NASA to assign it an important role in the construction and operation of the initial space station flotilla. Its tasks will include transporting material from the shuttle to the station site and helping assemble and position the various station elements. Eventually, the OMV will be berthed at the manned habitat and will support satellite viewing, and perform spacecraft and payload placement and retrieval. It will also provide logistics and maintenance support of the free-flying platforms containing scientific and/or materials-processing experiments.

Although the initial development will lack the full dexterity and feedback characteristics of future telepresence systems, the OMV will nevertheless blaze a pioneering path towards that goal. As the first vehicle to be operated as a free flier, it will initiate teleoperation and will gradually be equipped with front-end attachment kits that, over time, because of increasing sophistication, will enable it to act as a telepresence. All of the technology required to develop and fly the OMV now exists. Initial flights are projected for 1990.

The OMV (Figure 3) is a shuttle-

Weight summary, lb	
Structure	1241
Avionics	457
Elect power (AgZn)	413
Thermal control	181
Propulsion (main and RCS)	1115
Docking kit	423
Subtotal	3830
Contingency (15%)	575
Expendables	7063
Main (usable) 6713	
RCS 350	
Total	11,468

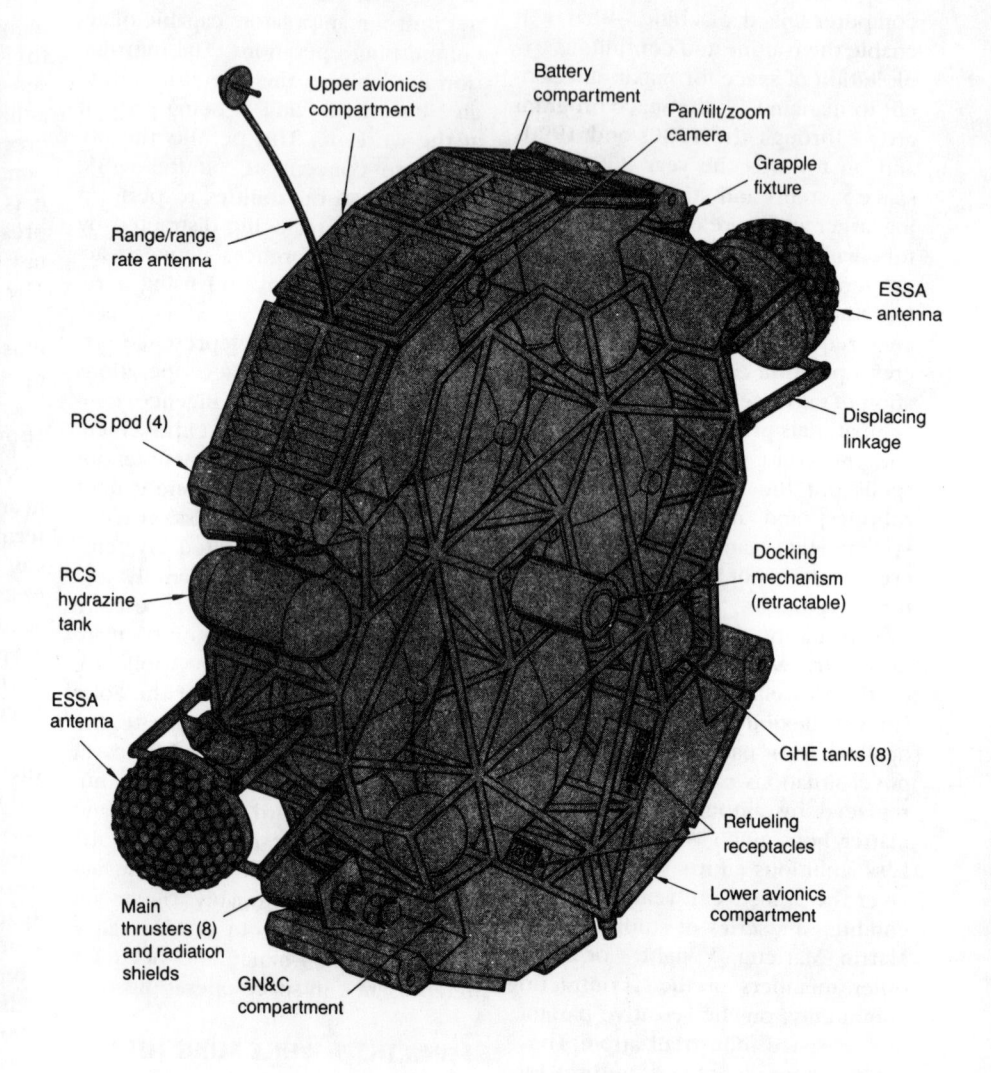

Figure 3. The orbital maneuvering vehicle: a conceptual view.

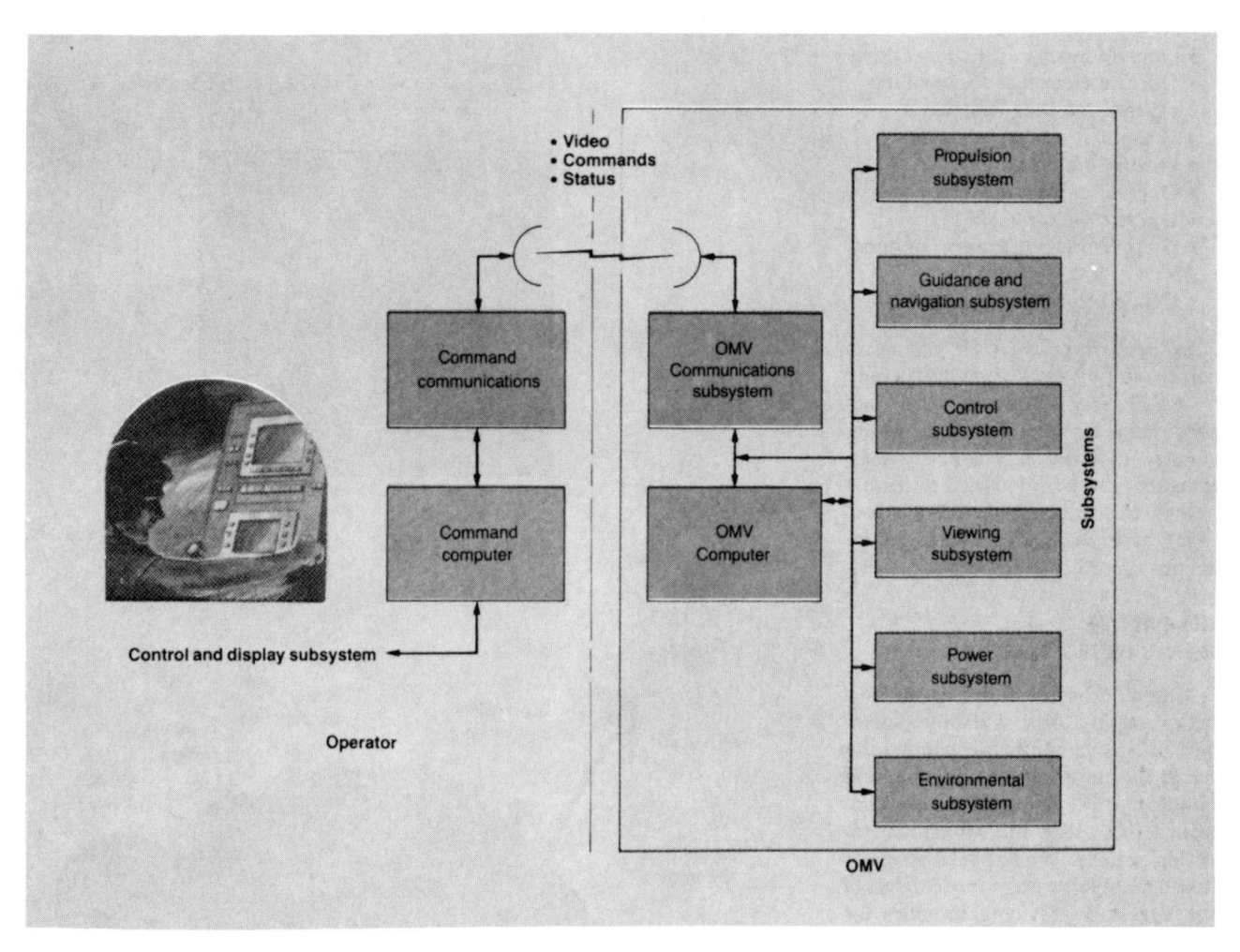

Figure 4. OMV systems.

launched, reusable, and remotely controlled propulsive stage that can either fly preprogrammed trajectories or be controlled or reprogrammed from a ground control station. Approximately 20 million pound-seconds of total impulse energy are available from N_2O_4/MMH biopropellant, which makes up approximately six percent of the 12,000-pound weight of the OMV. The vehicle is designed for installation directly into the space shuttle–orbiter payload-bay longeron and keel fittings for transport from the test range.

The OMV subsystems are shown in the block diagram of Figure 4. The vehicle will probably be controlled by an operator on the ground during the early years of flight, with orbiter aft flight deck control being implemented, if required, before the control and display station is installed on the space station. The OMV contains communications systems to transmit or receive status reports, commands, and video information between the orbiter and itself, as well as with ground stations, either directly or via the tracking and data relay satellite system (TDRSS).

OMV APPLICATIONS

A number of OMV configurations with special-purpose front-end kits for particular uses appropriate to selected categories of missions are identified in Figure 5.

Benefit studies have shown substantial earth-to-orbit shuttle-transportation cost savings using the OMV to help retrieve and service large space observatories operating above the nominal 160 nmi altitude for the orbiter. It will be possible to retrieve observatories such as the AXAF from operational altitudes and return them to the space shuttle–orbiter payload bay for EVA-supported in-bay servicing prior to redeployment to mission altitude. The OMV front-end kits for these retrieval/redeployment missions are expected to consist of mechanical grapple or latching devices (Figure 6) which adapt to existing interfaces between the observatories and the space shuttle orbiter, from which they were originally deployed. Other grappling devices are shown in Figures 7 and 8. Once the station is in operation, the observatories will probably be returned to the station for longer periods of maintenance or upgrading activities than can be economically supported by the shuttle orbiter, since the orbiters will be in demand for repetitive earth-to-orbit transportation flights.

SUPPORT OF MATERIALS-PROCESSING PLATFORMS

Commercial and government materials-processing platforms are microgravity vehicles that are expected to co-orbit in the vicinity of the space station. The OMV (Figure 9) will be used in flights to these deployed platforms to bring products back to the

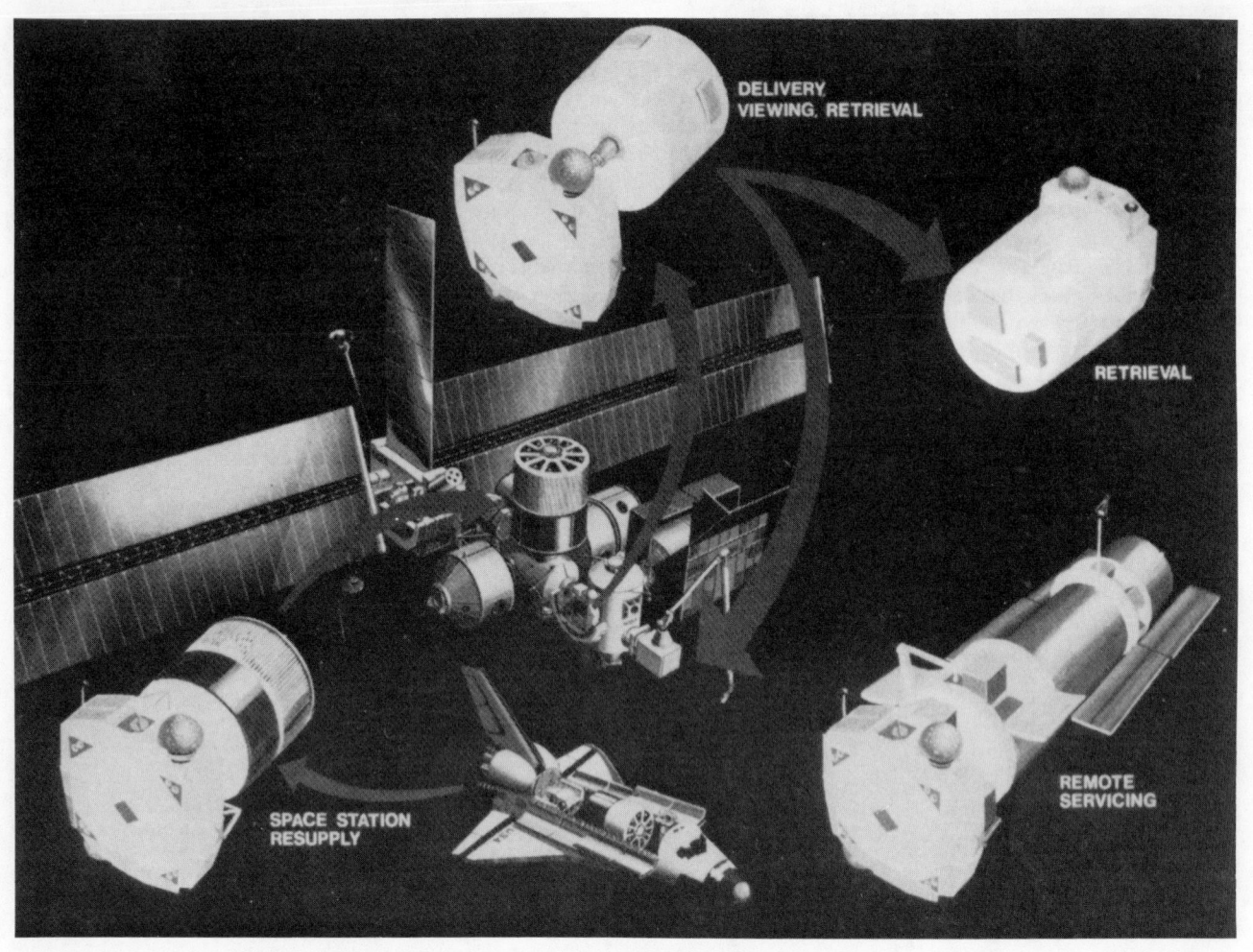

Figure 5. Artist's concept of spacecraft services performed by the OMV during the Space Station era. It is berthed at the manned habitat where it may sally forth for such duties as space station resupply, satellite viewing, spacecraft and/or payload placement and retrieval, and remote servicing.

space station. It will also be used to shepherd the platforms when aerodynamic drag characteristics different from those of the station, and deployment altitudes designed for specific mission durations cause the platforms to separate from each other at various distances from the station.

The OMV will be used for contingency operations at first from the shuttle orbiter, and subsequently when flying from the space station. Missions of the solar maximum repair type will be facilitated by dispatching the OMV from the station to retrieve such satellites for repair in pressurized or unpressurized shielded hangars on the station.

TELEPRESENCE TECHNOLOGY

The primary technology requirements for a near-term (1990-1992) telepresence system are:

• A stereo-optic vision system (preferably color) that will correspond to operator's head position.

• A head-mounted vision display system.

• Two seven-degree-of-freedom (DOF) manipulator arms with force control.

• Two grapple arms or one docking device.

• Interchangeable end-effectors.

• Force-indicating hand controllers or exoskeletal arms for control.

Vision. Telepresence will require a stereo-optic vision system that will provide depth perception and a sense of three-dimensional imaging similar to that provided by the human binocular vision system. In order to provide the capacity to slave the cameras to the operator's head position, the video displays should be helmet mounted. This allows the display screen to be always in view, regardless of the operator's head position. It also permits a separate image to be presented to each eye (necessary for true stereo vision) without requiring complex or expensive optics, which can restrict operator movement and cause discomfort. The use of color is desirable because it aids in scene recognition and understanding for both humans and machines.

Space-qualified video cameras have been in use since the 1960s. The technology for the required kind of vision system is well advanced, and a black-and-white stereo helmet-mounted video system has been developed and tested by the Naval Ocean Systems Center (NOSC) in Hawaii. The addition of color should present little problem.

Manipulator arms. Manipulator arms with seven DOF are desirable because they are similar to human arms and so are easily controlled by a master-slave control system. In addition, seven DOF are needed to be able to reach around objects or into confined spaces. Two arms are required because some space operations will need more than one arm in order to be completed. In addition, the human operator is probably more comfortable with controlling two seven-DOF manipulators than one seven-DOF arm and one arm with fewer than seven DOF. NOSC Hawaii has built and tested a system with two master-slave manipulator arms, and Martin Marietta has built a seven-DOF manipulator arm for the Marshall Space Flight Center that can easily be adapted for space use. MIT is building a manipulator system for neutral buoyancy simulation of space structural assembly and for testing telepresence control technology.

Force control of the manipulator arm is necessary because of the very high stress loads that can be accidentally applied without some limit on manipulator force. This control can include both total force limits that the manipulator will not exceed, and the ability to apply a force specified by the operator. Force feedback (sending the force data to the control site and allowing the operator to sense the force and limit it) is probably the most desirable technique, but time delays in the communications system could prevent the operator from sensing excessive force in time to prevent damage. Experiments have been performed with force-limited manipulators, but further research is necessary before this control technology becomes operational.

A telepresence sytem working on a satellite or a construction site must be able to supply forces and torques to nearby spacecraft and components. During these operations the servicer (telepresence system) must hold its position relative to the worksite or it will drift away and be unable to continue to apply force. Holding position by rocket thrust is difficult, wastes fuel, and may be impossible because the engines may not generate enough thrust to overcome the force applied by the manipulator arms.

Spacecraft docking has been performed since the 1960s and is a viable option for telepresence, but the telepresence system may have difficulty reaching the necessary locations at a worksite if it is fixed to one contact point. A solution is to use a second set of manipulator arms to grapple "hardpoints" (structural members, booster casings, EVA handrails, remote manipulator system fixtures, etc.). Since this would be their only purpose, this second set of arms need not be as sophisticated as the main arms. And since manipulator arms are a prerequisite for a telepresence system, the development of the less advanced grapple arms should not present any problems.

Figure 6. OMV (center) approaches a satellite for servicing. It carries a docking probe for latching onto satellite for either on-site repair or returning satellite to shuttle bay where more extensive repairs can be carried out prior to redeployment. Satellite can also be returned to earth aboard shuttle, should that be necessary.

End-effectors. The grappling of various hardpoints, the manipulation of objects, and the ability to use tools, are requirements that a near-term telepresence system must meet. A mechanical hand or hand-analogue is an option which, in theory, could perform these tasks. Such a device, however, would require a significant developmental effort, and it is unclear that the device would be easily controllable in an environment with a communications time delay. Interchangeable end-effectors have been demonstrated in the laboratory and can accomplish all near-term telepresence tasks. Since they are specialized, many of these end-effectors could perform better than a mechanical hand. The mechanical hand offers the advantage of high versatility, but at present such versatility is not necessary. More advanced telepresence systems (post 1995) will probably need some form of mechanical hand to perform complex tasks.

Control. The two most promising techniques for operator control of the manipulator arms are force-indicating hand controllers and exoskeletal master arms. A force-indicating hand controller is a multi-DOF "stick". The operator grasps the stick and applies force that moves the manipulator at a velocity proportional to the applied force. If the manipulator is in contact with a spacecraft or component, it applies the same (or proportional) force to the object with which it is in contact as is indicated by the operator. The operator applies forces to the hand controller to "fly" the end of the manipulator to the desired location.

The other attractive option is a master arm that monitors the position of the operator's arm and commands the telepresence manipulator to a similar position. Direct force control is more difficult with this system than with a hand controller because the master arm responds to an applied force by moving, thus the operator is not as aware of the forces being applied as when using the rigid hand controller. These exoskeletal controllers can use preset force limits instead of continuous operator commands.

The nuclear industry has used a third technique that is essentially a hybrid of these two control methods. The operator grasps a hand controller which commands the grippers or end-effector of the arm. The hand controller is attached to the end of a master arm which moves in response to forces applied to the hand controller by the operator. The actual manipulator arm follows the movement of the master arm. Both master-arm methods would benefit from force feedback, but communications time delays make this a questionable option.

All of these approaches are within present technological capabilities and are effective means of controlling a manipulator. The force-indicating hand controller is probably the best choice for a near-term telepresence system, but comparative experimental testing of these techniques is necessary before a final determination can be made.

Sensors. Proximity and force sensors for a manipulator arm control are necessary to provide information to the operator and control system. Proximity sensors are a well developed technology and are planned for use with the remote manipulator system. Force and torque sensors of various designs are available. Adapting them for space use should present no problems.

Communications. Communication with the telepresence unit is required for its operation. It can be accomplished using the K band single-access links provided by the TDRSS space-

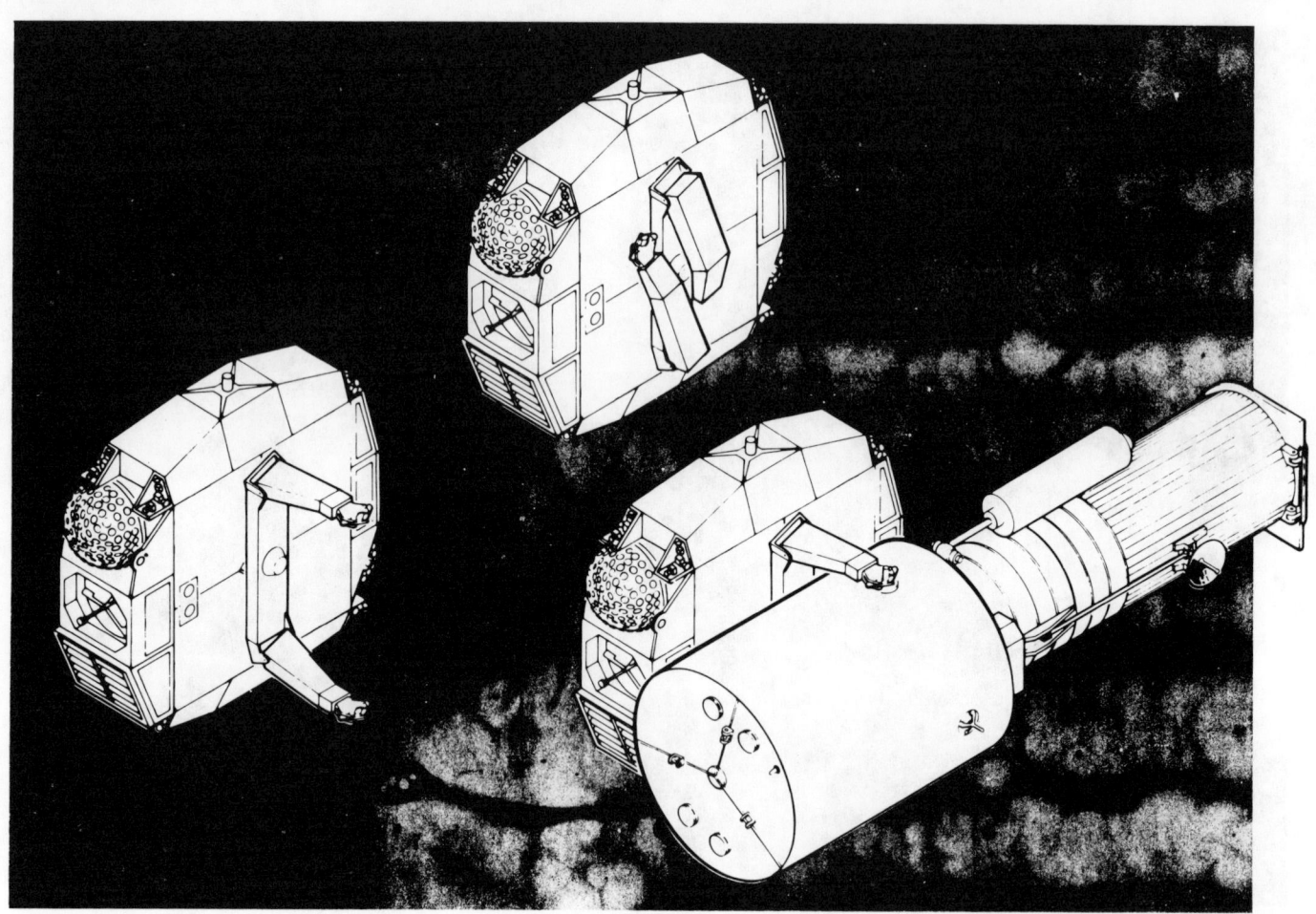

Figure 7. Space telescope retrieval.

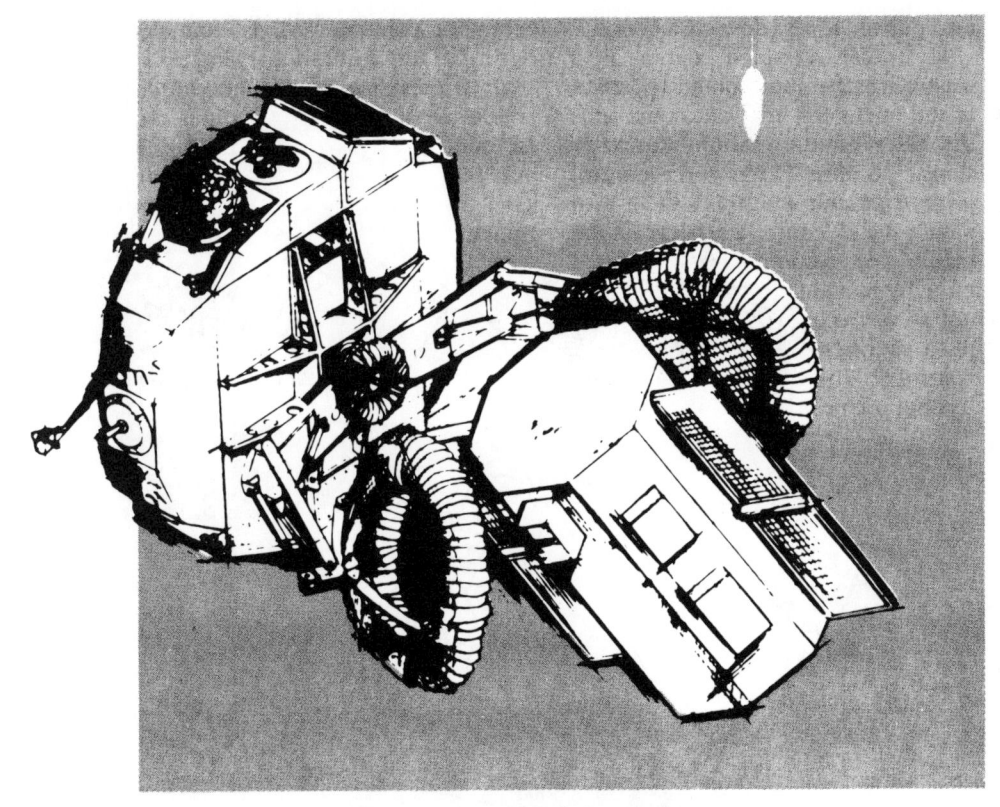

Figure 8. Debris capture.

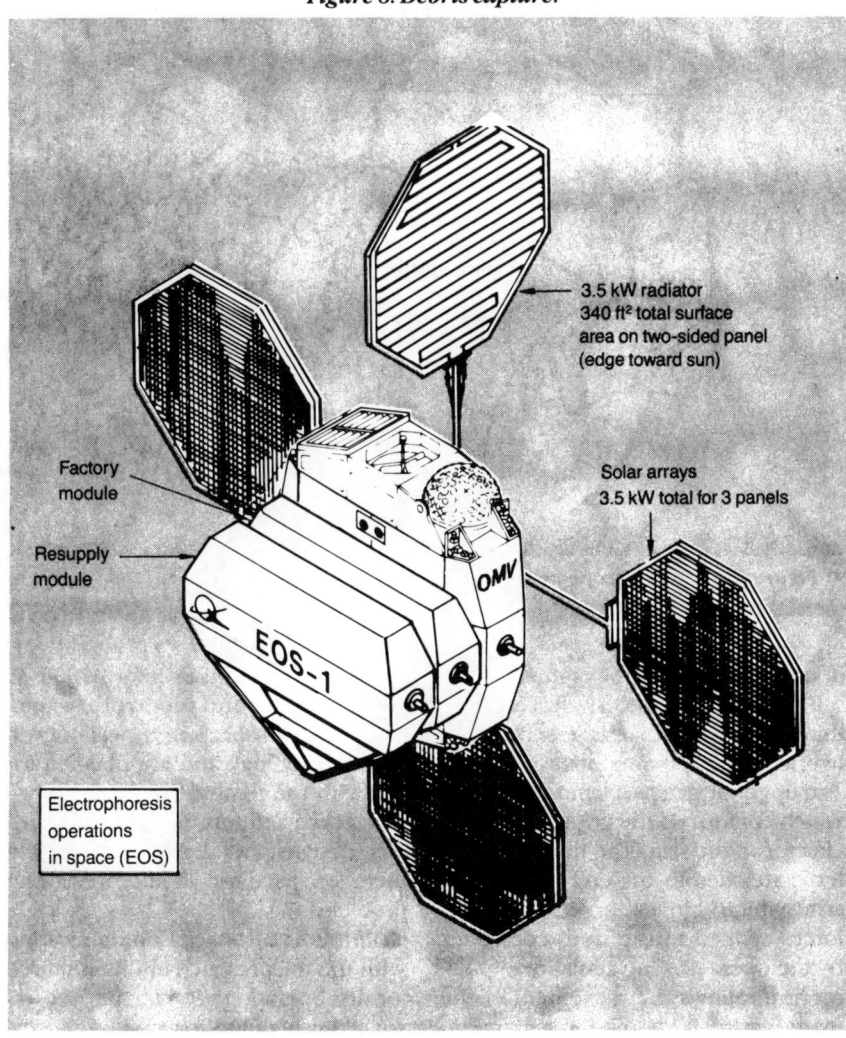

Figure 9. Materials processing support.

craft. Unfortunately, the minimum communications time delay for low-earth orbit (LEO) is 0.5 second. The delay can increase to 2.0 seconds if the control station must communicate with TDRSS via the NASCOM system. Since time delays degrade performance, the MIT study group recommends that every effort be made to minimize the communications time delay. This may require placing the telepresence control station at White Sands, New Mexico, where the TDRSS control center is located.

Predictive displays. Since the time delay cannot be completely eliminated from the communications system, predictive display technology should be investigated. Recent advances in computer-aided modeling make predictive displays a potential method of eliminating many of the restrictions imposed by time delays. For example, a computer could store a model of a spacecraft that would be updated and modified as the structure is altered by servicing. As the operator moves the manipulator, the computer would immediately show the operator where the manipulator links and end-effector are positioned in relation to the spacecraft, even though the video response from the spacecraft has not yet been received. In this manner, many of the problems caused by the "move-and-wait" techniques usually employed in dealing with time delays are reduced. Predictive display technology has the potential to be very useful for telepresence systems, but several years of development will be necessary before a usable system can be produced.

TELEPRESENCE PROJECT APPLICATION

The Marshall Space Flight Center has selected five projects for study:

- The space telescope (ST)
- The advanced X-ray astrophysics facility (AXAF)
- The very large space telescope (VLST)
- The coherent optical system of modular imaging collectors (COSMIC)
- The 100-m thinned aperture telescope (TAT)

These space projects were chosen to span the years 1985-2000, with ST representing a relatively near-term potential telepresence application, AXAF a mid-term application, and

127

VLST, COSMIC, and TAT long-term applications. These latter involve increased complexity and require technology well beyond the current state of the art. Together, the space projects cover a wide spectrum of tasks such as spacecraft servicing, resupply, rendezvous and docking, and in-orbit assembly. At present, the ST is the only project that is certain to be implemented, although there is a high probability that AXAF will also receive a go-ahead. Even if none of the three long-term space projects received full funding and development, it is felt that the telepresence technologies and capabilities they imply will be necessary in the late 1990s.

SPACE PROJECT TELEPRESENCE TASK ANALYSIS

Each of the five space projects has been analyzed by the MIT study group to determine, to the extent that is currently possible, the nature of the activities that an in-orbit telepresence system should be able to accomplish. Documents supplied by NASA were used as a basis for evaluations. For the ST, the physical parameters of the structure are known in detail; this task therefore consisted of analyzing, at a nuts-and-bolts level, each of the tasks necessary to perform ST servicing and maintenance. For AXAF, for which there are several tentative designs containing less detail than is available for the ST, this task consisted of evaluating anticipated telepresence requirements, and recommending modifications for the spacecraft to make these more easily met. Finally, for the advanced space telescope applications, telepresence requirements were evaluated at a very general level to determine appropriate areas for further research and development.

ST SERVICING TASKS

Present plans call for the ST to be deployed and inserted directly into orbit by the space shuttle. Further, current plans are to have pressure-suited astronauts perform ST servicing outside the vehicle. The ST has a design life of ten years, but this could be significantly extended with in-orbit maintenance and ground refurbishment. The ST configuration has undergone extensive testing through the use of neutral buoyancy simulations, which have clearly delineated the steps necessary to maintain, refurbish, and perform selected planned

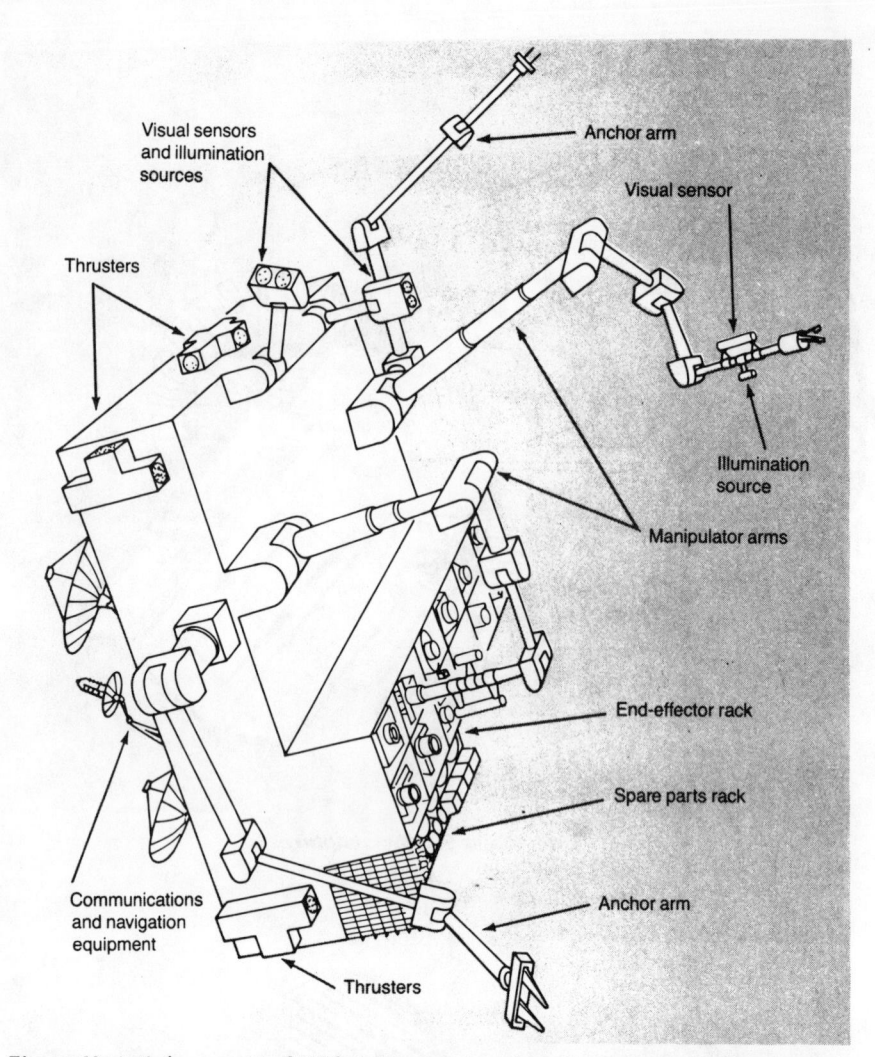

Figure 10. Artist's concept of MIT's telepresence robot. It carries tools, spare parts, and a variety of sensors for diagnosing and repairing faults. It is shown preparing to remove a defective module from a spacecraft and replace it with a new one.

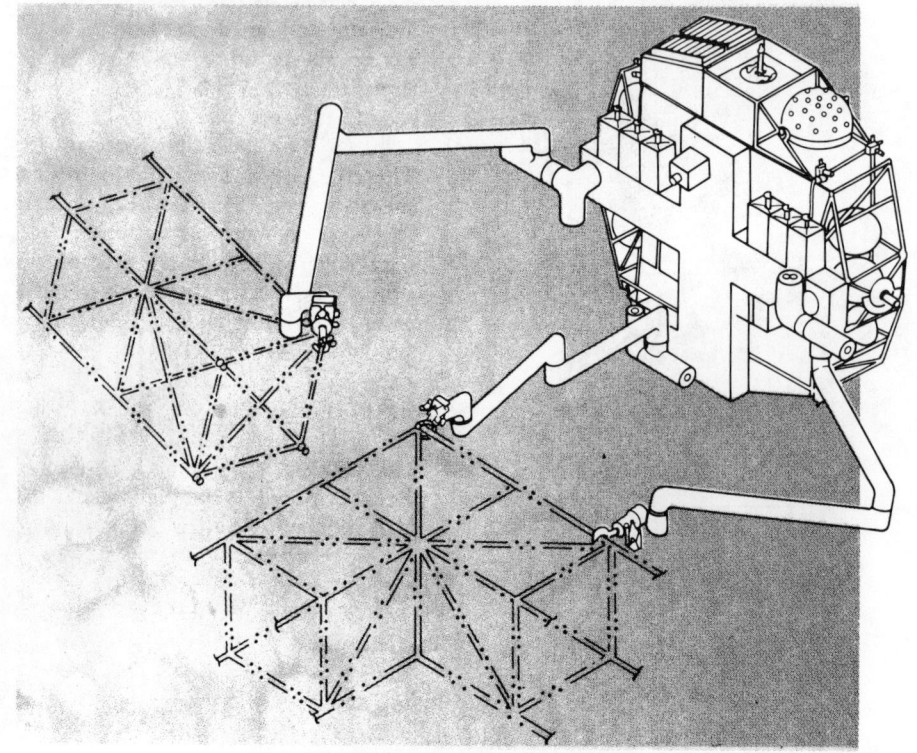

Figure 11. Robot-aided structural assembly.

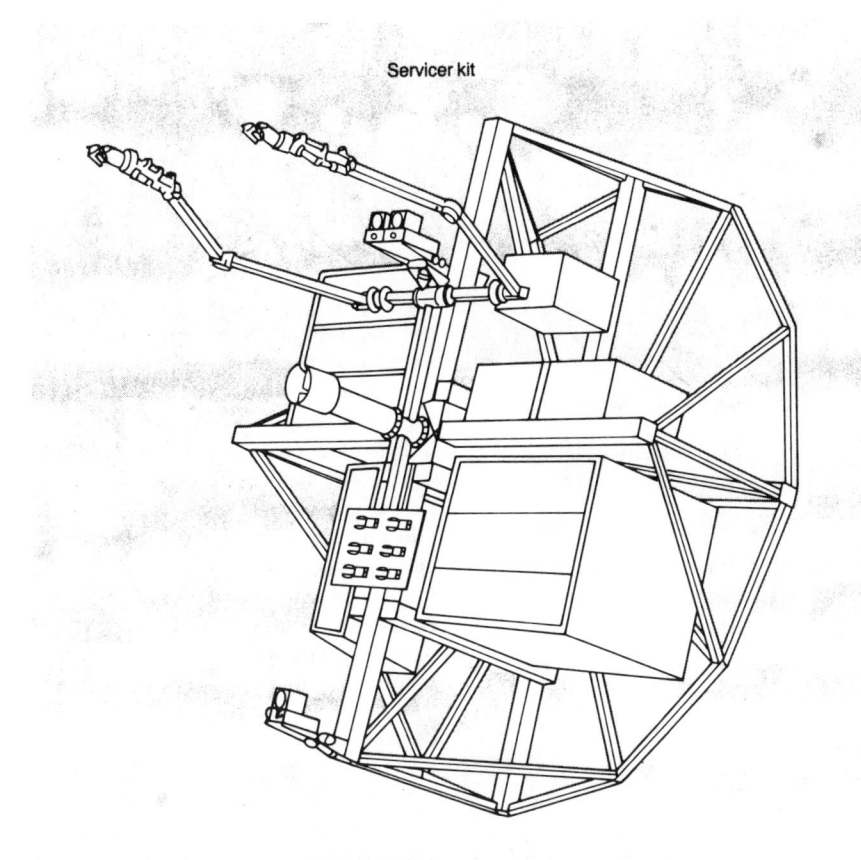

Figure 12. Remote orbital servicing system.

and contingency operations in EVA. These simulations determined the type and location of crew aids that have been integrated into ST to facilitate EVA servicing of the spacecraft. The methods developed and the crew aids devised, are being used as starting points for future efforts in ensuring spacecraft serviceability.

Orbital maintenance is expected to require 23 orbital units aboard the ST. These consist of five scientific instruments, three fine-guidance sensors, the science instrument control and data-handling unit, three rate sensor units, three rate gyro electronics units, three fine-guidance electronics units, and five batteries. Further, in-orbit overide of certain malfunctioning ST mechanisms (such as would be required on a contingency basis by faulty solar array deployment) has been designed. It is estimated that ST will require orbital maintenance anywhere from two and one half to five years after initial deployment.

Telepresence is potentially capable of handling all orbital maintenance activities, as well as the tasks of reboosting and orbital deployment from and retrieval to the space shuttle (with assistance from the OMV). While EVA activities are the currently planned method for performing orbital maintenance functions, the implementation of telepresence could potentially reduce costs of maintenance operations, free the shuttle and crew for other tasks, and offer other additional advantages. The cost reduction potential can be calculated by spreading the nonrecurring costs of a telepresence servicer over all the spacecraft it will service, rather than assigning the cost to a single space project.

OPERATIONAL ANALYSIS

Some preliminary work has been done to design a remote servicer that would be compatible with several spacecraft, and capable of performing servicing to the same extent as EVA. The free-flying hybrid teleoperator (Figure 10), a true telepresence robot, was conceptualized at MIT to be capable of propelling itself to a repair site; attaching itself to a structure; carrying tools, spare parts, and a variety of sensors; diagnosing and repairing faults; performing various structural activities (Figure 11); and communicating with human supervisors. The remote orbital servicing system (ROSS) (Figure 12) was conceptualized by Martin Marietta Aerospace to be capable of servicing *in situ* the ST, the solar maximum mission, and the long-duration exposure facility, using current state-of-the-art technology.

LONG-TERM PLANS AND GOALS

Unlike the technological developments necessary for near-term telepresence, much of the long-term (post 1995) development will be performed by research in artificial intelligence and supervisory control.

The most important long-term goals are increased system dexterity and the ability to handle contingency operations. As space operations become space industry, and the construction, modification, and repair of orbital systems become routine, on-site, high-dexterity manipulation will be mandatory. Equipment shipped from earth will not be preassembled as it is today, but will arrive as spare parts and components for orbital construction and assembly. Some of the components will probably require high-dexterity assembly. More important, the need to replace damaged and failed components, particularly in intricate mechanical devices or complex systems, will require dexterity

Comparison of Teleoperator and Telepresence

Teleoperator	Telepresence
Mobility base	
• Rotational and Translational Maneuvering	
• Large mobility base	• Small mobility base
• Best suited for external module change-out	• Best suited for internal service access
Manipulators	
• Primarily useful for pre-planned activities	• Interchangeable end-effectors
• Limited capability for general grasping and manipulation	• Ultimately dexterous, anthropomorphic end-effectors
• Insufficient for most contingency operations	• Generally EVA equivalent capabilities, ultimately exceeding these
Vision Systems	
• Video cameras for approach and docking	• Binocular stereo vision
• General worksite views	• High resolution
• Dedicated cameras for close-ups on end-effectors	• Wide-field color images
	• Cameras with four degrees of freedom (pan, tilt, fore-aft, side-to-side)
Communications	
Little difference	
Controls and Displays	
• Fairly traditional control-room consoles	• Master/slave control of a dexterous end-effector
• Limited degrees of freedom by tying vehicle to target prior to operations	• Wide-field binocular helmet-mounted displays
	• Camera position slaved to operator's head position
	• Possible multiple operators or multi-band control routes (e.g. voice recognition)
Sensors	
• None	• To be determined
	• Tactile sensors
	• Force sensors
	• Proximity sensors

simply to gain access to the repair site. An example is the modification or repair of a wiring harness. Despite clever design and much effort, there will still be places requiring hand dexterity, where wiring will need to be guided through a harness that is difficult to reach.

The potential size and scope of future space operations will prohibit the extreme caution and highly detailed planning that accompany present space missions. Since commercial space missions will be commonplace, industrial accidents will undoubtedly occur. The failure of a large materials-processing furnace or a high-pressure fuel line imply the need for crew rescue and versatile repair tasks. Tasks of this nature necessitate the ability to deal with nonfunctional and severely damaged equipment in an environment that may be unsuitable for EVA. The probability of successful advanced contingency operations is improved greatly by the availability of high-dexterity telepresence.

Increased system autonomy in the future is desirable, partly because of the scope of future operations and partly because transmission time delays may degrade dexterity. Many future tasks could be repetitive and boring; high-level supervisory control for these tasks would relieve operator fatigue and improve reliability. In regions of obscured communications, an autonomous operation capability is necessary. Transmission time delays may make remote high-dexterity control difficult or impossible, and cause even some otherwise mundane tasks to require supervisory control or autonomy.

Because of the cost of space vehicles, improvements to the telepresence system should be evolutionary, so that a new spacecraft is not required for each system upgrade. As spacecraft technology improves, the maneuvering system and telepresence unit may be replaced, but manipulator or computer system upgrades, for example, should not require replacing the entire spacecraft. The most radical advances in telepresence technology will occur in computer hardware and software, manipulators, and end-effectors. Once a high-dexterity manipulator is developed and installed, most system changes will be in software, and will be performed remotely from the ground or from space station control centers.

Presented at the American Nuclear Society 25th Conference
on Remote Systems Technology, 1977

MARS VIKING SURFACE SAMPLER SUBSYSTEM

DONALD S. CROUCH* *Martin Marietta Aerospace Denver Division, P.O. Box 179, Denver, Colorado 80201*

KEYWORDS: *manipulators, remote handling, remote vehicle*

ABSTRACT

A Surface Sampler Subsystem was developed for use on Viking Landers 1 and 2, which landed on the surface of the planet Mars in 1976. A major component of this subsystem is the Acquisition Assembly, which consists of a computer-controlled boom unit and collector head used for acquiring small samples of material from the Martian surface. The boom unit consists of an extendable/retractable furlable tube element capable of extending the tip of the collector head to a maximum of 3.45 m (136 in.), and an integral gimbal capable of 5 rad (288 deg) azimuth and 1.3-rad (74-deg) elevation movement. All Mars surface operations are performed automatically with periodic command uplinks and data downlinks through Deep Space Network antennae to the control center at the Jet Propulsion Laboratory in Pasadena, California.

INTRODUCTION

On August 20 and September 9, 1975, respectively, two lander/orbiter spacecraft were launched from Kennedy Space Center by Titan III Centaur rockets to begin a 10-month 6.4×10^8 km (400-million-mile) journey through space to the planet Mars. The missions of these spacecraft were to conduct both Mars orbital and landed surface scientific experiments primarily directed to answering the question of whether any level of life exists, or ever existed, on the planet. The spacecraft arrived safely and were put into planetary orbit to image the surface in search of safe landing sites within the preselected areas of interest. The landers were subsequently separated from the orbiters; each landed safely on the Martian surface on July 20 and September 3, 1976, respectively.

One of the major subsystems aboard each lander is the Surface Sampler (Fig. 1). This subsystem was designed to acquire, process, and deliver surface material samples to the Biology, Gas Chromatograph Mass Spectrometer (GCMS), and X-Ray Fluorescence (XRFS) experiments, and to provide support for the Surface Physical and Magnetic Properties investigations. The Surface Sampler consists of four major components:

1. Acquisition Assembly—acquires surface samples and delivers them to the required experiments
2. GCMS Processor—receives samples from the Acquisition Assembly, grinds the material to a <300 μm particle size, and delivers metered 1 cm^3 samples to the GCMS experiment
3. Biology Processor—accepts samples from the Acquisition Assembly, sieves the material to a <1500-μm particle size and delivers 7-cm^3 metered samples to the biology experiment
4. Control Assembly (SSCA)—receives digital commands from the spacecraft computer and provides appropriate control and data handling functions for each of the previously described electromechanical components.

The primary emphasis of this paper is a description of the Acquisition Assembly, a remote computer-controlled boom and collector head capable

*Invited.

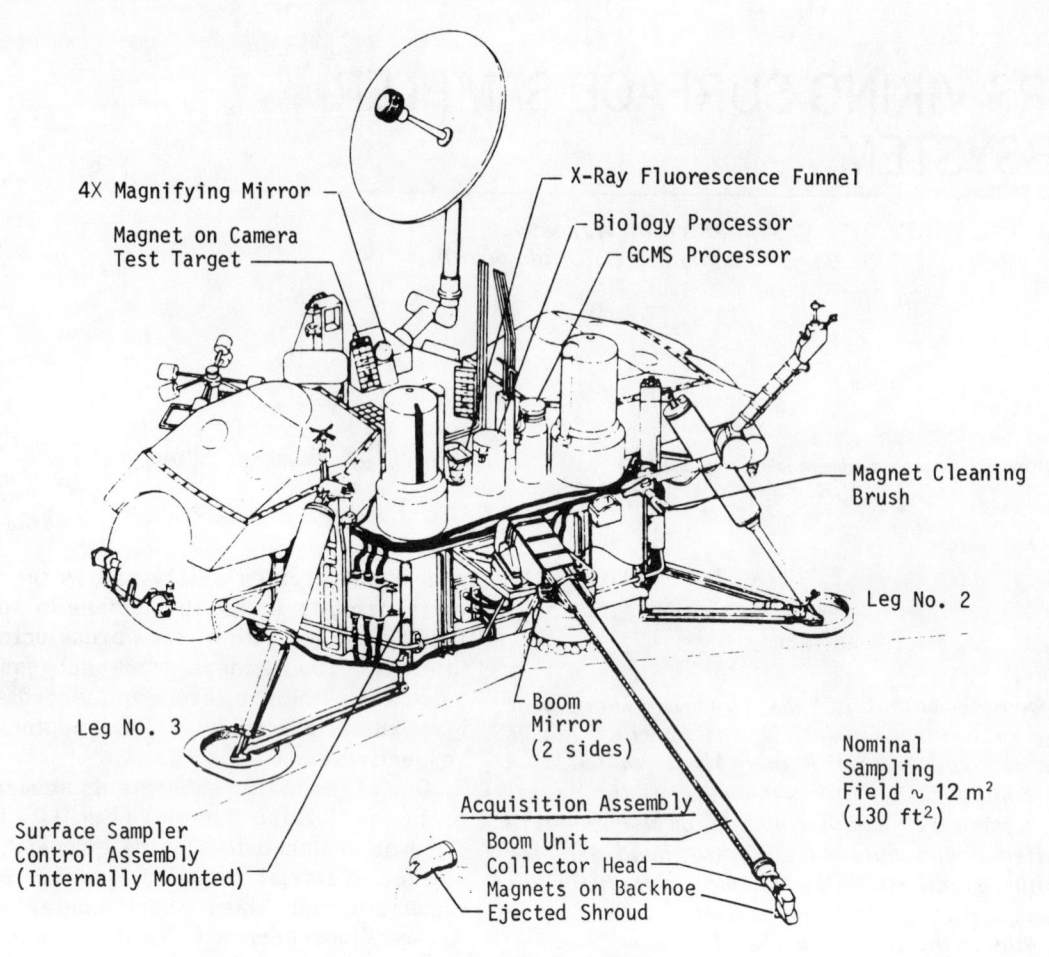

Fig. 1. Surface sampler components.

of automatically acquiring surface samples at any location within a prescribed 12-m² (130-ft²) semicircular area in front of the lander.

DESIGN REQUIREMENTS AND DEVELOPMENT

Design guidelines for the Surface Sampler Subsystem were severe regarding environmental requirements, and difficult to define regarding sampling requirements. The precise nature of the Mars surface material at the potential landing areas could be predicted only on the basis of Mariner orbital photography as related to earth and lunar analogs. Therefore, it was necessary to design, somewhat conservatively, for a surface material that may consist of one or more of the following materials: (a) noncohesive sand or pebbles, (b) lunar-like cohesive particulate, (c) hard-pan, or (d) any of the above material overlain with small- and medium-size rocks. In addition, the Surface Sampler was required to operate at any time during Martian day or night in the presence of moderate-to-heavy wind velocities.

A summary of the major environmental requirements for subsystem design includes the following:

1. prelaunch heat sterilization of +145°C (+293°F)

2. spacecraft level vibration, shock, acoustic noise, and acceleration

3. cruise temperatures of -18°C (0°F) to +77°C (+170°F) and pressure of 1.33×10^{-12} Pa (10^{-14} Torr)

4. Mars surface operating temperatures of -112°C (-170°F) to +49°C (+120°F), pressures of 266 to 2000 Pa (2 to 15 Torr), and wind velocities to 70 m/s (230 ft/s).

Development of the Surface Sampler was accomplished at Martin Marietta Aerospace in Denver, where the Titan III booster rocket and Viking Landers (VL) were produced. Design approach options were tested using a series of breadboard models. These breadboards evolved to a "design development" model that was subjected to environmental tests including heat sterilization, vibration, cruise and landed thermal vacuum, electromagnetic interference, and life operational tests. Additional tests were conducted using Air Force low-gravity aircraft, since operation of the Surface Sampler involves gravity-assist flow of surface sample material under $\frac{3}{8}$-g conditions.

The final configuration of the boom and collector head is illustrated in Fig. 2. The boom element is fabricated from two strips of Carpenter Custom 455 steel chemically milled to provide both a channel space for the integral flat cable, and a weight-saving longitudinal taper along the length of the boom. Two halves of the boom element are subsequently seam-welded together, sprocket holes are punched, and the material is heat treated to the required "free-state" configuration. Following heat treat, the flat cable is installed and the entire assembly is flattened and stored on the drum assembly. A series of rollers and guides flatten the free-state boom as it is rolled up on the drum.

Drive assemblies for the extend/retract, azimuth, and elevation axes are powered by hermetically sealed, solid-film lubricated, permanent magnet dc motors. The hermetic seal mechanical feedthrough drive is accomplished by samarium cobalt drive rotors that develop magnetic torque through a nonmagnetic membrane of the motor housing. A cermet element position potentiometer is coupled to each of the three drive assemblies. Sand and dust ingestion into the internal mechanisms is precluded by a silicone-coated nomex cloth bellows surrounding the gimbal area, and spring-loaded nomex cloth wipers that surround the flattened boom element located near the front of the boom housing.

Construction materials for the boom unit included aluminum alloy, titanium, and steel.

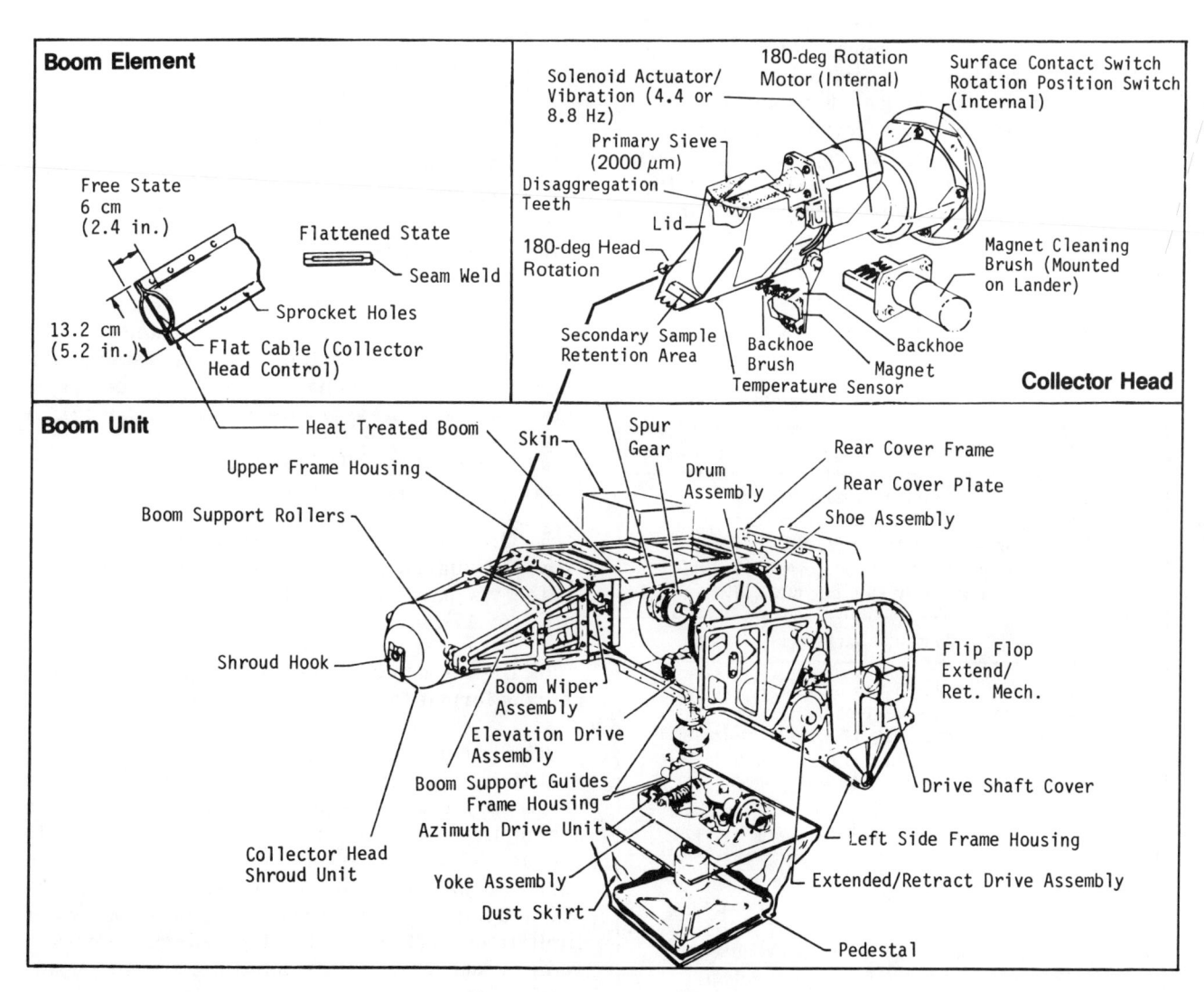

Fig. 2. Acquisition assembly.

Weight of the unit is ~10.4 kg (23 lb). A listing of major boom operating parameters is tabulated in Table I.

The collector head (Fig. 2) consists of a lower stationary jaw for digging into the surface, and a solenoid-actuated upper jaw for retaining the sample. It can direct-deliver the acquired bulk sample to the appropriate experiment in the upright position, or it can be rotated 3.14 rad (180 deg) and the solenoid-actuated upper lid (in the inverted position) can be vibrated at an 8-Hz rate to sieve the acquired sample through a 2000-μm sieve. Other features of the collector include a sample temperature sensor, a backhoe for trenching operations, and a magnet array used in conjunction with a lander-mounted magnet cleaning brush. A pressurized contamination control shroud is used to ensure maintenance of the nanograde cleanliness level of the collector head through all phases of prelaunch checkout, launch, cruise, descent, and landing on Mars.

Development of controls electronics proceeded in parallel with development of the mechanical components; initial integration evaluation was conducted on the design development model. Qualification test model hardware was fabricated and subjected to a rigorous repeat of the previously described environmental tests. Three flight model articles were subsequently fabricated to support the two Viking flights.

TABLE I

Room Unit Operating Parameters

Extend/Retract Axis
Maximum extension—3.45 m (136 in.)
Operating rate—2.5 cm/s (1 in./s)
Operating force—133 N (30 lb)
Motor current—170 to 600 mA @ 30 V dc
Control sensitivity—0.6 cm (0.25 in.)

Azimuth Axis
Maximum angle—5 rad (288 deg)
Operating rate—0.031 rad/s (1.8 deg/s)
Operating torque—10.8 N-m (8.0 ft-lb)
Motor current—13 to 50 mA @ 30 V dc
Control sensitivity—0.01 rad (0.6 deg)

Elevation Axis
Maximum angle—1.3 rad (74 deg)
Operating rate—0.026 rad/s (1.5 deg/s)
Operating torque—63.8 N-m (47 ft-lb)
Motor current—100 to 500 mA @ 30 V dc
Control sensitivity—0.01 rad (0.6 deg)

LANDER OPERATION

As previously described, the Surface Sampler Subsystem basically consists of three major electromechanical components and a controls assembly. Total operation of the subsystem requires use of the VL receiver, computer, data handling processor, transmitter, and Viking Orbiter relay receiver and transmitter. The following major steps are required between generation of a particular sequence at the Jet Propulsion Laboratory (JPL) and receipt of data indicating successful execution of the sequence on Mars:

1. Develop sequence (typically 30 to 80 commands) at JPL, verify on operational mockup, computer-check for conflicts with other subsystem commands, and integrate with total spacecraft command load (days 1 to 10).

2. Transmit commands to the operating Deep Space Network (DSN) antenna (Goldstone, Madrid, or Canberra) via hardline, satellite, and microwave, which, in turn, transmits through 4×10^8 km (250 million miles) of space to the lander S-band antenna and stores in the computer (day 11). (Note: This process is also reversed; i.e., memory readout to ground-verify that the commands are stored properly in the computer memory before execution.)

3. Computer sequentially issues each digital command of the sequence to the Surface Sampler Subsystem and the resultant data generated by the execution of each command is stored in the lander data-handling processor tape recorder (day 15).

4. Lander transmits stored data to the Orbiter, where it is stored on a tape recorder (day 16).

5. Orbiter transmits stored data to the operating DSN antenna, where it is subsequently transmitted via hardline, satellite, and microwave to JPL (day 17).

6. Digital data are computer-decoded into engineering format for analysis (day 18).

The time base indicated is typical and, in reality, can be somewhat longer or shorter. Short emergency corrective action sequences have been accomplished through all the above steps in 3 to 4 days. This represents an absolute minimum based on practical considerations such as earth-Mars-Orbiter transmitting geometry, power limitations, computer loading, 25-min one-way radio-frequency transmission time, and man-loading factors.

SURFACE SAMPLER CONTROL

There are 12 available commands relating to the Surface Sampler Acquisition Assembly which include the following: extend/retract, ± elevation, cw/ccw azimuth, cw/ccw collector head rotation, collector head open, 4.4/8.8 Hz collector head vibration, and collector head close. Each surface sampler command is encoded as a 24-bit digital binary word containing motor address, direction of rotation, and quantitative position data. Figure 3 illustrates a typical control loop for a boom movement. The 24-bit command word is issued to the SSCA input register, decoded, and the appropriate motor is switched to operate in the correct direction. Position of the boom is tracked by the potentiometer with its output digitized by an analog-to-digital converter. These data are routed to a comparator, where it is continuously compared with the input command. When the digitized output of the potentiometer matches the input command, the motor drive power is terminated. Operation of solenoids is similar to the motors, except no feedback loop is involved. Digitized data are accumulated with each command execution, which includes position achieved, operating currents, temperatures, switch positions, etc.

Each sequence to be executed involves the following lander computer activities:

1. Mission Sequence Event Word—sets the local lander SOL[a] (day) and time (hour, minute, second) for sequence execution and establishes logic parameters for the sequence. Logic parameters include items such as experiment to be serviced, number of sampling attempts if the preceding attempt provided an inadequate sample, accept or bypass No-Gos, etc.

2. Control Logic—controls the general flow of command table execution.

3. Fourteen Command Tables—contain specific commands and delta times allowed for execution;

---
[a] A Mars SOL = 24.62 earth hours.

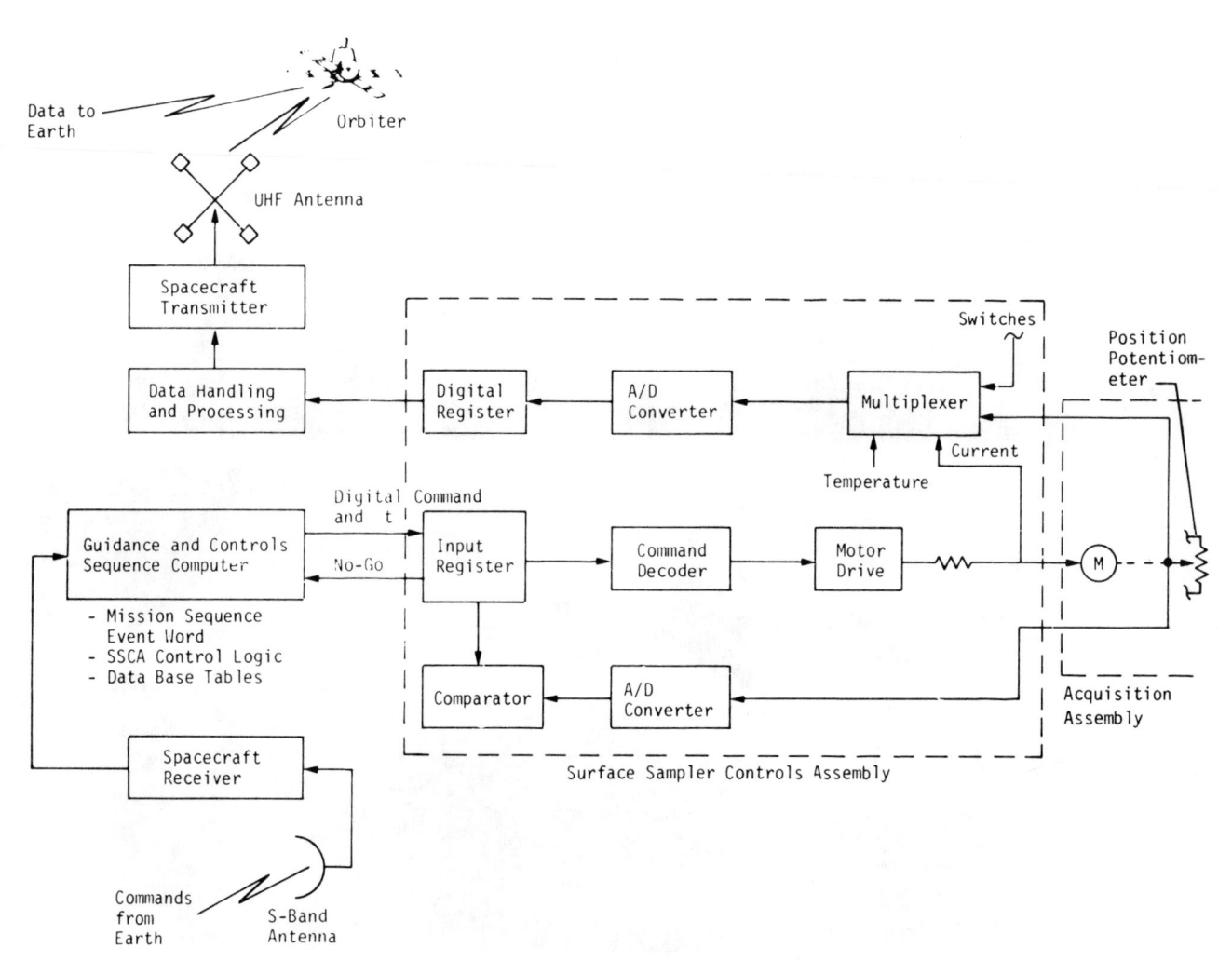

Fig. 3. Simplified boom control loop.

a maximum of 226 commands can be stored in these tables.

When a sequence is scheduled for execution, the lander computer automatically commands the lander to power-up the SSCA for a 10-s stabilization. The computer then begins to sequentially issue the 24-bit commands to the SSCA in accordance with established control logic and command table values. The computer waits a specific delta time between the issue of each command, which has been precomputed to allow sufficient time for command execution and data compilation. If the command is not successfully completed, or a No-Go is generated by an unsafe operation; the computer will automatically power-down the surface sampler and terminate all further commands until corrective actions are transmitted from earth. The computer continues to issue the commands until all scheduled operations of the particular sequence are completed, and the surface sampler is powered-down.

MARS SURFACE OPERATIONS

During the period between the landing of VL-1 on July 20, 1976 and June 13, 1977, a total of 82 surface sampler sequences involving over 3800 commands were performed by VL-1 and VL-2. During this period of operation, five hardware anomalies were encountered and diagnosed. Corrective actions were taken, which resulted in the continued operation of both systems. A description of these sequences and anomalies is presented below.

A preprogrammed sampling sequence was stored in both Viking computers before landing to

Fig. 4. Viking Lander-1 landing area panograph as viewed by camera no. 2.

provide a limited sampling capability in the event of an uplink commanding system failure. Arbitrary sampling site coordinates were established for the biology, GCMS, and XRFS sampling sequence. Both VL-1 and VL-2 landing areas were extremely rocky; "Murphy's Law" resulted in a large rock residing at the preprogrammed sampling site in both areas. A typical scene from the VL-1 site is illustrated in Fig. 4.

A computer program is available at JPL for producing topographic profile maps of sampling sites using stereographic data from the two lander cameras. Therefore, it was a relatively simple task to determine new sampling coordinates in a rock-free area and uplink site coordinate changes for the surface sampler before the first scheduled sampling sequence on SOL 8. The accuracy of the sample site coordinate determination program is on the order of 1 to 2 cm (0.4 to 0.8 in.).

All sequences planned for execution on the landers are developed and verified on the Viking Science Test Lander (Fig. 5) at JPL. Typical types of sequences that have been verified on the Science Test Lander and executed on Mars include the following:

1. delivery of 2000-μm sieved surface samples to the biology, GCMS, and XRFS experiments

2. delivery of small rock and pebble samples to the XRFS experiment

3. pushing medium-size rocks to permit sampling material that has been protected from ultraviolet radiation

4. magnified observation of surface particulate accomplished by positioning the collector head with a full sample in front of the magnifying mirror for camera imaging

Fig. 5. Science Test Lander at JPL, Pasadena, California.

5. collection of magnetic fraction surface material on the collector head backhoe magnet array and observation by direct imaging or indirectly through the magnifying mirror
(Note: The collector head backhoe magnet array can be cleaned of magnetic material by rotating the backhoe through the lander-mounted cleaning brush.)

6. support of surface imaging under the lander by proper positioning of boom-mounted mirrors

7. support of the meteorology investigation by acquiring temperature profile data at various increments between surface level and 3-m (118-in.) above the surface

8. support of surface thermal properties investigation by extending the collector head and integral temperature sensor several centimeters under the surface and monitoring diurnal temperature variations throughout a Martian day-night cycle

9. support of surface physical properties investigation by indirect measurement of surface bearing strength

10. support of surface physical properties investigation by depositing a sample of material on the lander grid for imaging a "splash" pattern

11. support of lander imaging investigations by positioning the boom to occult the sun

12. backhoeing a 23-cm (9-in.) hole for special Biology Gas Exchange Experiment investigation.

Some of the typical Mars surface sampler operations were photographed by lander cameras. Figure 6 illustrates a typical sampling sequence where a scientifically interesting caliche crust sample was acquired from a relatively tight area between rocks at the VL-2 site. Figure 7 illustrates pushing "Badger Rock" at the VL-2 site to acquire a biology sample from material that has been shaded from ultraviolet radiation for millions of years. One of the most interesting sequences performed during VL-1 operations was digging a 23-cm (9-in.)-deep hole (Fig. 8) in support of special biology and XRFS sampling requirements. Since the surface sampler was not designed for digging deep holes, a total of 688 commands were executed to backhoe the hole in a somewhat inefficient manner. Tables II and III summarize the results of all sequences performed on both landers as of June 13, 1977.

During the performance of surface sampler operations on both landers, a total of five hardware anomalies were encountered. A summary of these anomalies and corrective actions is presented below:

1. The VL-1 SOL 2 boom launch and landing restraint pin jam—during the shroud eject sequence, the pin failed to drop free to the surface when the boom was extended. During the retraction command, the pin jammed precluding retraction to the commanded position, and a No-Go resulted. The potential failure was duplicated in the laboratory, and a successful corrective action was executed, which involved extending the boom

Fig. 6. Viking Lander-2 sampling sequence for caliche crust sample.

TABLE II

Viking Lander-1 Surface Sampler Performance Summary

Operation	Local Lander Time (LLT)	UTC (GMT)	Duration (h)	Number of Commands	Results
		1976			
VL-1 landing	0/16:13:12	202/11:53:06	---	---	Successful
Biol PDA lid deploy	0/16:19:08	202/11:59:02	---	---	Successful
CH shroud eject	2/10:20:50	204/07:19:54	0.25	13	Shroud ejected, pin jam, No-Go
GCMS PDA lid deploy	3/14:00:00	205/11:38:39	---	---	Successful
Pin jam recovery	5/10:40:00	207/09:37:49	0.21	12	Successful
Biol sampling	8/06:54:28	210/08:05:27	1.89	58	Successful
GCMS sampling	8/08:52:50	210/10:03:49	1.15	76	Partial success, no "level full"
XRFS sampling	8/10:35:50	210/11:49:49	0.72	38	Successful
XRFS sampling	8/11:22:50	210/12:33:49	0.72	38	Successful
GCMS sampling	14/06:25:00	216/11:19:07	0.28	13	Acquired sample, no delivery, No-Go
E/R diagnostic sequence	18/14:00:00	220/21:32:48	0.10	29	Successful
GCMS sample delivery	22/12:00:54	224/22:11:43	0.48	29	Successful
Dump GCMS sample	31/10:00:00	234/02:07:06	0.07	11	Successful
GCMS sampling	31/10:40:00	234/02:47:06	0.75	49	Successful
XRFS sampling	34/10:10:00	237/04:15:52	0.66	41	Successful
XRFS sampling	34/11:10:00	237/05:15:52	0.66	41	Successful
Biol sampling	36/11:10:00	239/06:35:02	2.34	60	Successful
XRFS sampling	40/10:27:50	243/08:31:13	0.95	47	Successful
XRFS sampling	40/12:07:50	243/10:11:13	0.95	47	Successful
Phys/Mag Properties	41/15:30:00	244/14:12:58	0.97	40	Successful
Biology sampling	91/07:00:00	295/14:42:21	1.92	61	Successful
Temporary Termination of Surface Sampler Activities for Solar Conjunction					
		1977			
Met temperature profile	177/11:47:30	18/04:14:24	1.99	80	Successful
XRFS sampling	177/14:30:00	18/06:56:54	0.47	31	Successful
XRFS sampling	178/14:40:00	19/07:46:29	0.47	31	Successful
XRFS sampling	179/10:30:00	20/04:16:04	0.47	31	Successful
XRFS sampling	180/10:30:00	21/04:55:39	0.47	31	Successful
Deep hole sequence	202/14:00:00	43/22:56:35	1.58	165	Successful
Deep hole sequence	204/11:53:00	45/22:08:45	1.78	179	Successful
Deep hole sequence	218/12:00:00	60/07:29:59	1.58	165	Successful
Deep hole sequence	219/11:53:00	61/08:02:34	1.78	179	Successful
XRFS sampling	229/08:00:00	71/10:45:26	0.82	55	Successful
Biol sampling	250/08:00:00	93/00:36:47	2.32	80	Successful
XRFS sampling	250/11:00:00	93/03:36:47	0.81	55	Successful
Magnetic abundance	270/13:00:00	113/18:48:31	1.78	104	Successful
Physical Properties	280/10:46:40	123/23:11:04	1.69	44	Successful
XRFS skim sampling	285/11:00:00	129/02:42:19	0.40	35	Successful
XRFS skim sampling	286/11:00:00	130/03:21:55	0.38	35	Successful
Physical Properties	296/09:00:00	140/07:57:48	1.16	39	Partial success, early sequence term.
Recovery sequence	297/15:00:00	141/14:37:24	0.22	12	Successful
XRFS rock push	302/09:00:00	146/11:55:20	0.53	30	Successful
XRFS sampling	311/10:59:27	155/19:51:04	0.84	42	Successful
XRFS sampling	312/10:59:27	156/20:30:40	0.84	42	Successful
Totals			37.45	2168	

NOTE: VL-1 Surface Sampler Operations continuing normally as of June 13, 1977, ~11 months after landing on Mars.

TABLE III

Viking Lander-2 Surface Sampler Performance Summary

Operation	Local Lander Time (LLT)	UTC (GMT)	Duration (h)	Number of Commands	Results
		1976			
VL-2 landing	0/09:49:05	247/22:37:50	---	---	Successful
Biol PDA lid deploy	0/09:55:53	247/22:44:38	---	---	Successful
CH shroud eject	1/10:44:50	249/10:05:32	0.36	19	Successful
GCMS PDA lid deploy	3/14:00:00	251/14:39:53	---	---	Successful
Biol sampling	8/16:00:00	256/10:05:27	1.66	50	Partial success; XRFS No-Go
CH diagnostic sequence	10/15:30:00	258/10:54:37	0.02	2	Successful
XRFS delivery	13/16:45:00	261/14:08:22	0.06	8	Partial success; insufficient rocks[a]
GCMS sampling	21/10:00:00	269/12:40:04	1.06	67	Successful
Biol sampling	28/16:00:00	276/23:17:11	1.95	64	Successful
XRFS sampling	29/13:30:00	277/21:26:46	0.50	42	Successful
XRFS sampling	30/10:30:00	278/19:06:22	0.50	42	Successful
GCMS rock nudge	30/11:20:00	278/19:56:22	0.37	19	Successful
GCMS rock push	34/10:35:00	282/21:49:43	0.36	23	Successful
GCMS rock push	37/10:00:00	285/23:13:28	0.17	18	Successful
GCMS sampling	37/16:00:00	286/05:13:28	1.29	97	Successful
GCMS sample delivery	40/15:50:00	289/07:02:14	0.34	23	Successful
Biol 2-rock nudge	45/10:00:00	294/04:30:10	0.49	36	Successful
XRFS sampling	46/13:00:00	295/08:09:46	0.61	39	Partial success; insufficient rocks[a]
XRFS sampling	46/13:40:00	295/08:49:46	0.61	39	Partial success; insufficient rocks[a]
XRFS sampling	47/13:00:00	296/08:49:21	0.61	39	Partial success; insufficient rocks[a]
XRFS sampling	47/13:40:00	296/09:29:21	0.61	39	Partial success; insufficient rocks[a]
Biol rock push	51/06:15:00	300/04:42:42	0.19	19	Successful
Biol sampling	51/06:40:00	300/05:07:42	2.58	79	Successful
Phys/Mag Properties	56/14:00:00	305 15:45:38	1.24	59	Successful
Phys/Mag Properties	57/06:43:00	306/09:08:13	0.96	43	Successful
XRFS sampling	57/08:00:00	306/10:25:13	0.45	39	Partial success; insufficient rocks[a]
XRFS sampling	57/08:45:00	306/11:10:13	0.45	39	Partial success; insufficient rocks[a]
XRFS sampling	58/08:00:00	307/11:04:49	0.45	39	Partial success; insufficient rocks[a]
XRFS sampling	58/08:45:00	307/11:49:49	0.45	39	Partial success; insufficient rocks[a]
Temporary Termination of Surface Sampler Activities for Solar Conjunction					
		1977			
Met temperature profile	131/10:35:00	16/13:49:43	0.72	45	Successful
Biol sampling	145/09:00:00	30/21:28:57	1.73	63	Successful
XRFS rock pile	145/13:30:00	31/01:58:56	0.63	39	Partial success; insufficient rocks[a]
XRFS rock pile	145/14:30:00	31/02:58:56	0.64	39	Partial success; insufficient rocks[a]
XRFS rock pile	145/15:30:00	31/03:58:56	0.63	39	Partial success; insufficient rocks[a]
XRFS rock pile	145/16:30:00	31/04:58:56	0.64	39	Partial success; insufficient rocks[a]
Phys/Mag Properties	154/16:05:02	40/10:30:15	0.25	11	Successful
XRFS sampling	161/11:10:00	47/10:12:20	0.70	50	Successful
XRFS sampling	161/11:55:00	47/10:57:20	0.70	50	Successful
XRFS rock pile	172/12:00:00	58/18:17:48	0.63	39	Partial success; insufficient rocks[a]
XRFS rock pile	172/12:40:00	58/18:47:48	0.63	39	Partial success; insufficient rocks[a]
XRFS rock pile	172/13:20:00	58/19:37:48	0.63	39	Partial success; insufficient rocks[a]
XRFS rock pile	172/14:00:00	58/20:17:48	0.63	39	Partial success; insufficient rocks[a]
XRFS sampling	185/14:00:00	72/04:52:27	0.74	50	Successful
XRFS sampling	186/13:00:00	73/04:32:01	0.74	50	Successful
Biology sampling	195 09:45:00	82/07:13:19	0.14	3	No-Go; low temperature azimuth stall
Boom diagnostic sequence	205 12:00:00	92 16:04:12	0.04	10	Successful
Totals			29.16	1666	

NOTE: VL-2 Surface Sampler Operations temporarily suspended April 2, 1977 due to the extremely low temperatures encountered during the Mars "winter" months. Operations scheduled to resume in September 1977 during the Mars "spring."

[a] The partial successes were due to the nature of the Martian's surface materials rather than a problem with the surface sampler.

Fig. 7. Viking Lander-2 surface sampler pushing "Badger" rock.

Fig. 8. Viking Lander-1 23-cm-deep hole dug by the surface sampler.

and vibrating the collector head such that the pin-jam was freed.

2. Failure of the VL-1 SOL 14 boom to retract—during a GCMS sampling sequence, the boom failed to respond to a retract command. Investigation revealed that a combination of consecutive retract commands, which excessively tighten the furlable tube on its storage drum, combined with low-temperature effects, can cause the motor starting torque to be exceeded. All subsequent sequences were revised to preclude consecutive extend or retract commands; the problem has not recurred.

3. Control of VL-2 SOL 8 collector head rotation—during delivery of an XRFS sample, the collector head discrete position feedback switch failed to operate and terminate collector head rotation at an intermediate position. All subsequent sequences were revised to "time" control rotation of the collector head based on knowledge of lander bus voltage and rotation motor operating characteristics.

4. Failure of the VL-2 SOL 195 boom azimuth—during an azimuth command, the boom ceased operation after rotating ~0.68 rad (39 deg) of a commanded 3.5 rad (201 deg). The -112°C (-170°F) temperature at the time of the anomaly was lower than during any previous operation. The operation was subsequently repeated at a higher temperature of -84°C (-120°F) with no problems encountered. We presume that low temperature combined with gear-tooth lubricant wear and a 0.14-rad (8-deg) incline of the lander may have resulted in a rotation torque that exceeded the motor stall torque. All subsequent sequences will be performed at temperatures greater than -101°C (-150°F).

5. Shutdown of the VL-1 SOL 296 boom extend—during a boom command to extend to 28 cm (110 in.), a No-Go was encountered and the Surface Sampler Sequence was automatically terminated. Subsequent analysis of imagery data revealed the boom had attained the commanded position. Subsequent Surface Sampler Operations were performed successfully, and analyses of all data indicated all components were performing within specification. We assumed the automatic shutdown was probably caused by a spurious electrical transient.

CONCLUSIONS

The Viking Surface Samplers were a success. Martian surface materials were delivered to all experiment instruments and special sequences were performed in support of other scientific investigations.

Thus, the general design concepts employed in the Viking surface sampler have proven to be adequate as a result of interplanetary space environment exposure for 11 months, and successful operation on the surface of Mars for 10.8 months for VL-1 and 9.3 months for VL-2, as of June 13, 1977. Design concerns such as solid-film lubricant degradation, low-temperature operation of motors, and low-temperature stressing of the furlable tube boom element were alleviated. The anomalies encountered to date are understood and corrective action sequences and operating limitations have been imposed. The V-Ls are still operating nominally, and it is anticipated that the surface sampler subsystems will continue to perform satisfactorily through 1978.

A LARGE-SCALE MANIPULATOR FOR SPACE SHUTTLE PAYLOAD HANDLING—THE SHUTTLE REMOTE MANIPULATOR SYSTEM

H. J. TAYLOR *Spar Aerospace Ltd., 825 Caledonia Road
Toronto, Ontario, Canada M6B 3X8*

KEYWORDS: *manipulators, remote handling, orbiting systems*

ABSTRACT

The design and manufacture of an engineering model space shuttle remote manipulator system are now complete. The concept embodied in the manipulator and its control system is designed to meet National Aeronautics and Space Administration-generated specifications for a payload handling system. Many new techniques are used to achieve hardware that meets the exacting requirements of the space environment and repeated launch and reentry.

INTRODUCTION

The shuttle remote manipulator system (SRMS), the subject of a Canadian/U.S. agreement between the National Research Council of Canada (NRCC) and the National Aeronautics and Space Administration (NASA), is described. The SRMS is a joint undertaking between NRCC and NASA, in which Canada provides this major orbiter subsystem. Spar Aerospace Ltd. is the prime contractor responsible to NRCC for the design, development, and manufacture of the first flight system in response to requirements developed at NASA's Lyndon B. Johnson Space Center.

SYSTEM FUNCTION

The SRMS is part of the overall space shuttle in-flight payload handling system. It is primarily used to deploy payloads from the orbiter payload bay for release into orbit and to retrieve payloads from orbit and stow them in the payload bay for servicing or return to earth. Figure 1 depicts the shuttle on orbit with the manipulator arm deployed to handle a satellite payload.

The basic remote manipulator system (RMS) consists of two manipulator arms plus supporting equipment. When only one manipulator arm is required, normally it will be installed on the port side of the orbiter payload bay.

ELEMENTS OF THE SYSTEM

The SRMS system, shown in Fig. 2, is made up of a 6-deg-of-freedom manipulator arm that is

Fig. 1. The SRMS.

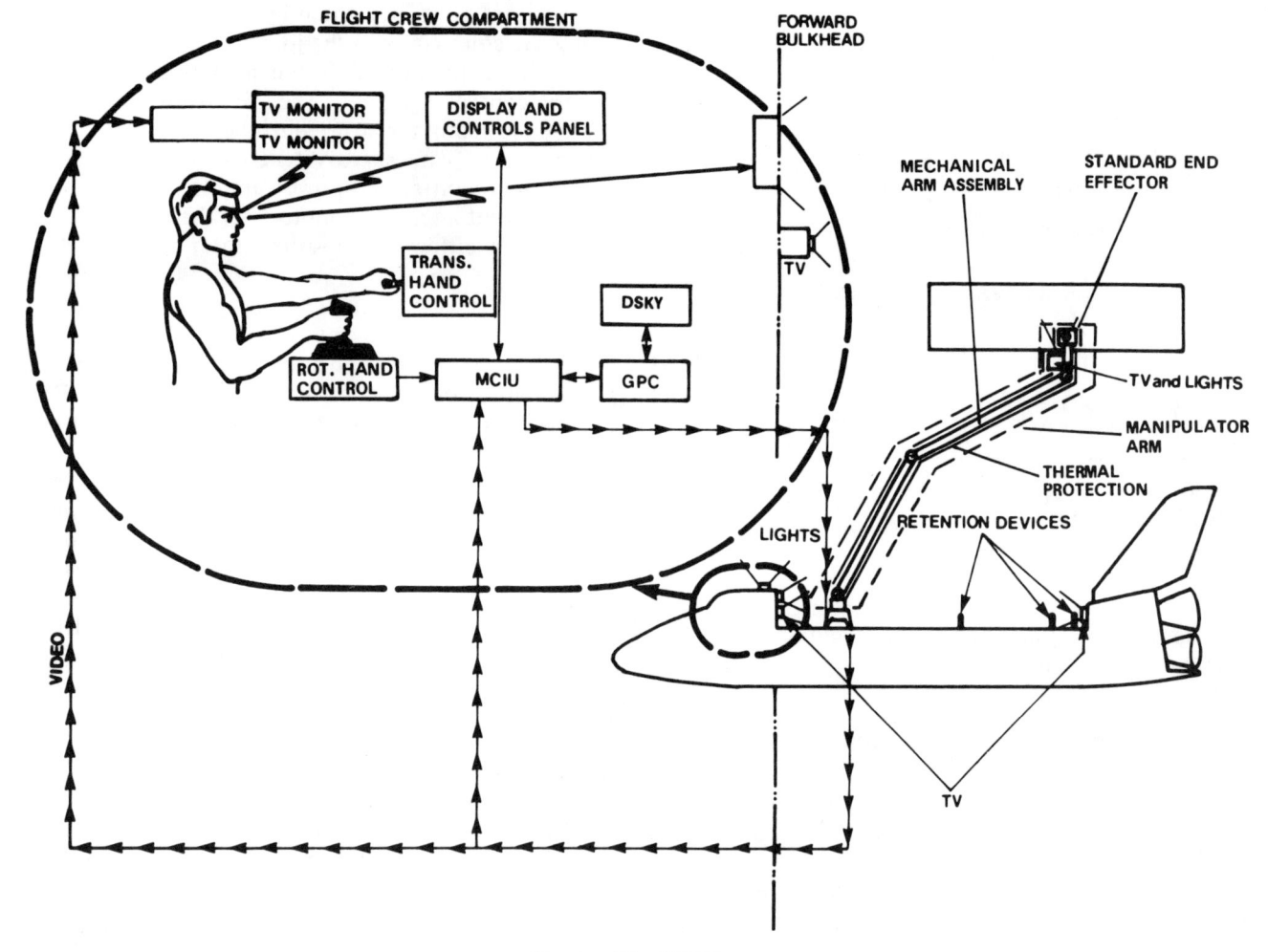

Fig. 2. The SRMS.

operated by a payload specialist located in the aft area of the shuttle's crew compartment. The operator has direct vision from two windows looking aft into the shuttle cargo bay and also from two windows directly above him. He also monitors RMS activity on two television screens located adjacent to his control panel. These receive signals from cameras located on the forward and aft payload bay bulkheads and also on the wrist of the RMS. A control panel includes displays that provide him with the status of the RMS as it performs its functions. The operator controls the arm by utilizing two 3-deg-of-freedom hand controllers. The left-hand controller provides translational motions of the manipulator, while the right-hand controller gives rotational motions. Commands from the controls are routed through a manipulator controller interface unit (MCIU), which is the main interface between the orbiter's general-purpose computers, the control and display panel, and the manipulator arm and its end effector.

PERFORMANCE REQUIREMENTS

The manipulator system was designed to meet the following basic requirements:

1. The end effector and the arm control system was designed to allow capture of a payload which may have an initial misalignment with respect to the end effector of ±15 deg or ±10 cm after capture. It is capable of releasing payloads with attitude errors of <5 deg.

2. The arm itself is designed so that it can achieve a maximum tip speed of 0.06 m/s with a 14 500-kg payload attached and 0.6 m/s with no payload. It is designed so that the stall torque in the drive system can provide a 7-kg force at the tip of the arm in all axes.

3. The RMS will be able to position payloads relative to orbiter axes with considerable precision.

TABLE I

Shoulder Joint Performance Parameters

	Shoulder Yaw	Shoulder Pitch
Stall torque, minimum (Nm)	1050	1050
Joint rate, maximum (deg/s) (no payload)	2.29	2.29
Joint rate, maximum (deg/s)[a]	0.229	0.229
Single joint drive tip speed (m/s)[a,b]	0.06	0.06
Rigid body stop distance of tip (m)[c]	0.57	0.57

[a]Up to 14 500-kg payloads.
[b]Straight-arm configuration.
[c]Baseline 14 500-kg payload.

Manipulator joint performance requirements are summarized in Table I.

MECHANICAL ARM SUBSYSTEM

The commands from the operator, when suitably transformed by the MCIU, are delivered to the mechanical arm subsystem (MAS), shown in Fig. 3. This subsystem is a 15.3-m-long 6-deg-of-freedom manipulator arm. The arm has three segments separated by electromechanically driven joints. The joint at the shoulder has shoulder pitch and shoulder yaw motions. A 6.4-m arm segment then terminates in the elbow joint, which provides elbow pitch. A further 7-m arm segment terminates in a wrist joint that provides wrist motions in pitch and yaw. A further 1.8-m segment of the arm includes a joint that provides continuous wrist roll motions. In turn, this segment is attached to an end effector, which is the device that actually interfaces with the payload.

CONTROL SYSTEM

The control system consists of joint servos and control algorithms. The primary function of the joint servos is to control joint angles and rates to satisfy specified commands. The commands can be provided either by the operator directly from control input devices on the displays and controls (D&C) panel, as in the case of the "joint-by-joint" control (direct drive), or through control algorithms. Control algorithms receive commands from the operator through suitable control input devices in terms of position and orientation of end effector or translational and rotational velocities of the end effector and attempt to satisfy these by computing and specifying the commands to the joint servos.

Each of the joints in the manipulator contains a servocontrol system. The servosystem receives commands from the arm controller or directly from the operator. The servosystem consists of a servomotor, servocontroller, power amplifier, and sensors to sense joint angles and motor rate. The servo can be operated in either the position, rate, or direct drive mode.

In the position mode, the position feedback signal obtained from the encoder in the joint is compared with the position input command to provide an error signal to drive the servomotor. Rate feedback from a tachometer on the motor shaft provides damping.

In the rate mode of operation, the position feedback signal is switched out and the rate input command is compared with a scaled signal from the motor tachometer to provide a joint rate error signal to drive the motor.

In the direct drive mode, the motor is driven "open-loop", bypassing the control electronics. This is not a servomode, but is only a configuration suitable for applying inputs to the motor.

Any one of the above operating modes can be selected by mode select logic, which configures the system as required.

To perform useful tasks, the end effector attached to the tip of the arm must be controlled in a specified manner. The state-of-the-arm (and hence, the end effector) can be specified by the position and orientation of the end effector and its translational and rotational velocities. The arm can be controlled only through the joint servos. The arm control algorithms basically generate commands to the joint servos. The arm control algorithms are based on a kinematic model of the arm. The arm can be controlled in position or rate. In position control, the position and orientation of the end effector are controlled, whereas in rate control, the translational and rotational velocities of the end effector are controlled.

DESIGN DESCRIPTION

The engineering model shoulder joint is described in some detail, as it is representative of all the joints in the manipulator arm subsystem in basic structure and mechanism. Figure 4 is a cutaway illustration showing the main elements of the assembly.

Shoulder Joint

The shoulder joint interfaces mechanically with the arm swing-out mechanism at its lower flange and with the upper arm boom at the electronic housing flange.

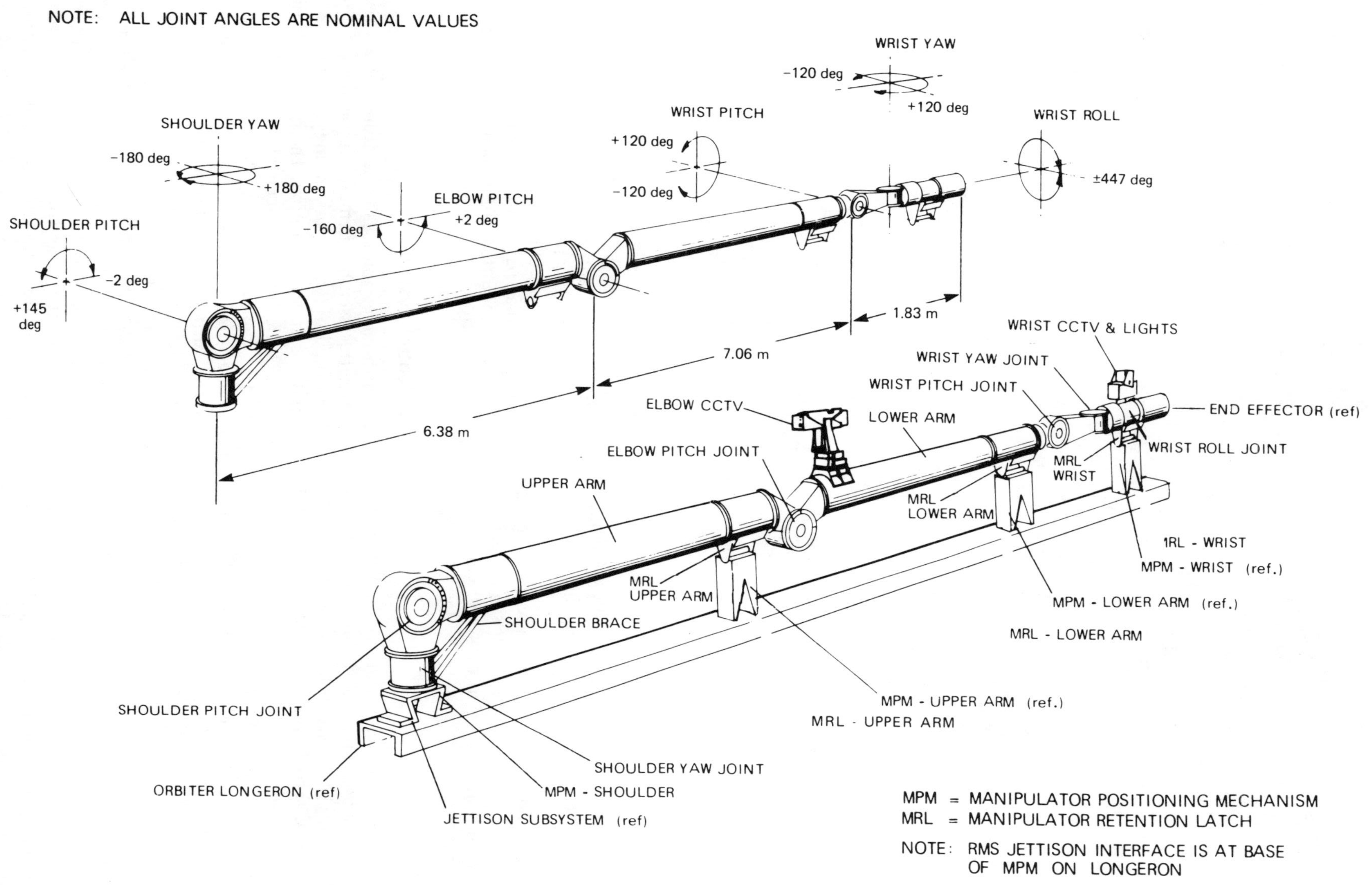
NOTE: ALL JOINT ANGLES ARE NOMINAL VALUES
WRIST YAW
-120 deg
+120 deg
SHOULDER YAW
-180 deg
+180 deg
WRIST PITCH
+120 deg
-120 deg
WRIST ROLL
±447 deg
ELBOW PITCH
-160 deg
+2 deg
SHOULDER PITCH
-2 deg
+145 deg
1.83 m
7.06 m
6.38 m
WRIST CCTV & LIGHTS
WRIST YAW JOINT
WRIST PITCH JOINT
END EFFECTOR (ref)
ELBOW CCTV
LOWER ARM
ELBOW PITCH JOINT
WRIST ROLL JOINT
MRL WRIST
UPPER ARM
MRL LOWER ARM
1RL - WRIST
MPM - WRIST (ref.)
MRL UPPER ARM
SHOULDER BRACE
MPM - LOWER ARM (ref.)
MRL - LOWER ARM
SHOULDER PITCH JOINT
MPM - UPPER ARM (ref.)
MRL - UPPER ARM
SHOULDER YAW JOINT
ORBITER LONGERON (ref)
MPM - SHOULDER
JETTISON SUBSYSTEM (ref)
MPM = MANIPULATOR POSITIONING MECHANISM
MRL = MANIPULATOR RETENTION LATCH
NOTE: RMS JETTISON INTERFACE IS AT BASE OF MPM ON LONGERON

Fig. 3. The MAS.

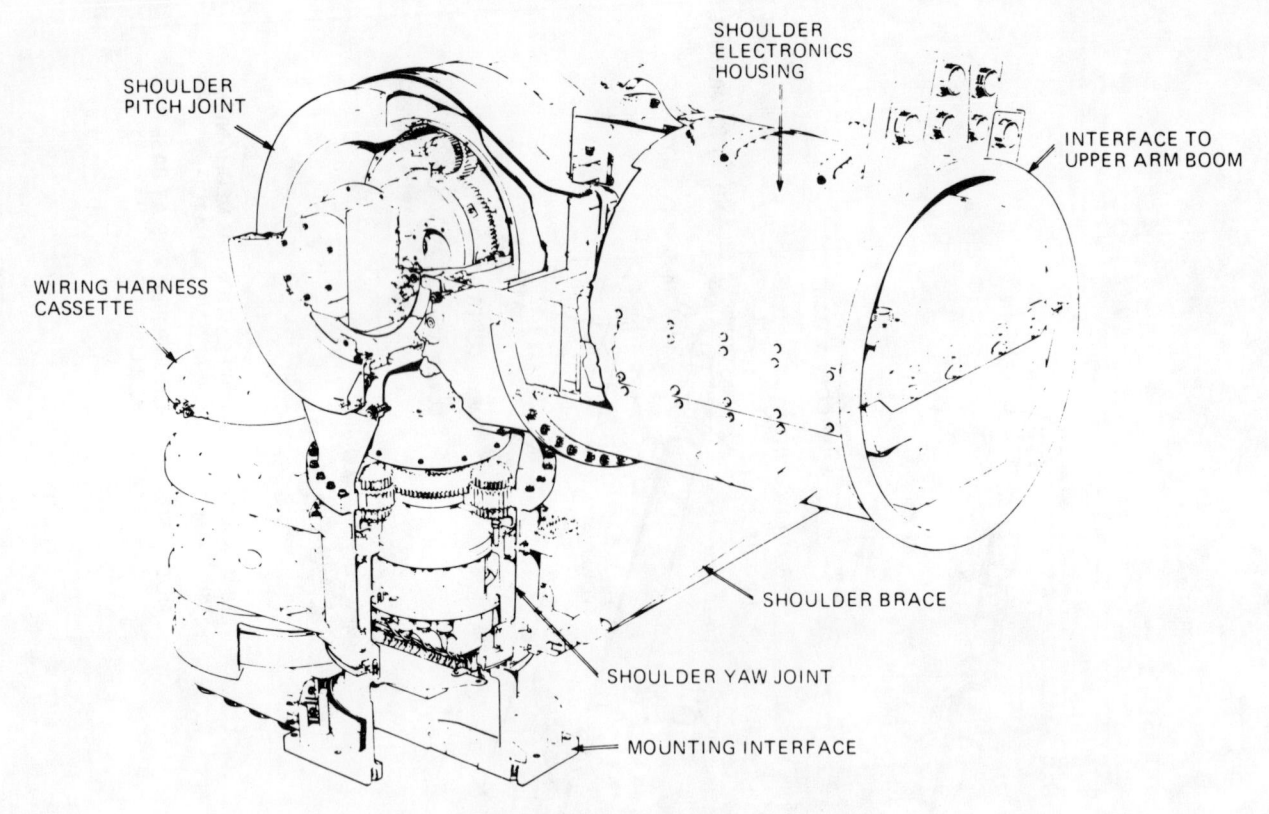

Fig. 4. Overall configuration of the shoulder joint.

Yaw Joint

The yaw joint structure uses a fully machined inner sleeve as the mounting flange and support pedestal for the complete arm. The yaw outer casing is carried by the inner sleeve through a pair of angular contact bearings mounted at the upper sleeve position and a cylindrical roller bearing at the lower end.

Transition Piece

The yaw transition piece connects the yaw outer sleeve to the shoulder pitch joint through a circular flange and a pair of crescent-shaped flanges at the pitch joint inner toroid. All three flanges use bolt and dowel fastening.

The pitch joint inner toroid and the upper arm transition together form a clevis-type connection coupled by an arrangement of angular contact ball bearings and cylindrical roller bearings as in the yaw joint.

Electronics Compartment

This assembly has a one-piece cylindrical outer casing with bolt and dowel flanges at each end which interface the shoulder pitch joint and the upper arm boom.

The shoulder brace saddle is riveted to the underside of this outer casing. A tray spans the inner cylindrical space and supports electronics packages. The connectors for the brace actuator and the main electrical cable are bulkhead types mounted to a doubler plate that permits removal of the electronics tray, the tray-mounted components, and connectors as one assembly.

Cable Control System

The shoulder joint cable is stored in a cassette. The cassette assembly is mounted on the aft outside surface of the yaw joint casing and moves with manipulator arm yaw motion. The internal configuration is a helical tunnel open at each end. The flat conductor cable bundle is threaded through the helix, the lower opening accepts the cable slack generated during yaw motion by moving the cassette-stored cable from the helix minor diameter toward the major diameter. The pitch-generated slack is handled in the same manner, but through a 90-deg chute into the upper helix opening. The cassette absorbs the effects of pitch-and-yaw motion simultaneously without adding torque load to the pitch-and-yaw drive systems. The arm cable harness terminates at the shoulder in a connector box mounted at the lower end of the yaw pedestal.

Shoulder Brace

The shoulder brace is designed to prevent pitch axis rotational excitation of the shoulder and, thus, avoid gear tooth damage during launch vibration. The twin strut brace cradles the electronics compartment at the upper end and terminates at the yaw casing in an expandable collet actuated by a motor-driven ball screw.

The collet attachment is released on command after launch and prior to arm deployment, and is not reengaged during the less severe reentry and landing phase.

STRUCTURAL MECHANICS

The shoulder joint is stiffness critical. Lack of stiffness amplified by the total arm length affects arm control performance. The shoulder and all other major components have been through a computer optimization process to arrive at the materials selected, mainly based on the constraints of weight, stiffness, and strength. Table II summarizes shoulder stiffness.

The steel yaw-to-pitch transition piece and aluminum pitch-to-electronics-compartment transition piece are both castings. Castings were selected as the most stiffness-efficient structural design; stresses are generally low in the shoulder joint and so lower strength properties are acceptable. The interface with the orbiter is through the stiffness-critical inner yaw shell.

INTERNAL LOADS/DEFLECTIONS

The major shell assemblies were analyzed for unit loads using the finite element technique to determine stiffness contributions. Stress analyses were conducted later using actual load case predictions.

FRACTURE CONTROL

All components that experience cyclic loading have been analyzed for crack instability with use against a desired mission life. If crack growth (using standard nondestructive evaluation crack size) appears unstable at or more than the desired mission life, the design is deemed to be not fracture critical.

VENTING

Thermal blankets and structural components are vented to accommodate pressure changes during orbiter ascent and descent.

TABLE II
Shoulder Joint Stiffness
(Straight-Arm Configuration)

Moment Direction	Total Stiffness (Nm/rad × 10^5)
Vertical	10.8
Inboard	11.7
Aft	12.6

GEAR DESIGN

The RMS joint drives are required to drive or be driven by both very high and very low inertia masses attached to the end effector.

After several trade-off studies, a basic gearing package was derived which consists of a low-speed gear group coupled to a high-speed gear group. The high-speed grouping consists of two reductions of parallel axis gearing, referred to as "G-1," while the low-speed grouping consists of two reductions of planetary-type gears referred to as "G-2." All gearing is spur type. The schematic diagram for the shoulder gear train is shown in Fig. 5.

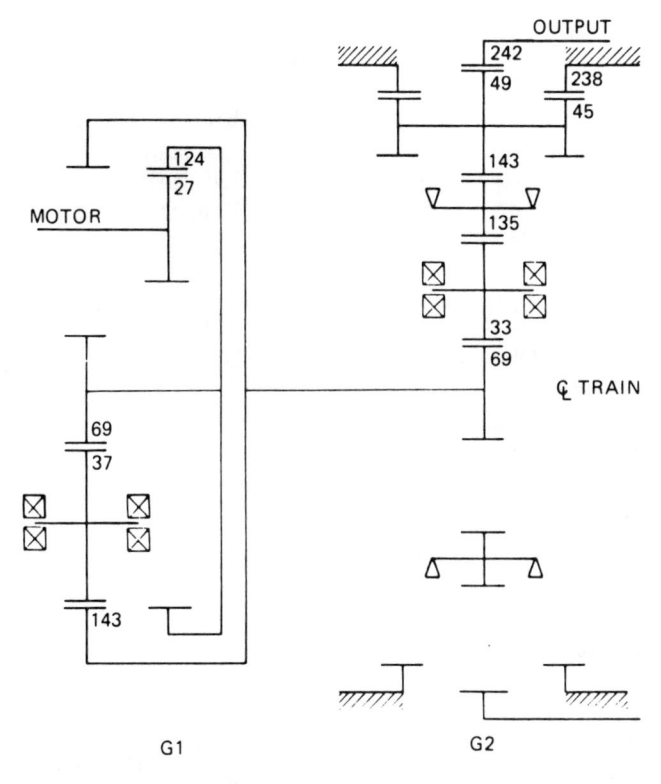

Fig. 5. Schematic of the shoulder joint gear.

Differential Planetary Final Reduction

Attention was given to the characteristics of epicyclic trains in which a disparity exists between torque multiplication ratio and a velocity ratio. Torque multiplication begins to level out at ratios in excess of 37:1. Therefore, higher planetary ratios of 50:1 or more could not be used without a serious loss of efficiency and a loss in the ability to backdrive. Eleven equally spaced planet pinions maximize the number of load paths.

Gear Ratio Selection

The SRMS joint gear ratio was selected to match the speed/torque characteristics of available motors. A planetary final reduction ratio of 39:67 was used on all joints. Overall ratio adjustments to suit individual joint requirements were accomplished in the high-speed G-1 train.

Other design drivers affecting gear train selection were:

1. forward drive efficiency
2. backdrive efficiency
3. minimum backlash
4. backdrive inertia
5. forward drive motor/load inertia match.

Backdriveability and Efficiency

Gear train efficiency was a basic consideration in designing the joint mechanism since backdriveability was a factor. By proper selection of planet-to-ring gear ratios, it was possible to predict a differential planetary efficiency in excess of 76%. An actual efficiency of 85% was recorded during the test. Backdriveability is assured at all efficiency levels in excess of 50%.

Gear Train Backlash

Backlash at individual meshes is targeted at 0.05/0.1 mm to give the workshop an achievable level of composite tolerance, tooth thickness, and other factors.

Contact Ratio

The tooth-mesh contact ratio is one of the gear parameters subject to manipulation and control. All gearings for the MAS were held to a maximum contact ratio of 1:5.

Gear Stresses

All gears in the joint gear trains were designed to transmit maximum power within the limitations of envelope and weight. Each mesh of the entire train was analyzed to withstand the worst case of bending and contact stresses developed under maximum load conditions.

Contact (hertzian) stresses have been maintained below the 6900×10^5 N/m² load capacity limit specified for Lubeco 905 dry lubricant under fatigue conditions. The material used for the MAS gearing will sustain an "uncorrected" allowable gear-tooth bending stress of $\sim 2750 \times 10^5$ N/m². Individual gear elements of the system have been sized to experience between 1240×10^5 and 1650×10^5 N/m².

G-2 Gear Reduction

Figure 6 shows a typical G-2 low-speed gear group. The centrally located star gear reduction can be seen in mesh with the second-stage reduction sun gear. The planetary system of 11 gear clusters is without conventional bearing support; the radial component of the output and reaction gear contact forces is countered by bearing rings with controlled flexibility to provide a minutely variable operating center distance. This unique Spar development provides a means of reducing backlash at low load levels.

G-2 Manufacture

Manufacturing the G-2 group of the reduction gearbox presents some unique shop problems. To function, the outer gear elements must be the same size over pins and their active profile must be precisely indexed relative to each other. The optimized design of G-2 demands a uniform load

Fig. 6. The low-speed gear group.

distribution among these planets. Therefore, all planet gear elements are selected for dimensional uniformity. The requirements for precision are met in practice by making the planet gear elements as separate pieces, through hardened, finish ground, and stress relieved. The pieces are matched for size and mounted concentrically in a fixture designed to provide indexing and alignment. Electron beam welding joins the cluster.

THERMAL DESIGN

The thermal design ensures acceptable operating times for the shoulder joint under the most severe environmental conditions. Thermistors are located on the most thermally critical components. Thermostatically controlled electrical resistance heaters maintain acceptable temperature levels under nonoperational cold case environmental conditions.

ELECTRICAL DESIGN

The shoulder joint assembly contains two motor-drive modules with the following:

1. *Servopower amplifier (SPA).* An SPA is provided for each motor module and contains a motor-drive amplifier, velocity feedback, input and output data buss circuits, commutator scanner and position encoder circuits, and tachometer-generator excitation.

2. *Joint power conditioner (JPC).* The JPC receives orbiter 28-V dc main power and generates various regulated voltages for SPA use. The shoulder joint JPC supplies power to the shoulder yaw, shoulder pitch, and elbow electrical equipment.

3. *Backup drive amplifier (BDA).* The BDA drives any one of the 6 deg of freedom in a backup mode on command from the orbiter.

4. *Position encoder.* An optical position encoder is provided for each degree of freedom to provide a digital position signal to the appropriate SPA, which allows transmission of the angle signal through the data buss.

5. *Cable harness.* Electrical receptacles on the connector box located at the pedestal of the shoulder yaw assembly provide the orbiter/arm electrical interface. The cable harness is attached along the external arm surface, inside the thermal control blanket.

MATERIALS AND PROCESSES

Materials used in SRMS construction have been selected within the constraints of the NASA/JSC materials selection list, and the limitations imposed by the predicted environmental conditions of humidity, salt fog, vibration, vacuum, ultraviolet radiation, penetrating radiation, and temperature excursions.

Candidate materials not included in the materials selection documents were subject to a comprehensive range of tests including flammability, outgassing, volatile condensation, and toxicity prior to acceptance or rejection.

Thermal Control Materials

Goldized Kapton filing, separated by Dacron mesh, is used for thermal insulation blankets. Surface coatings are limited to thermal control paint, anodizing, and black oxide.

Solid Lubrication System

Lubeco 905 was selected for all MAS gears and bearings.

TESTING

Each mechanical arm joint is subject to a functional test program performed using specially designed test rigs capable of exercising the degree of freedom and monitoring the performance under simulated loading. Figure 7 shows a partially assembled MAS mounted on the system test rig, which is supported on air bearings and permits arm exercise in the plane of the air-bearing floor with close to "zero G" conditions. Joints with axes normal to those joints under test are exercised subsequently by axially rotating the mechanical arm in the test rig until the degrees of freedom of the remaining joints are parallel to the floor plane.

SIMULATION FACILITY (SIMFAC)

Figure 8 shows part of the Spar facility installed to provide real-time simulation of the behavior of complex electromechanical systems through mathematical modeling and computer-generated scenes. Simfac is particularly useful for man-in-the-loop systems and has proved very effective in determining the operational characteristics of the SRMS. The orbiter-type controls and displays are located in a crew station patterned after the orbiter flight deck. The dynamics of the mechanical arm are math-modeled and displayed in four cathode ray tube

Fig. 7. The mechanical arm on the system test rig.

Fig. 8. The shuttle orbiter complex with part of the simulation subsystem in the background.

monitors. Two of these monitors simulate orbiter cargo bay camera views. The other two simulate direct views as seen by the arm operator.

CONCLUSION

A brief examination of the technical solutions to a unique manipulator problem, that of a complex mechanism in the space environment, has been considered. A summary of the results of a 4-yr concentrated effort by a team of Canadian engineers and scientists has also been presented.

ACKNOWLEDGMENTS

The author wishes to thank Spar Aerospace Ltd. and the National Research Council of Canada for permission to present and publish this paper.

CHAPTER 5
ROBOTIC ADAPTATIONS TO HOSTILE ENVIRONMENTS

Presented at the SME 5th International Symposium
on Industrial Robots, September 1975

Application of High Performance Heavy Duty Industrial Manipulators

by Henry G. Krsnak
and
Michael J. Howe
General Electric Company

ABSTRACT

The motivation to greater automation has been and continues to be improved productivity along with reduction in hazard to personnel and compliance with safety regulations. Among the last areas to be mechanized are assembly and materials handling. Robots and manipulators are now being applied in assembly and material transfer operations.

This paper discussed one approach to mechanization utilizing large (9.3m extension maximum), heavy duty (2700 kg. payload maximum) high precision manipulator systems. A generalized system description is presented along with specific industrial applications from the foundry, automotive and forging industries.

INTRODUCTION

The traditional motivation toward automation of industrial processes in order to increase productivity has been supplemented by new incentives in recent years. The environmental constraints of the Occupational Safety & Health Act (OSHA), the rapidly increasing cost of labor and the problem of hiring and retaining help for hot, dirty or physically punishing jobs have all helped to intensify interest in industrial automation.

Automation of a task or process frequently requires a considerable investment over and above the cost of the handling equipment. In many cases, plant and process changes are required. Parts handled by automatic machinery must be fairly uniform in size, shape and position. The system requires sensors for detecting the presence (or absence) of parts, for determining the orientation of the parts, and for monitoring the anomalies in the process. Quite frequently human judgement and agility are required in order to make the decisions and to perform the physical tasks (e.g., clearing equipment log jams, orienting parts, etc.) that are beyond the machine's capabilities. There are many material handling tasks in industry for which automation is desirable but, for technical or economic reasons, are not automated. Productivity and cost effectiveness of such tasks can most efficiently be improved by some balanced use of man and machine. A man operated, high load, dextrous manipulator is such a system.

MAN AMPLIFIER MANIPULATOR SYSTEMS

A man-amplifier is a piece of equipment that provides performance analogous to that of man with magnification in strength and reach. Such hardware allows relatively complex tasks to be performed in a mechanized manner while retaining the unique capabilities of human senses and judgement as an integral part of the control loop.

In manufacturing, the problem best served by man-amplifiers is a handling process or task which may not be precisely repetitive or for which adequate process sensory information required by fully automated control systems is not available. It is in this realm that the unequalled adaptability of human judgement and action can be used to best advantage.

Typical applications are those where parts entering the work area are random in size, shape and orientation. Retention of a man in the system permits the machine to adapt to the position of the part; the man can readily make decisions on sorting, passing defective parts and accommodating perturbations in process input and output. Man's capabilities can also be used for the operation of auxiliary equipment (e.g., presses, conveyors, furnace doors, etc.).

Manned manipulators do not compete with fully automatic systems (robots). Processes that have the inherent uniformity (of product shape, feeding and positioning) to permit economical automation should be automated. However, lines that cannot be economically automated full time (due to intermittent short runs, mixed products, or product overlaps during changeovers requiring a frequent manual supplement) may qualify for a hybrid system. In a hybrid configuration the system is operated through either computer control for uniform product operations or manual control for random or rapidly changing product operations. With such an arrangement the manual mode also serves as a back-up for the computer mode.

FUNCTIONAL DESCRIPTION

The MAN-MATE® Industrial Manipulator (see Figure 1) is a man amplifier in which the movements and forces applied by the operator to the master arm are amplified and duplicated by the slave arm (boom). The two arms (similar in form and function to human arms) are interconnected through a bilateral servo system which provides two-way communication between the operator and the work. Displacement of the replica master by the operator creates an error signal in the servo system; the system responds to correct the error by moving the boom a proportional amount in the same direction (see Figure 2). The

® Registered Trademark of the General Electric Company.

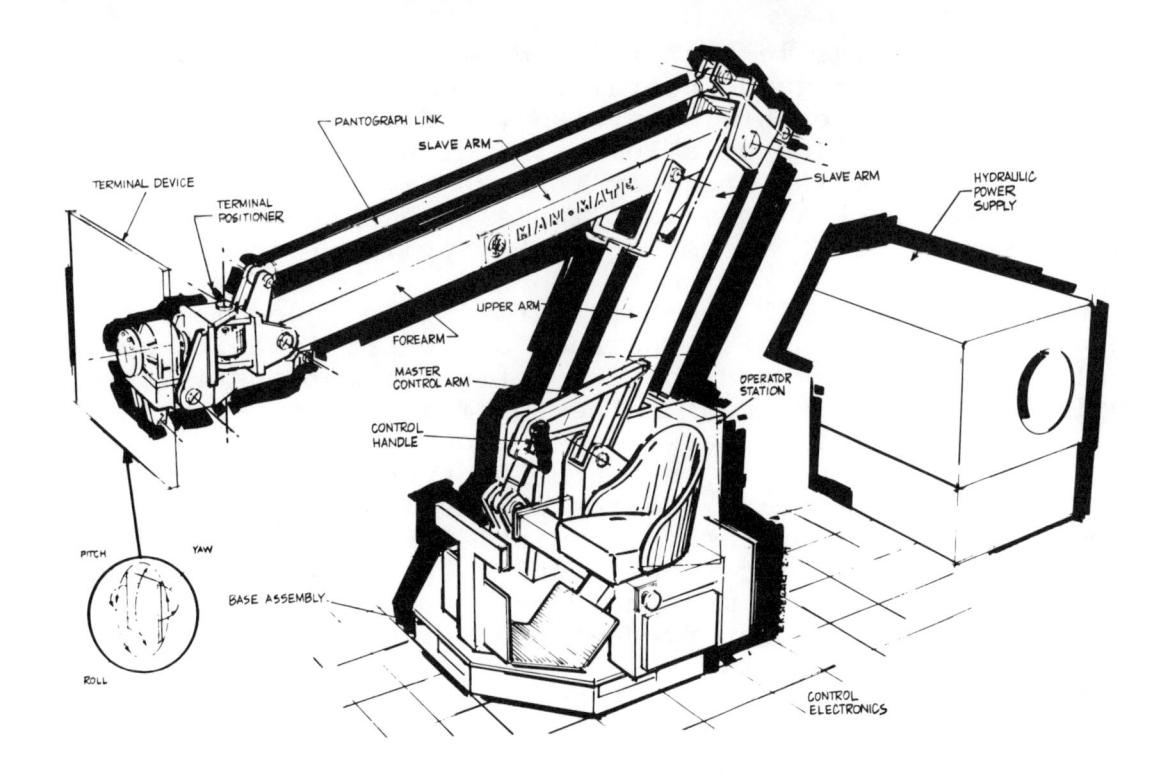

Figure 1. Manipulator configuration.

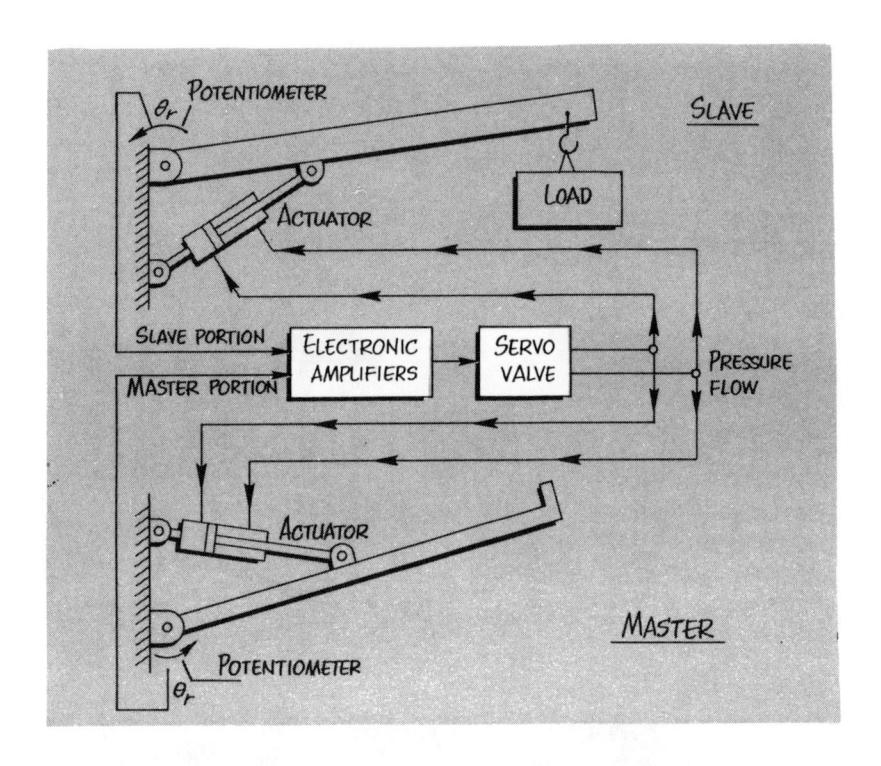

Figure 2. Physical representation of a servo loop.

system is responsive to both displacement and rate of change in the master, giving the operator a broad range of control in a single control element. Conversely, if the boom is displaced by the addition of a load, the displacement and magnitude are sensed by the operator through his master control. This force feedback allows compliance with forces of constraint as well as the sensing of the applied forces.

The correspondence with the human arm is continued into the outboard end of the boom. The boom can be fitted with any or all of the pitch, yaw and roll end effector motions which jointly function as a wrist. The terminal device serves as a hand. The wrist motions are used to position the terminal device for picking up the part and for proper part orientation at drop off. The terminal device is the interface between the manipulator and the process. Terminal devices take many forms: vacuum cups for handling sheet stock; opposed claws for bar stock or castings; a simple hook or a specialized device designed for a particular product (see Figure 3). The controls for the end effector and terminal device are integrated into the master control arm at the control grip. A series of pushbuttons controls the hydraulic actuators to operate the wrist and hand motions. A variety of applications has shown this configuration of three replica motions and up to three grip-integrated motions to be cost effective and a prudent baseline of standardization.

Figure 3.   Special device for handling V-8 and V-12 engine heads.

An operator, comfortably seated, exerting a maximum of 3.4 kg
(7½ lb.) force on the master arm can control a load of 2700 kg
(6000 lb.) through 270° rotation, 3.7m to 9.3m (12 ft. to
30 ft.) of extension and 6.3m (20 ft.) in lift (see Figure 4)
with six degrees of freedom. The master counterbalance is
adjusted so that with no payload virtually no force is
required on the master to maintain slave position. With rated
load on the slave, a force of 3.4 kg (7½ lb) is required at
the master to maintain the position of the load. The force
feedback enables the operator to sense weight and inertia
forces of the load; it supplements his depth perception by
enabling him to feel forces on the boom when placing, picking
up and releasing loads.

Figure 4.   Manipulator performing appliance handling task.

COMPUTER CONTROL

The Man-Mate manipulator capability can be extended to provide
full computer controlled operation without sacrificing manned
capability. It can be converted from manual to computer
operation merely by control switching. As the operator performs the transfer task, the coordinates of the sequence are
recorded and the operation is programmed. The manipulator can
then repeat the task independently of the operator. This
feature of the Man-Mate system has been demonstrated in development tests and is available for selected applications.

TRAINING

Operators are trained in the basic machine functions in a
matter of minutes. An operator with normal physical dexterity

and depth perception can become totally familiar with the operation within one eight-hour shift. The training time is short because of the instinctive control technique which employs both spatial correspondence and force feedback. The ease of operation has been demonstrated consistently; people who are unfamiliar with the machine have become reasonably proficient operators in 10 to 15 minutes.

MAINTAINABILITY

User data shows that based upon 4,500 production hours per year, the system "availability" is approximately 98.5% (i.e., production time lost due to Man-Mate repairs is about 50 hours per year). Maintenance can be performed by plant electrical and hydraulic technicians. Established troubleshooting routines can be performed with standard test equipment; no extensive training in solid state electronics or servo theory is required.

APPLICATIONS

The cases described below illustrate some applications of the Man-Mate industrial manipulator to material handling tasks in industry. Additionally, they show the variety of tasks that may be performed by the operator and the flexibility retained in the process line by using a manned manipulator.

In most applications, the manipulator may be installed in existing production lines with little or no modification to the line. The flexibility of maintaining production without the manipulator is retained; frequently adequate space is available for men (working safely outside the machine's envelope) to supplement the manipulator tasks (e.g., during peak production or changeover where the variety of parts exceeds the number of sorting bins within the machine's envelope).

The prime motivations for introduction of the manipulators in these applications are:

- o to improve a material handling task.
- o to get men out of a hostile environment.
- o to improve productivity.
- o to reduce breakage and scrap.
- o to retain human decision-making capability as a vital element in the process.

In addition to discharging his material handling duties, the manipulator operator, through a control station placed on the manipulator, can perform operating tasks formerly done by the line crew.

Several installations have been developed with the manned manipulator as an integral part of the system. In these installations, man's decision-making capability was a key factor in the selection of the chosen system. The system parameters (e.g., part or process variability) made the choice of man as a sensing and control element an economically viable one.

Figure 5 shows the manipulator in the center of the most advanced forging system in the world. The system produces, in a single strike of the forging press, a precision turbine blade up to 167 cm (66 in.) long. The system consists of one of the world's largest 11 x 10^6 kg (12,000 ton) mechanical forging presses, a 6.4 m (21 ft.) diameter rotary hearth electric furnace, a 450 x 10^3 kg (500 ton) straight side trim press, and a Man-Mate industrial manipulator from whose console the entire operation is controlled.

Figure 5. Manipulator installation in turbine blade forging process.

The system operator, seated safely and comfortably away from the furnace and the presses, has a pushbutton panel (ref. Figure 5) within easy reach for controlling the furnace doors, the forging press, and the trim press. The operator opens the furnace door, removes the billet, and closes the door as he transfers the billet to the forge press. He locates the billet in the die of the press, releases it and initiates the press stroke. He then retrieves the formed blade and transfers it to the trim press, cycles that press and places the finished part on a pallet. Occasionally, the blade sticks in the upper die; the operator senses and responds to the anomaly routinely, with no significant delay in the operating cycle.

The manipulator is equipped with pitch, yaw, and roll positioner motions and a special stainless steel gripper (Figure 6) for precise positioning of the blades in the forge and the trim presses. The unit is skid mounted so it can be lifted out of the way with a bridge crane for die changing.

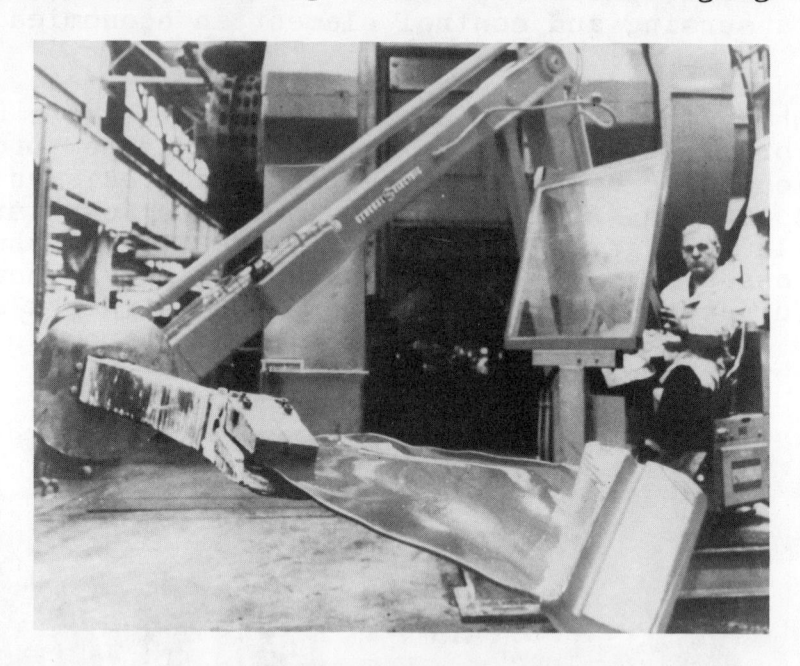

Figure 6. Manipulator handling large turbine blade.

In another forging shop, a manipulator system was inserted into an existing production line (Figures 7 and 8). The

Figure 7.
Manual handling
of axle forging.

Figure 8.
Man-amplified handling
of axle forging.

objective here was to relieve men from the job of handling truck axle housings weighing nearly 180 kg (400 lb.). When large housings were forged, two men were utilized. Work on this line had to be stopped during the summer when it became too hot. As the figures show, the manipulator was simply placed on the line as a man amplifier. The man is now well back from the press, comfortably seated, using one hand to perform a task formerly taxing to two men. Production is now maintained year round.

A foundry application, where the manipulator was added to an existing facility is shown in Figure 9. The task is to remove sprues and runners from a foundry shakeout table. A shakeout area is one of the worst working environments in industry. Workers are subject to high noise levels, abundant airborne dust and sand grit as well as intense heat from red hot castings. Previously, two men (with 100% relief) stood adjacent to the shakeout, snagged sprues and runners with hooks and transferred them to scrap boxes manually.

Figure 9. Manipulator in foundry shakeout application.

Because the prime incentive for change here was to remove men from an inhospitable environment, the master station was separated from the boom and placed in an air and sound conditioned room. The electrical, electronic and hydraulic controls were also placed in the room thus providing a clean atmosphere for maintenance. In this installation, space considerations prohibited placing the manipulator adjacent to the shakeout where the man stood formerly. The machine and operator's enclosure were placed over the shakeout in previously unused space. No valuable floor space was sacrificed; no process line modifications were required; the operator's view of the line was improved; and in the event that the manipulator is down, the job can still be done manually from the former work station. One man in total comfort now performs a job that previously required four men working in conditions that are no longer acceptable. This is a good example of an application that would be unfeasible to automate. The scrap pieces removed (i.e., sprues and runners) are randomly mixed with the castings. Mechanical separation is impractical because scrap and parts are nearly the same size. Human judgement is needed to differentiate between good parts and scrap.

CONCLUSION

The outstanding feature of such a system is that it fills the void between dull, punishing manual labor and sophisticated automatic machinery. It can handle heavy, hot, irregular loads with speed and precision; man's dexterity and senses are retained but his strength and endurance are multiplied many times over. It will perform in an environment unbearable to humans while the operator comfortably controls an entire production process through minimal physical effort.

BIBLIOGRAPHY

BANGS, S. "No 'Loafing' with New Forge Press", *Precision Metal*, November, 1974.

BRUST, H.A. "For Versatile, Heavy Handling, Put a Man in the Loop", *Automation*, December, 1973.

GEORGE, R. L. "Materials Handler Has Hydraulic Muscle", *Hydraulics & Pneumatics*, August, 1972.

GRAY, W. E. "A Role of Mechanization in Improving Industrial Working Conditions", Proceedings: *NATO Symposium on Working Place Safety*, Bad Grund, West Germany, July, 1974.

WINSHIP, J. J. "Forging Begins a New Era", *American Machinist*, November, 1972.

ZELDMAN, M. A. "The Business Case for Robotics", *1975 IEEE Intercon Conference Record: Session 13*, New York, New York, April, 1975.

Presented at the RI/SME Robots 8 Conference, June 1984

Remote Demilitarization of Chemical Munitions

by Robert H. Leary
James M. McNair
and
John F. Follin
GA Technologies, Incorporated

This paper describes work in progress at GA Technologies, Inc. under U.S. Army contract DACA87-C-0031, "Mechanical Process Development Project - Chemical Agent/Munitions Demilitarization Program." GA is currently constructing a prototype facility to develop and demonstrate novel process concepts for the demilitarization of obsolete chemical agent munitions. Part of this contractual effort involves the design, development, and demonstration of remote automated manufacturing technology for material handling and process control. Here we discuss the design and implementation of a multi-robot material handling system and its associated minicomputer-based control system for a particularly promising disposal process that will be demonstrated at the prototype facility.

Project Goals and Design Criteria

The major project goal is to develop and demonstrate a process technology that has the potential for major (e.g. 30%) reductions, relative to current methods, in the demilitarization costs of existing inventories of obsolete chemical munitions. This goal, together with technical and safety constraints, forms the basis for the major design criteria for the prototype facility.

Under normal operating conditions, manned entry into process areas is prohibitively disruptive and expensive. The presence of lethal chemical agents requires the use of efficiency-limiting protective suits and extensive manpower and equipment for backup support. Thus the first major design criterion for the process is remote operation and maintenance under normal circumstances.

The second major design criterion arises from economic considerations. Economic analysis of existing demilitarization technology reveals a large manpower component in the operating costs. Manpower reduction appears to have the most potential cost saving impact in view of the overall project goal. This leads to the design criterion that the process should require a minimal number of operating personnel and ideally should run autonomously under computer control. Indeed, a major goal of the prototype facility is to demonstrate the feasibility of autonomous process operation under computer control, including the capability to recover from many process upset conditions without operator intervention.

The third major design criterion also arises from economic considerations in the form of throughput rates. Target single line processing rates have been established for each munition type with typical values on the order of one to two munitions per minute. Supporting criteria for system availability and maintenance have been set for consistency with this project goal.

While the prototype facility at GA will operate using relatively small numbers of non-explosively configured munitions filled with simulated agents, most of the operational constraints imposed by remote operation with actual munition inventories are addressed in the prototype design. Thus the robot and computer control technology described below will be directly applicable to a production facility.

Process Overview

The novel process selected for demonstration at the prototype facility is called <u>cryo-fracture</u> as it is characterized by cryogenic embrittlement followed by fracturing of the munitions. Figure 1 shows a perspective of an anticipated demilitarization production facility based on the cryo-fracture concept. The feed material for this disposal process consists of munitions of various types such as projectiles (155-mm, 105-mm, and 8-inch diameter), 4.2-inch mortars, and 105-mm cartridges, all containing highly toxic chemical agents. Several munition types also contain propellants and high explosive bursters designed to disperse the agent in military use. The disposal process thus must safely and efficiently handle several types of hazardous materials.

The basic process under development involves three stages as shown in Figure 2. In the first or unpack stage the munitions are depalletized, singularized, and placed in holding racks in a staging area. The material handling operations in this stage vary in complexity with munition type and initial packaging but in general require several distinct task sequences.

In the second or cryogenic treatment stage, the munition holding racks are placed in a bath of liquid nitrogen until the structural metal components (typically low carbon steel) of the munitions are embrittled. After an appropriate amount of immersion time, the racks are removed from the bath and transported to an airlock for transfer of the munitions to the fracturing and disposal area.

In the final cryo-fracture stage, the embrittled munitions are loaded into a hydraulic fracturing machine where each munition is fractured under controlled conditions. The fragments and contents fall into a high temperature furnace which burns the chemical agent, high explosive, and propellant and also decontaminates the residual metal pieces.

II. REMOTE OPERATIONS AND MAINTENANCE

Remote Technology Selection

An initial study on the selection and use of remote technology for the cryo-fracture process considered both robotics and the manual manipulator technology developed for the nuclear industry. It was concluded that nuclear manipulators are too slow and labor-intensive to meet the required production rates and manpower criteria. However, these manipulators are useful for certain remote maintenance operations. One the other hand, the capabilities of commercially available robots were found to be consistent with our remote

FIGURE 1
ANTICIPATED CRYO-FRACTURE PRODUCTION FACILITY

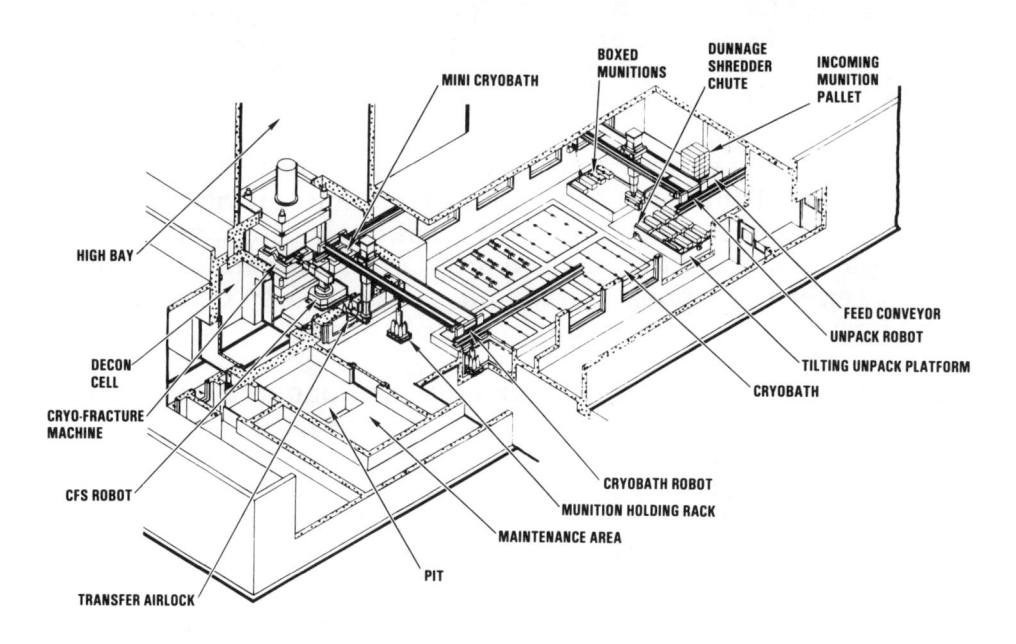

FIGURE 2

CRYO-FRACTURE PROCESS FLOW DIAGRAM

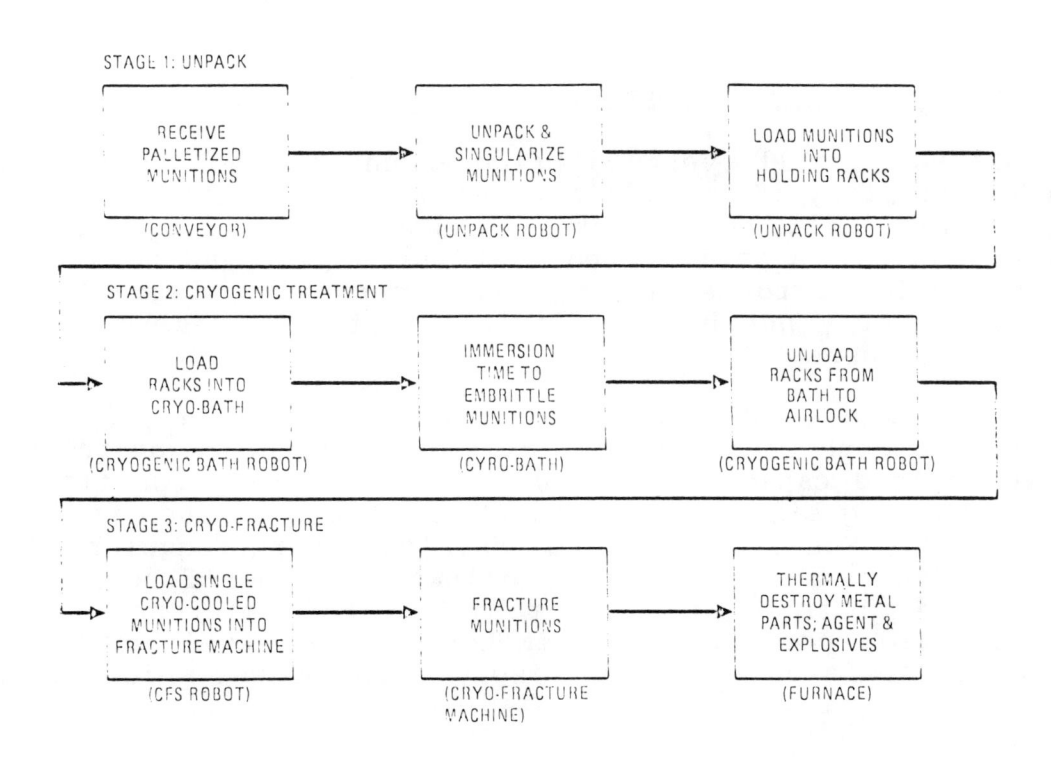

165

operation, production rate, and autonomous operation goals if the robots could be integrated into a overall minicomputer-based process control system. Robots were also selected due to their flexibility in processing the different munition types. Therefore the process concept was developed ultilizing robot material handling and computer control technology.

The prototype facility incorporates two robot systems, one for unpacking munition pallets and transporting munitions to and from the cryogenic treatment pool and one for loading the cryo-fracture machine. Although these systems are quite different in function and mechanical requirements, they share a number of common general specifications derived from the design criteria:

1. In order to facilitate remote autonomous process operation, the robots shall be capable of running under control of an external supervisory minicomputer. This implies the availability of an interface and communication protocol between each robot controller and the minicomputer. The robot controllers shall themselves be programmable in a high-level language which supports this interface.

2. The robots shall be capable of highly flexible operation in terms of working envelope and degrees of freedom, even if such flexibility exceeds normal process operation requirements. This will allow maximal capability to respond to off-normal and process upset conditions and to demonstrate robotic operations for remote maintenance under both manual and automatic control.

3. End effectors and special tooling shall be designed for fast, automatic connect and disconnect under computer control. Materials shall be resistant to caustic (sodium hydroxide solution) decontamination sprays and thermal cycling from room to liquid nitrogen temperatures.

Unpack and Cryogenic Bath Robot

A GCA/PAR XR6100 overhead bridge-mounted robot is used in the prototype facility (Figure 3) both to unpack munitions and to service the cryogenic bath. In an actual production facility such as shown in Figure 1, it is anticipated that the unpack and bath servicing tasks would be performed by separate overhead systems; however, a single heavy duty but highly flexible system serves both roles for prototype purposes.

The overhead robot has six degrees of freedom (x,y,z, yaw, pitch, and roll) to cope with the complexity of the unpack operation, while a heavy payload capability (1500 lb) is necessary for transferring racks containing six of the heaviest munitions (8-inch projectiles at 200 lb each). The large working envelope (a rectangular volume with dimensions 50-ft x 16-ft x 8-ft) allows coverage of both the unpack and cryogenic bath areas. The robot's CIMROC microprocessor-based controller is programmable in CIMPLER, a high-level robot language. The CIMROC is linked to a supervisory minicomputer which can download

FIGURE 3
ISOMETRIC OF GA TECHNOLOGIES PROTOTYPE FACILITY

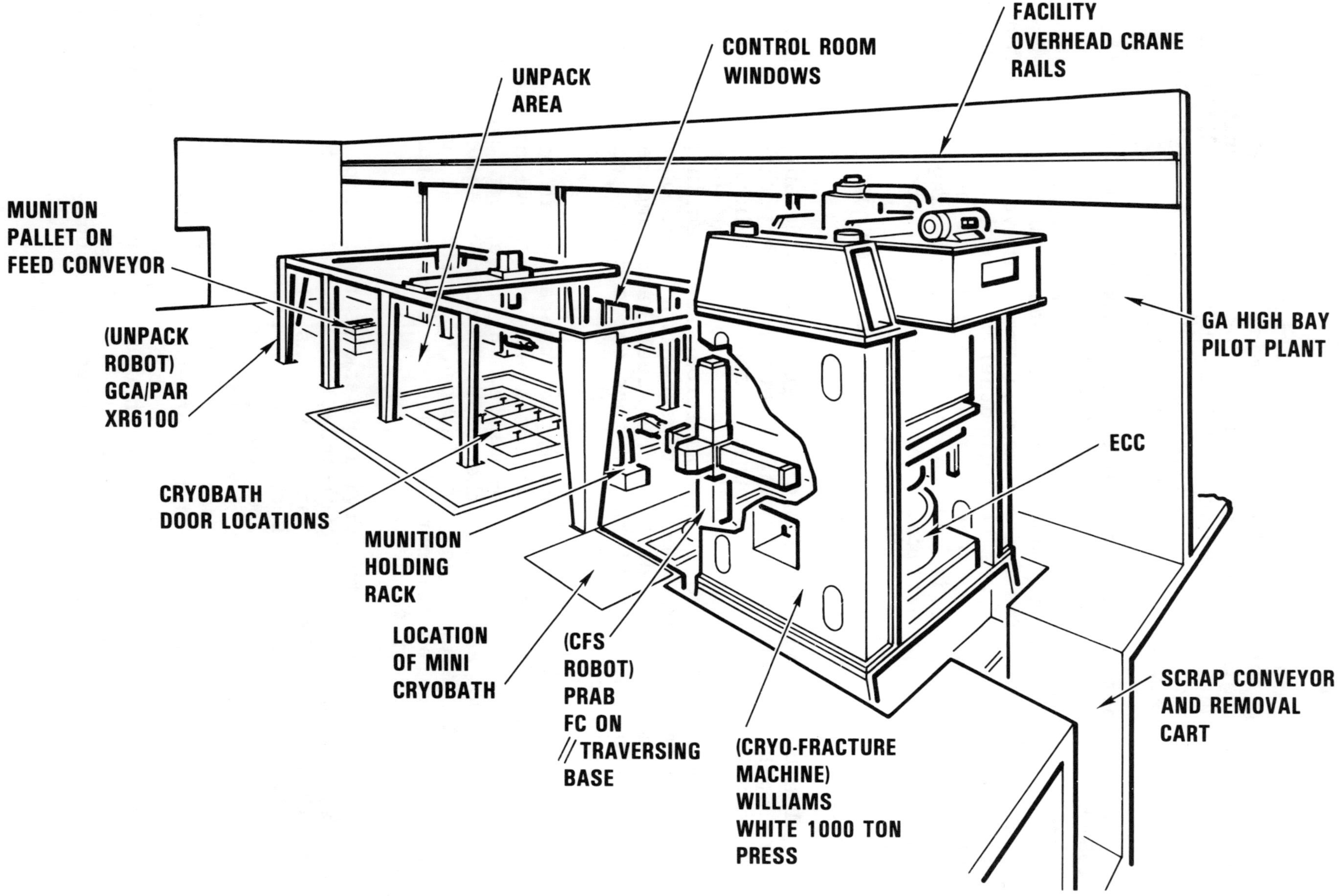

167

single instructions or entire programs, thus allowing the overhead robot operations to be easily coordinated with the rest of the computer-controlled process line.

As shown in Figure 4, unpack functions for the various munition types can be categorized into two major tasking sequences based on initial packaging configurations of the received munitions. Mortars and cartridges are boxed and must be handled by a different unpack procedure than the various types of projectiles which are palletized in a relatively open configuration.

The unpack tasks for the boxed munitions, 4.2-in mortars and 155-mm cartridges, are perhaps the most complex of the automated material handling operations. These munitions are packaged individually in tubed containers and sealed inside wooden boxes, two rounds per box. The boxes are palletized in various configurations of from 12 to 48 boxes held together by steel bands. During the unpack operation the steel bands are removed and the boxes are singularized and oriented correctly for opening. A special multi-function pneumatically actuated end effector is then used to open each box, remove the packing material, and finally extract the tubed munitions for transfer to a munition holding rack for processing in the cryogenic cooling operation.

The unpacking of the projectiles is considerably simpler due to a more open initial pallet configuration. The projectiles arrive in pallets containing 6 to 8 munitions arranged in a rectangular array with each individual munition oriented in a vertical standing position (nose upward). The munitions are held in position by wooden top and bottom pieces bound together by steel bands. The unpack sequence consists of cutting the steel bands and removing the wooden pallet top. The munitions are thus exposed for pairwise transfer to the munition racks.

Detailed analysis of the unpack operation in both cases results in a complex series of individual steps. Complicating the task further is the fact that the pallet may have been packed as long as 30 years ago and while the general configuration is known, significant initial and age-induced geometrical variations are possible. The control methods used to cope with this situation are described in section III below.

The cryogenic bath servicing task is much simpler than the unpack tasks, amounting to a pick-and-place operation under controlled conditions for the overhead robot. As illustrated in Figure 5, the cryogenic bath is a pool of liquid nitrogen covered by insulating doors which swing to an open position for munition rack immersion and retrieval operations. A rack holding six munitions is transferred over an unoccupied door location of the pool and lowered through the doors covering that location. After a suitable immersion time (30 minutes to 2 hours, depending on munition type), the rack is lifted by its handle (which protrudes through the doors) and removed from the pool for transfer through a simulated airlock to the cryo-fracture system. The supervisory minicomputer keeps track of

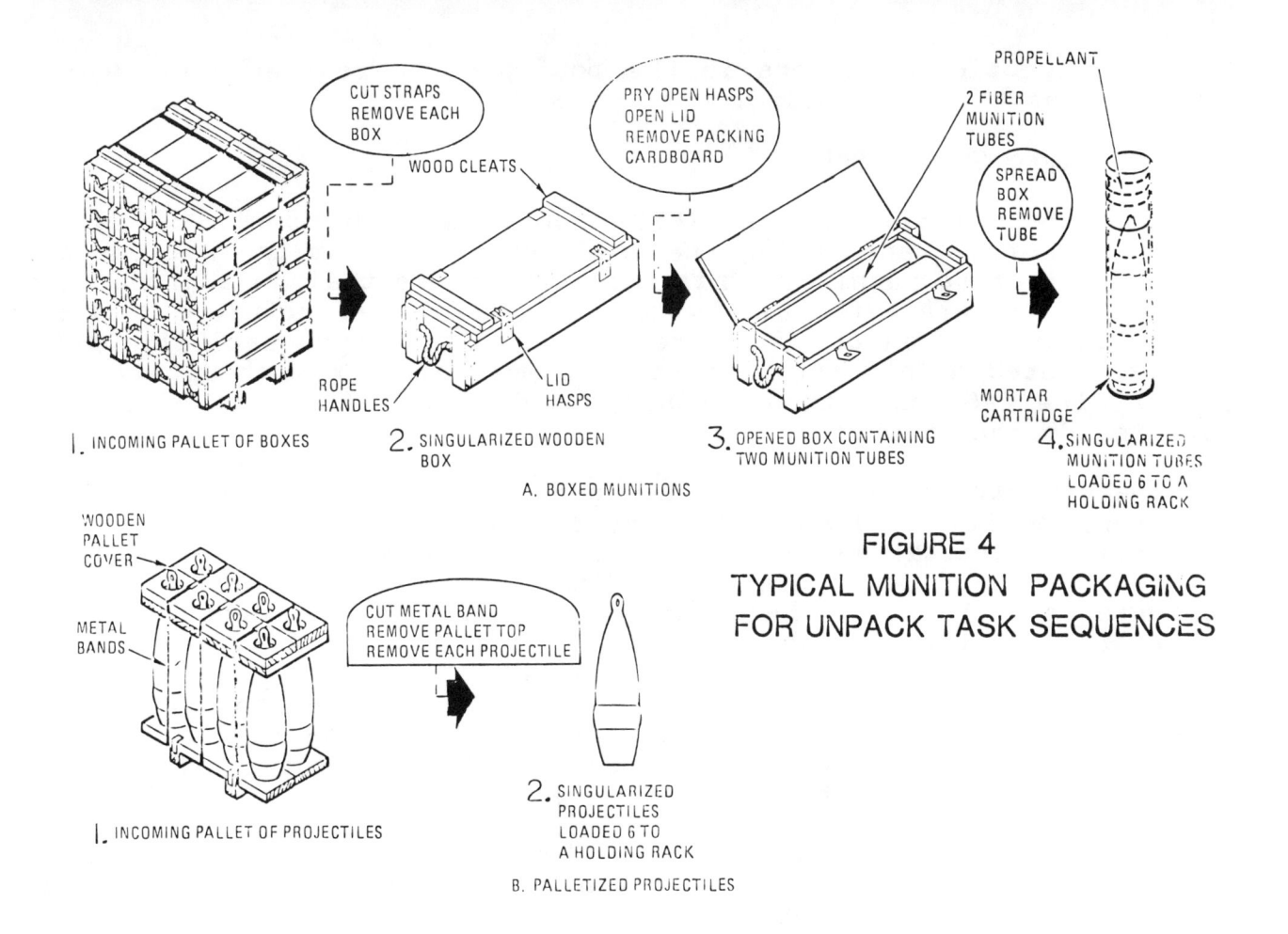

FIGURE 4
TYPICAL MUNITION PACKAGING
FOR UNPACK TASK SEQUENCES

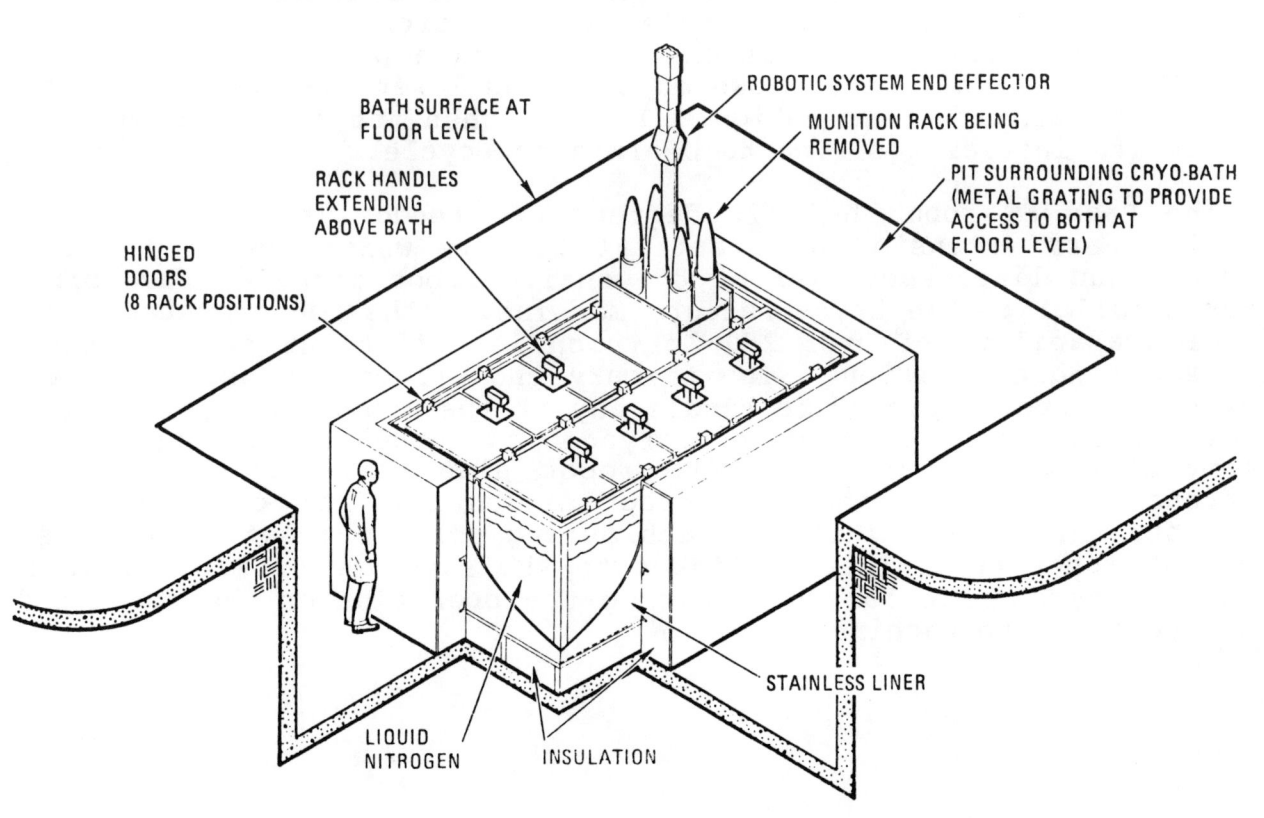

FIGURE 5 ROBOTIC SYSTEM UNLOADING MUNITION RACK OUT OF PROTOTYPE CRYO-BATH

free and occupied locations in the pool and the residency times of the racks as they are cooled.

Cryo-fracture System Robot

The cryo-fracture system (CFS) robot, a Prab model FC, is responsible for transferring the cold munitions from the airlock to the cryo-fracture machine. (This airlock is essential in a production facility to separate the highly contaminated cryo-fracture cell from the relatively clean cryobath area. In the prototype facility where only simulated agents are present, a simple airlock mock-up will be used.) The CFS robot then loads the individual munitions onto die sets inside the fracturing machine.

The cryo-fracture machine, as shown in Figure 3, is essentially a 1000-ton, custom Williams White hydraulic press. The machine executes a vertical stroke to fracture the embrittled munitions between upper and lower die sets. The fractured munitions and their contents fall through a chute which leads to a removal cart and scrap conveyor. (In the production facility the fragments will fall directly into a high temperature disposal furnace.) The munition fracturing occurs inside an cylindrical explosive containment chamber (ECC) which surrounds the cryo-fracture machine. The ECC is designed to mitigate blast and fragment effects in the unlikely event of a munition burster detonation.

A typical loading cycle starts with the munitions standing in a vertical orientation at the airlock station. The CFS robot selects and grasps a munition, rotates it into a horizontal orientation, and sweeps through a 180 degree arc to bring the munition directly in front of the fracturing machine. The robot arm then elevates the munition to the proper height for entry through a narrow loading window in the open ECC. The arm extends horizontally through this 22-inch high window and positions the munition over the lower die set. The munition is then lowered onto the lower die set by a wrist bend and/or arm lowering motion. The arm then retracts and rotates back to its initial position to begin a new cycle.

The Prab FC robot has six degrees of freedom (pedestal base rotation, arm extension and vertical travel, wrist rotation and bending, and linear movement of the entire robot along a traversing track parallel to the cryo-fracture machine.) The robot system has a payload capability of over 250 lb which is sufficient to load the heaviest single munition. The primary end effector consists of a univeral gripper designed to handle any of the munition types as well as withstand the effects of cryogenic temperatures and decomtamination fluids. The CFS robot also performs specific maintenance tasks on the cryo-fracture machine and can cycle munitions through a "mini" cryobath during certain upset conditions. The Prab robot controller is linked to the supervisory minicomputer which is responsible for coordinating the operation of the robot and the cryo-fracture machine.

Remote Maintenance

The systems used in the prototype facility in general will not be capable of being remotely maintained. Since remote maintenance is an important element in the overall maintenance plan for a production facility, the capabilities of the unpack and CFS robots for such operations will be investigated and demonstrated in the prototype facility. A maintenance manual is being developed that outlines general remote maintenance principles and their application to the cryo-fracture process. This manual and the experience gained from the prototype facility will be used to develop a general maintenance strategy for a production facility.

III. COMPUTER CONTROL SYSTEM

Control Philosophy

In order to achieve the goal of remote autonomous operation, high-level process control has been centralized in a single, relatively powerful minicomputer. There are several reasons for this decision.

1. Autonomous operation of complex processes is inherently computationally demanding. The control system must be able to cope with minor and perhaps major process variations using adaptive methods and feedback from a variety of sensors located throughout the work area. In the event of off-normal or process upset conditions, the software must be capable of diagnosing the problem and planning and executing an effective recovery procedure. Most current stand-alone robot controllers are based on microprocessors which are too limited in memory, computational power, and operating system sophistication to deal with such sensor integration, adaptive control, and error recovery requirements in real time.

2. The multi-stage cryo-fracture process is semi-continuous with closely coupled stages and minimal material buffers at the stage interfaces. Thus the simultaneous activities in each stage must be closely coordinated and integrated. This is most easily accomplished by centralizing high-level process control in a single powerful computer which supports concurrent running of intercommunicating real-time control programs.

3. The software development effort for the remote autonomous control system is complex and extensive. Software development is greatly facilitated by the sophisticated operating systems, large memory and disk storage capacities, high-level languages, and utility programs such as text editors and on-line debuggers that are available on modern minicomputer systems.

Control System Components and Functions

The control system hierarchy is illustrated in Figure 6. The supervisory minicomputer, a Data General MV/4000 running under the

FIGURE 6
PROTOTYPE PROCESS CONTROL SYSTEM

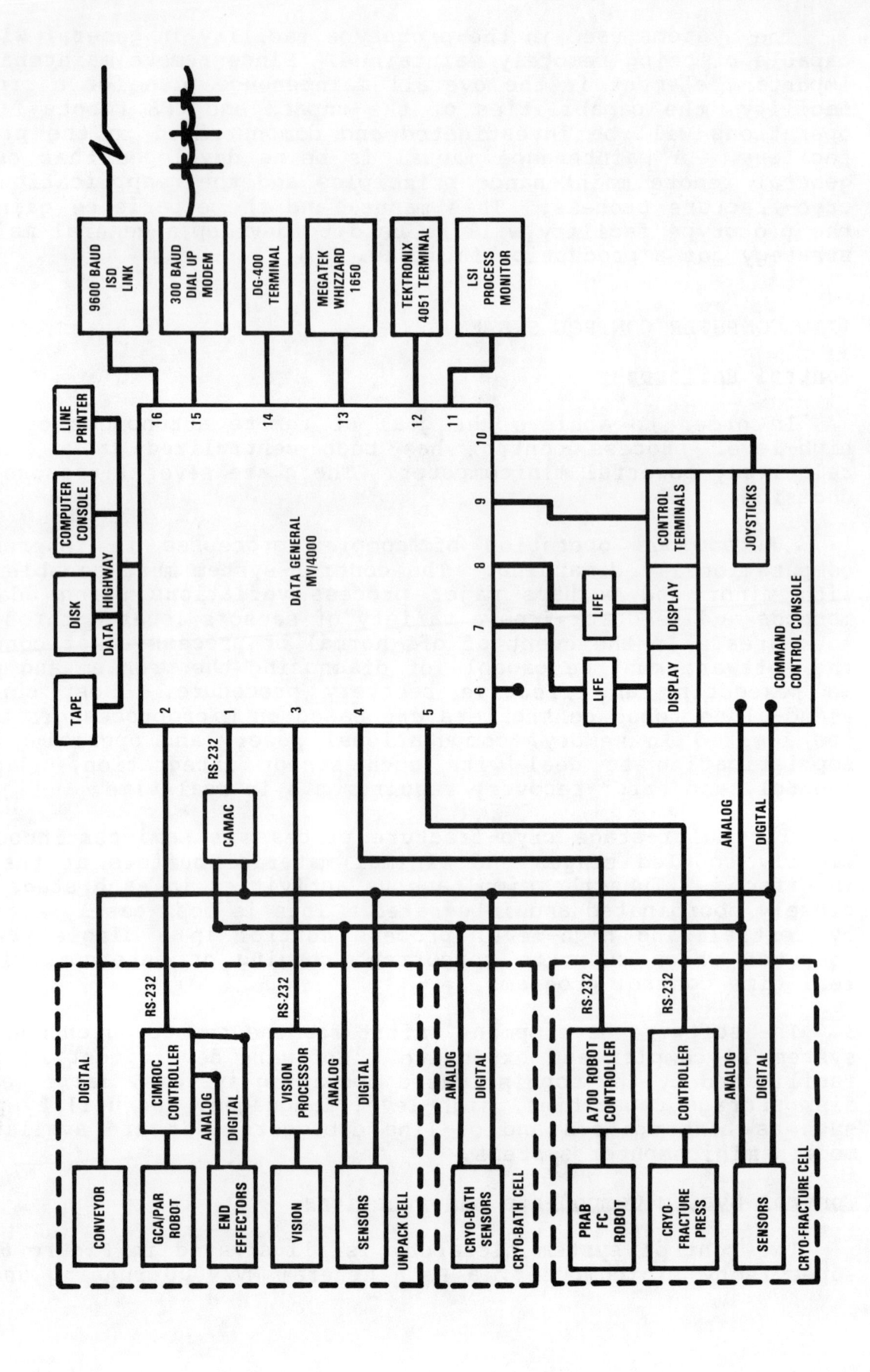

AOS/VS operating system, directly supports standard devices such as a disk, tape drive, terminals, and line printer as well as a sophisticated graphics system with animation capability whose function is described below. The minicomputer is interfaced to each robot controller via a RS-232C communications link which uses a protocol based on ASCII character strings. Complex pieces of equipment with dedicated local controllers such as the vision system are handled in the same manner. Simple sensors and instruments which provide temperature, safety, and process status information over analog or digital lines are interfaced either to one of the robot controllers, directly to the supervisory minicomputer, or to an intermediate CAMAC (Computer Automated Measurement And Control) controller which provides high speed I/O capability to the minicomputer. Finally, a manual control capability for the robots is provided by a joystick linked indirectly to the robot controllers through the minicomputer.

The major functions of the high-level control system software implemented on the minicomputer are tasking (detailed sequencing and execution control of individual process steps for a given operation such as unpacking pallets of 155-mm projectiles), process line monitoring and status display, diagnosing, planning and execution of recovery procedures in the event of process upsets (error recovery), and operator support in the event a condition arises that requires operator intervention.

The tasking function can be illustrated by the interaction of the minicomputer and the overhead bridge-mounted robot controller in an unpack operation. Local controller software provided by the robot vendor consists of an operating system, a high level robot language, and kinematic subroutines for converting specific robot language commands into physical motions and operations. User-supplied software at the controller level consists of a series of individual programs written in the robot language that accomplish subtasks pertinent to the overall unpack operation. Various unpack operations for different munition types may share similar subtasks that differ only in parameters such as location or gripping force. Thus the tasking software at the minicomputer level for a given unpack operation involves a very straightforward FORTRAN 77 program that generates the parameter values and command strings that initiate the proper subtasks in the proper order. As each subtask is completed, the minicomputer program verifies success or failure through the sensor network before initiating the next subtask or the appropriate corrective action.

Process line monitoring and status display involves the interaction of a graphics display program and various sensor polling programs. The polling programs acquire process and equipment status data and store it in a designated area of memory which is continually being updated as new data comes in. The graphics program accesses the data through this shared memory area and displays it in a form easily and quickly assimilated by an operator on a schematic representation of the process line. Out-of-range or unexpected parameter values are localized and flagged for attention by blinking the appropriate display fields containing the deviant values.

The error recovery function is based on the triangle table formalism developed by the artificial intelligence community to characterize the operation planning process. In simplest terms, a list of essential precursor conditions for the successful execution of each operation step is maintained as well as addition and deletion lists of conditions that are changed by the successful execution of the step. These lists, combined over all sequential steps, form the "triangle table". Given these tables, the individual task programs can easily monitor their own progress, verifying successful completion of the current step before triggering execution of the next. In the event of an inconsistency between the table entries and the conditions observed through sensor inputs, an error recovery program plans a sequence of actions which allows reentry into the triangle table at an appropriate step. Particulary extensive or severe errors are referred to the operator for recovery under manual control.

The control software supports manual operation and recovery in two ways. First, the manual controls such as the joystick for operating the unpack robot, operate through minicomputer software. Control movements are monitored by the software and translated into the appropriate command strings which are then communicated to the robot controller. This approach has the advantage of allowing software safety censoring of inappropriate or potentially damaging manual control actions.

The second category of operator support for manual operations involves the graphics system, a Megatek Whizzard 1650 intelligent color graphics display which supports animation. Graphics programs are being developed which display a real time animated representation of operations in the robot work cell using position and velocity data from the robot controller and sensors. This provides a critical visual feedback to the operator during manual control in addition to that from a set of closed circuit television monitors. Perhaps more importantly, the system can operate in a simulation mode in which commands from the minicomputer are "executed" on the display but not passed to the robot controller for physical execution. This allows particularly difficult manual manuevers to be practiced before being performed with the robot. This simulation mode can also be used for operator training as well as testing and verification of the tasking software as it is developed without exposing the robot to hardware damage from catastrophic software bugs that are likely to be present during development.

IV. CONCLUSION

The robot and computer control technology discussed in this paper is being developed to demonstrate the feasibility of remote autonomous operation of a cryo-fracture process line for demilitarization of obsolete chemical munitions. It promises to show significant cost, labor, availability and safety advantages with respect to current demilitarization technology. Additional development work remains, particularly in the area of remote maintenance, before the goal of remote autonomous operation can be achieved in a production facility.

Presented at the RI/SME Robots 9 Conference, June 1985

Applying Robots in Nuclear Applications
by John J. Fisher
E.I. du Pont de Nemours & Company

INTRODUCTION

Nuclear workcells offer obvious requirements for remote handling technology and equipment. Radiation, hazardous chemicals, and particulate contamination require many process areas to be inaccessible to humans and hostile to all but selected equipment. During the 1950's and 60's, master/slave and remote-controlled power manipulators were developed. Evolved forms of these devices have been used for virtually all nuclear manipulator applications to date. Recent advances, however, in computer-assisted teleoperators and robotics offer new options for remote manipulators. This paper will focus on robotic systems in nuclear processes.

BACKGROUND

Proper selection of a robotic remote handling system requires a basic knowledge of the other types of manipulator systems. Therefore, these systems will be described before robot systems are discussed. Figure 1 shows four types of commercially available manipulator devices suitable for nuclear operations: mechanical master/slaves, power manipulators, servo master/slaves, and robots. Each manipulator type has advantages in different applications. Master/slave manipulators are operated through the walls of shielded cells to perform dexterous tasks in line-of-sight applications. The master and slave arms are mechanically coupled with steel tapes which allow forces to be transmitted. These systems remain unchallenged in price and performance in many existing facilities.

In large process areas, power manipulators have been mounted on large cartesian gantry cranes. The manipulators are controlled unilaterally with switches which move each joint of the arm individually. These units are best suited for heavy material handling and tasks where precision motions are not critical. Robotic controllers with resolved motion algorithms are greatly improving the capabilities of these devices (Equipment Selection section).

Computerized servo master/slave manipulators have been developed to expand the capabilities of mechanical master/slaves to remote process areas. Like mechanical master/slaves, they may be bilaterally controlled, meaning that position (and velocity) information is transmitted from the master arm to the slave arm and vice versa. Force reflection, based on position errors between the actual and commanded locations of the slave arm, has been developed. In general, servo master/slaves have light, low inertia linkages, and backdrivable gear trains and are intended for light duty applications (roughly that of a human operator). Thus, they are ideally suited for emergency response operations, maintenance functions, and tasks where the operator must make motion responses based on tactile information. Martin has presented a thorough summary of the history, development, and manufacturers of this equipment.[1]

This paper was prepared in connection with work under Contract No. DE-AC09-76SR00001 with the U.S. Department of Energy. By acceptance of this paper, the publisher and/or recipient acknowledges the U.S. Government's right to retain a nonexclusive, royalty-free license in and to any copyright covering this paper, along with the right to reproduce and to authorize others to reproduce all or part of the copyrighted paper.

Mechanical Master/Slave

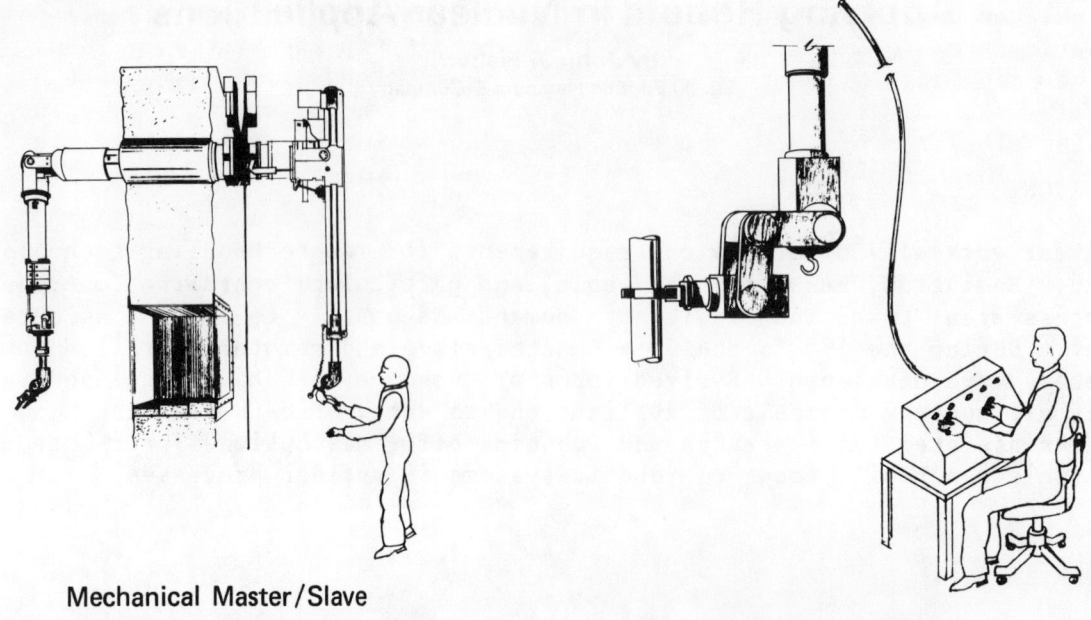

Power Manipulator

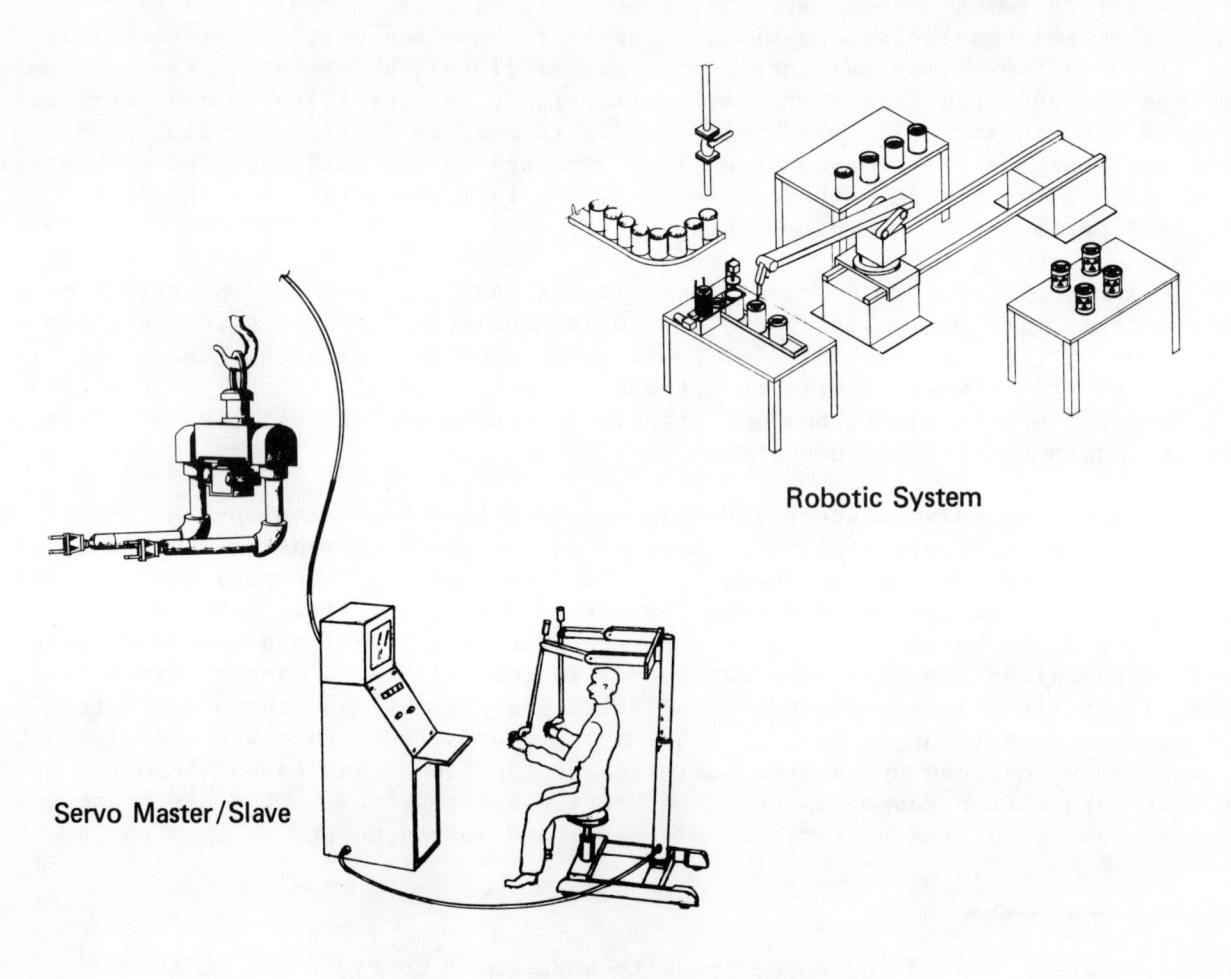

Robotic System

Servo Master/Slave

Figure 1. Four Types of Manipulators

Other special purpose computerized manipulator systems have been developed for selected nuclear operations. Asano describes a seventeen-degree-of-freedom, snake-like manipulator developed by Toshiba, Inc. which can pass through a small port hole to perform remote inspections.[2] The Westinghouse Energy System Service Division has developed a remotely operated service arm (ROSA). This device has been successfully used to perform eddy current inspections and manual tube plugging operations in steam generators. It has also been used to insert plugs into a reactor core barrel. The high cost of these devices, however, has limited their widespread use.

DISCUSSION

Industrial robots also offer remote handling alternatives in nuclear environments. They are generically very stiff, may have large lifting capacities, and are capable of high accuracy and repeatability. These benefits in conjunction with computer-based controls allow these devices to perform preprogrammed routines in structured environments. The wide availability and low prices of commercial robot arms and components make them attractive for many nuclear applications. Due to their advantages and low cost, interest in applying robots in nuclear applications is increasing.

E.I. du Pont de Nemours and Company, which operates the Savannah River Plant and Laboratory for the U.S. Department of Energy, is employing robotic systems in nuclear applications. The inherently limited versatility of industrial robots has been greatly enhanced by adopting current sensor technologies and supervisory, man-in-the-loop control schemes. Using these enhancements, robotic systems can offer viable, cost effective alternatives for remote handling applications. An identification of tasks, a choice of equipment, an analysis of availability factors, and a mockup testing program are essential to the success of the robot installation. This paper will develop these concepts. In addition, five case studies of robotic applications which explain equipment and control system implementations are reviewed.

TASK IDENTIFICATION AND ANALYSIS

The remote handling systems engineer can use robotic systems to solve three types of remote handling operations. In nuclear applications which contain repeated structured tasks, automation can be considered. Automated tasks are those which can be performed in a completely preprogrammed mode. They require a structured work environment. A second type of application contains specific tasks which can be performed automatically. These, however, require operating personnel to monitor and sequence their execution. A robot can perform these tasks through a manually supervised control system instead of a fully automatic one. The last type of application requires manual (teleoperated) control of the robot's motions in addition to programmed motions. This type of application requires the arm to function in a tele-robotic control mode. The three types of applications are discussed below.

Automation

Industrial robots can be used to automate routine and nonroutine tasks, which are repeated in structured nuclear applications. Routine process operations are the easiest to recognize. These are production tasks which include handling fuel slugs, performing radiation assays on parts, and preparing samples. These nuclear operations share many similarities with standard industrial robot applications, and a dedicated robot work cell may provide a successful automation alternative. Figure 2 shows a Zymark chemical automation system filling sample vials with a reagent for a nuclear analysis. Note the fixturing required and the location of machines and devices in relation to the robot.

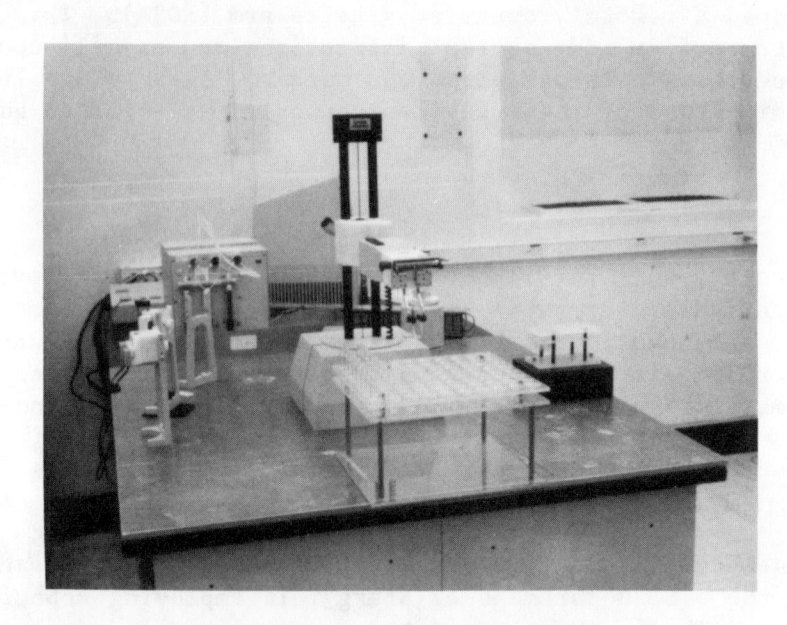

Figure 2. Zymark System

Other tasks suitable for robots, but not as easily recognized, are nonroutine but repeatable jobs which are awkward or potentially dangerous to perform. Examples of these tasks include: handling radioactive waste containers, changing filters, and performing selected (planned) maintenance functions. A dedicated workcell is not generally warranted in these cases, but a gantry or track-mounted robot performing several specific tasks may be justifiable. For example, a shielded cell transfer system is discussed in the Case Studies section. This system is being developed to transfer materials into nuclear hot cells, but will also make radiation measurements, contamination smears, and perform a source packaging operation.

Complicated or multiple step operations may also be candidates for automation. Robotic systems are not generally limited by the complexity of particular tasks since arms can be adapted with multiple tools and end effectors. Furthermore, the number of steps required in a operation is also not a limiting factor since robots can be adapted with virtually unlimited memory for stored routines. A factor which does govern the feasibility of automating applications is the degree to which a work area is structured.

"Task structuring" refers to the orderliness required in a work area for tasks to be preprogrammed and reproduced. The degree of structuring present determines the functions a robot can perform in a programmed mode. Highly structured operations require fewer external sensors than less structured ones. In addition, the need for enhanced control strategies is also reduced. The following questions should be addressed to determine if an existing process can be automated:

1) Can a robotic device be mounted in the process area or adapted to an overhead gantry system, floor conveyer, or mobile platform to access the parts and machines in the process?

2) What additional fixtures such as tool and parts racks, conveyers, and parts feeders are required?

3) How repeatable are the sequences and part locations? What tolerances are required to manipulate the parts or to access other devices?

4) Which portions of the operation require verification (i.e., sensory inputs)?

5) When and how often are operations modified with special procedures? Can these instances be recognized and alternate procedures be implemented automatically?

In new facilities or processes, the use of robots should always be considered. Workcells can be structured and fixtured to allow accessibility to components, and sensors can be hardwired into the facility, thus eliminating the sensory burden on the robot. Additionally, a robot mounting arrangement or positioning system can be designed to make the optimum use of the capabilities of the system.

Manual Supervisory Controlled Tasks

The feasibility of automating a task may be limited by a difficult inspection operation or by the need to control and verify a sequence of processes. Adding additional hardware such as an advanced computer vision system and a programmable controller might solve the technical problems but add economic ones. In such cases, a manual supervisory control scheme should be investigated. Manual supervisory control allows a system operator to communicate to the robot and process machines through the robot controller or a host computer, to gain information, sequence tasks, or issue new commands in a programmed operation. The system, however, must have the intelligence and sensory capabilities to perform the task automatically, once instructed. Operator-interactive control is a less structured form of manual supervisory control. In this case, the operator may perform inspections or control devices while the robotic system is in operation. In nuclear operations, the reduction in the number of personnel in a process area is generally based on potential exposure considerations and not economic ones. For this reason, manual supervisory control techniques, using an operator at a remote control console, are often used instead of fully automatic applications.

Teleoperated Tasks

In many applications, task structuring will not permit automation. In these operations, a separate teleoperated manipulator should assist the robotic device; otherwise manual control capabilities must be added to the robot.

A telerobot is a device which can perform both robotic and teleoperator functions. Robots can be operated in a telerobotic mode by adopting a technique to easily control the arm manually and by improving the executive control commands to simplify routine operations.

Standard robot teach pendants use push buttons to move the individual joints of the arm at several predetermined speeds. This works well for teaching points but is extremely awkward for controlling complicated manual motions. Various techniques may be used to enhance this control. The most common is to select a robot controller which can move the robot arm in straight lines corresponding to the three orthogonal axes of a selected coordinate reference frame such as "base x-y-z" or "tool center point (TCP) x-y-z," then to adapt a three-axis rate control joystick to move the arm in these directions. Straight line robot movements are referred to as resolved coordinate motions since all the joints of the manipulators are controlled simultaneously to move the arm in one direction. Using a joystick, operators can easily control the robot to move up & down, left & right, and forward & backward to simplify pick-and-place operations. A number of robot vendors will provide a joystick control package as an option. In addition to enhanced manual controls, executive control functions should be easy to use. For example, the control system should allow an operator to easily switch between the manual and automatic modes without typing several commands from a terminal. It should allow the maximum velocity of manual and programmed moves to be scaled to test a new program or perform a delicate manual movement. Another useful control modification enables manual control functions to proceed while the robot is executing a preprogrammed routine. This function allows the robot to automatically pick up tools and move to desired locations, but it can also stop to allow objects to be located and grasped under manual control. Obviously, the robot must keep track of its position during the manually controled movements.

In addition to control modifications, some mechanical changes may be required for telerobot control. Robot arms, using nonbackdrivable gear trains, may be damaged if they are moved against a rigid object. Damage may occur before the robot controller can respond to a motor current limit and shut off the motor. To reduce the possibility of damaging the arm, backdrivable gears, slip clutches, or some form of compliance is required. Supplementing manual control operations with force feedback from a force/moment sensor or motor current sensors also helps to protect the arm and can improve the manual capabilities of the unit. Spar Aerospace, Inc. is experimenting with a force moment/sensor and special joystick equipped with torque motors on each axis to provide force feedback sensory information to an operator. The forces generated by the motors may be scaled as necessary to permit both heavy and delicate operations.

CHOICE OF EQUIPMENT

In this section, the choice of equipment will be discussed including considerations of radiation environment, availability, and sensory requirements.

Radiation Considerations

Industrial robot controllers and sensory devices rely heavily on digital integrated circuit (IC) technology. The IC's in a robot or sensor are usually the limiting factor in the radiation hardness of the device. Consequently, an analysis of the expected radiation fields is required to determine if standard off-the-shelf robotic components are usable or if specially designed or selected devices are required.

Digital IC technologies [i.e., bipolar and metal oxide semiconductor (MOS)] have varying immunities to radiation types and fields. The radiation hardness of IC technologies also differs between venders due to different manufacturing techniques. For this reason, the radiation hardnesses of different technologies is usually given as a range. Table 1 lists the failure thresholds for a variety of IC families. Note that bipolar devices are inherently hardened to gamma radiation [10^6 rads (si)] while MOS technologies have high neutron tolerances. Also note the extremely low gamma tolerance of MOS UV EPROM's. Unfortunately, the vast majority of robots and sensory devices use EPROM components. Using the table as a guide, potential trouble areas may be located before a system is installed.

TABLE 1.

Radiation Hardness of Semiconductors*

Digital IC Technology	Neutron Hardness n/cm^2 (1 Mev)	Total Dose Hardness Commercial (Rad(si))	Total Dose Hardness Hardened (Rad(si))	Short Pulse Transient Upset Hardness (Rad(si)/sec)
Bipolar:				
TTL: LSI products	1x10^{14}	1x10^6	1x10^6	5x10^7
ECL: 100K series	1x10^{15}	1x10^7	1x10^7	3x10^8
IL: 9900 series	3x10^{13}	3x10^6	3x10^6	3x10^9
ISL	5x10^{14}	5x10^6	5x10^6	2x10^7
MOS:				
NMOS: LSI products	1x10^{15}	7x10^2 (EPROMS)	1x10^4	3x10^6
CMOS: LSI products	1x10^{15}	1x10^3	5x10^4	3x10^7
4000 series	1x10^{15}	5x10^3	1x10^6	3x10^7
CMOS/SOS:	1x10^{15}	1x10^3	1x10^7	5x10^9

* Taken in part from Long, D. M., IEEE Trans Nucl. Sci, NS-27(16):1674. 1980.

In general, robot arms are controlled from seperate control cabinets. Consequently, radiation effects on the robot may not be of great concern except for the SSI or MSI circuits used in the joint position transducers. Many sensory devices, however, contain integral VLSI signal conditioning circuits and microprocessors. These units may have to be shielded, or certain weak components may be interchanged. As an example, many UV EPROM's are pin for pin compatible with bipolar "fused link" PROM's manufactured by Harris and others. Similarly, SSI devices present on the robot itself can generally be replaced with hardened devices or, as in the case of optical encoders, another compatible type can be substituted.

In high radiation applications (exposures of 10^6 rads and higher), special radiation hardened IC's are required. Gover presents the advantages of hardened CMOS technology for these applications.[3] At this radiation level, the selection of other special components beside IC's might be in order; electrical insulation, elastomerics, and greases begin to degrade. A reference on radiation resistances of specific engineering materials are listed at the end of this report.

Availability

The availability of a robotic system depends on two factors: reliability and maintainability. Industrial robots have proven to be reliable in many commercial applications (mean time between failure >5000 hr for some units). When these systems fail, however, repairs may be complicated and labor intensive. Consequently, special maintenance considerations must be addressed before a system is installed in an area with high radiation fields or potentially harmful levels of nuclear contamination.

In high radiation applications, the robotic system and components should be installed so they can be easily transported to a "cool" regulated area. This greatly simplifies installation and maintenance operations and may also extend equipment life since radiation-sensitive components can be removed from the radiation zone when they are not in use. One design approach is to mount the robot on a sliding platform. The slide mounting provides a stable base for the robot and an easy means to transport the system. An alternate approach uses an overhead bridge crane to transfer the robot and system components. Lifting bails, alignment pins and fixtures must be integrated into the workcell to allow the robot to be removed and subsequently remounted.

Contaminated nuclear workcells present special problems to system maintenance. In many existing nuclear processes, hands on repairs, using maintenance personnel in plastic suits, are required. This practice, however, is avoided whenever possible to reduce radiation exposure and the possibility of nuclear material assimilations. A maintenance method preferred to operating through plastic suits is to remotely decontaminate components to a suitable level; then to transfer them to a "cool" regulated area for repairs. Decontamination chemicals used in nuclear processes include: nitric acid, caustic solutions, and "Freon." Unfortunately, very few commercial robot arms and components can be cleaned with these chemicals without damaging their internal components and wiring. Protective covers, boots and gauntlets, however, can accomplish the same function as mechanical sealing (Figure 3). Covers should be designed for easy removal and replacement so they can be changed in a glove box. Air purging is also desired to keep contaminated process air from entering the cover as it expands. Scoville describes a different method of protecting the robot for a finishing application. In his application, the robot operates in an enclosed box through a combination convoluted rubber boot and bellows arrangement.[5] As a result, the arm is protected but maintains the dexterity necessary to perform the required operation.

An alternative to decontamination is to employ a modular system design. System components such as end effectors, sensor devices, and arm sections can be removed remotely and replaced remotely via master/slaves, remote cranes, or through glove ports. Many times this approach is safer, easier, and more economical than attempting to clean a contaminated piece of process equipment. Depending on the accessibility of the equipment in a process, modularity may be an important design requirement. Few commercial robots use modular designs, but several venders will quote modular designs as special options. The cost for developing a special purpose

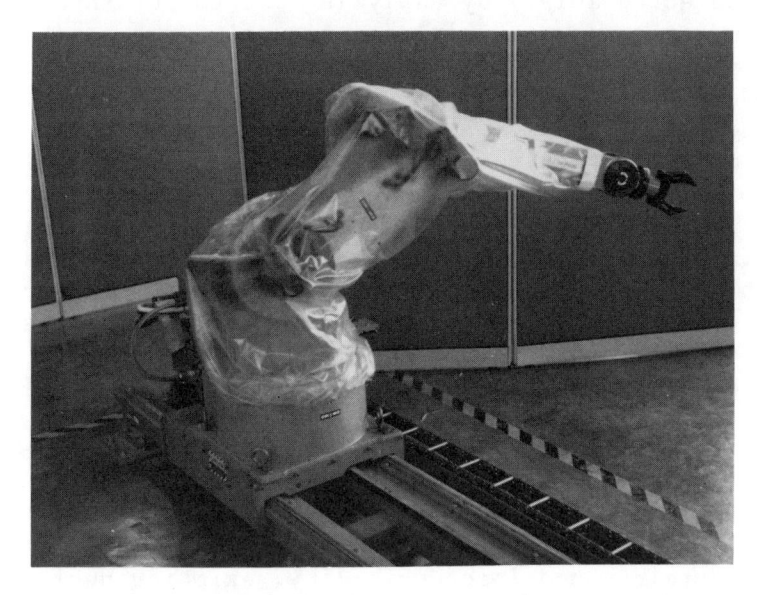

Figure 3. Robot with Protective Covering

modular robot may not be justified since the cost of replacing most commercial arms (less the controller) is low. A duty cycle analysis is required for this decision. TeleOperator System, Inc. (TOS) is developing a modular robot. The features of this unit are discussed in the Case Studies section.

Equipment Selection

Sensors are required both to verify and to control process steps. Nuclear processes typically require more sensors than similar industrial applications since operating personnel cannot make hands-on adjustments or corrections to the process. To avoid manual intervention, the robotic system should have the means to correct, or at least recognize, basic operational problems. These problems may be detected by placing limit switches and proximity sensors on tool racks, end effectors, and peripheral devices. The sensors may be wired through an I/O panel on the process or robot controller to stop the operation, warn a operator, or signal the robot controller to branch to a subroutine to correct the problem.

Another sensory device that has proved to be very useful for nuclear applications is the force and moment sensor. This device communicates to the robot controller through a dedicated processor. Using strain gages, the sensor generates digital or analog signals which represent the six principal forces and moments acting on the end effector. With proper programming, the sensor can determine the weights of parts, measure torque or friction (for threading or insertion operations), and detect collisions with unexpected obstacles. JR3, Inc. manufactures several models of six axis force/moment sensors. Unlike other similar commercial devices, the JR3 sensor contains no electronics in the sensor housing. This allows it to be used in radioactive and high temperature applications.

Frequently, robot systems require several end effectors to perform different process tasks. To this end, special flanges to allow automatic hand changes may be used. Several commercial venders, including Applied Robotics and EOA Systems, Inc. specialize in this technology and can supply specially designed devices.

Robots are manufactured in a variety of sizes and configurations. The kinematic design is selected solely by analyzing the constraints of the application. Whenever possible, investigate commercial sensors, fixtures, and robots since they are generally faster and more economical to implement than specially designed devices. Robot venders will generally allow one to perform experiments with their products to determine suitabilty for a process. If no commercial system is available, however, a special design may be in order. Off-the-self motion control packages are available which can control the position, acceleration, and velocity of several servomotors concurrently. They use point-to-point control approaches and can easily be interfaced to specially designed mechanical linkages and positioners to enable pick-and-place routines. Electric robots are better accepted than hydraulic systems in remote nuclear applications. Hydraulic systems must be designed to avoid contamination problems resulting from hydraulic leaks during operation and service. In addition, seals and fluids must be evaluated for radiation resistance.

MOCKUP TESTING

Due to space constraints and inaccessibility, existing nuclear processes are difficult to automate. With this in mind, it is advisable to mock up an application as closely as possible. This allows the robot and all critical components to be thoroughly tested before installation. Major engineering efforts may also be required to design special devices, develop appropriate sensors, and to solve interfacing and control problems. Of course it is impossible to account for all details, but it is the author's experience that hours in the laboratory save days during the actual installation.

Automated operations, as discussed previously, may be designed to operate autonomously or through manual supervisory control. In both cases, some form of human inspection and manual sequencing may be required for operational/safety, or security purposes. In these cases, an operator control console should be fabricated and tested to determine if it contains the feedback devices and control functions necessary to allow a process to be performed and monitored. Feedback medium include: closed circuit television monitors, audio feedback, and special operator interfaces. Control consoles may use indicator lights, CRT screens, terminals, function buttons, touch screens, and computerized voice modules. The interface components are connected to the system sensors, machines, and the robot controller. Operators can determine the status of the system components through visual inspections or sensor inputs and then execute the necessary procedures. In emergency operations, operators should have the means to overide the preprogrammed sequence and control the robot manually.

Robot controllers which use features like relative move commands and an offline programming language should be chosen over simple point-to-point controllers. These features enable application programs to be easily transferred from the mockup area to the production setting. In addition, they provide more capability for future upgrades or modifications. Adaptive control algorithms, discussed later, should also be investigated to speed the installation.

In practice, the cost of the robot hardware is small, typically 10 to 25%, when compared to the total hardware and development costs of a system. However, in structured environments where sensory and peripheral devices are low in cost, the robot system may be the major cost of the system.

CASE STUDIES

Automated Bag Out

A robotic system was developed and installed to remotely remove and bag one-gallon waste cans from a glove box in a californium source processing facility (Figure 4). Waste cans from the facility contain sources of radiation and contamination. The process was well structured for automation, but the area offered severe space constraints which dictated the selection of the robot and its peripheral components. The basic sequence of the automated bag-out operation is the same as that used for manual operation. No major structural modifications were made to the process area. A special bag port, however, was installed on the glove box to enable the automated operation.

The automated system consists of a GCA 300V industrial robot, a pneumatic end effector, and a specially designed bag gathering-clipping machine. A long vinyl sleeve, sealed at one end, is gathered on the bag port and used to make the bags for four cans. The robot reaches through the port and grips a waste container, through the plastic sleeve, from a holding fixture attached to an existing cell transfer conveyer. Two feet of the sleeve are pulled off the bag port as the arm reaches into the glove box. A slight negative pressure inside the glove box causes the sleeve to expand around the waste can as the robot performs the gripping operations. As the robot arm is withdrawn, the bag turns inside out and envelopes around the can forming a tight protective cover. The sleeve is sealed twice with aluminum clips from the bag gathering-clipping machine which seals one waste package and forms the bottom of the next bag. The bag is then cut between the clips. Packaged cans are deposited in a shielded drum accessible by a remote overhead crane. The robot and process components remain clean during the operation.

Figure 5 shows the control console for the waste removal system. Five television monitors, a PA system, radiation monitors, and indicator lights provide feedback to the operator. The waste removal program performs the process automatically but stops at various steps to allow the operator to inspect the process. Alternate routines, to stop the operation or to back the robot out of the glove box, are implemented by pushing buttons on the control console. These buttons trigger conditional subroutine branches from the main application program. Sixteen inputs, attached to system components and the control console, provide the necessary interrupts and system interlocks. Sixteen output signals are used to operate the external devices and to illuminate indicator lights on the control console. Manual motions are not allowed during normal operations due to space constraints.

The operator-interactive control philosophy requires operators to sequence and inspect critical steps in the operation. Operators can recognize potential problems, such as damaged cans or a conveyer misalignment, and correct the problem remotely. The radiation monitors are located outside the glove box to determine the exact quanity of material inside each can. Operators use this information to determine the number of cans that may be removed in a given operation. The operator-interactive control scheme greatly reduced system installation time. A fully automatic process would have required the existing cell conveyer to be replaced and several more sensors to be implemented to perform the necessary human inspections. The entire process was mocked up in a clean area before installation since the radiation background at the glove box was 200 mrads/hr. Consequently, the total personnel radiation exposure absorbed during installation was limited to 3 rems. Radiation exposure savings totaling over 20 rems during the next year will be realized with the system.

Figure 4. CPF Robot

Figure 5. Control Console for CPF Robot

Sample Aisle Robot

In this application, a robot replicates a manual radioactive sample collection process. A typical automated sequence consists of opening a sample box, placing a shielded lid on a sample container, removing the container from the sample box, and placing a new container inside. The containers are heavy shielded cylinders which hold vials containing the samples. Radiation fields at the sample boxes may exceed 2 rads/hr after the doors are opened. The sample collection process is duplicated in approximately twenty sample boxes in a sample aisle.

A robot system was designed to mount to an existing 400-foot I-beam runway that was previously used to support sample container lifting hoists (Figure 6). The robot is manually positioned at a given sample box and clamped to one side. The operator then stands away from the sampler and sequences the robot motions with a move pendant. The move pendant asks the operator which action he wants to perform, then performs it automatically. Separate routines, such as opening or closing the door, finding the container, getting and placing the shielded lid, and removing and replacing the sample container, are executed by pressing designated buttons on the pendant (Figure 7). After a routine is completed, the arm is moved to a safe location and awaits further instructions. Interlocks are built into the program so the operator cannot execute consecutive operations which are potentially hazardous, i.e., attempting to place two containers in the same sampler. A radiation detector is mounted to the robot to to make a gross radiation reading on the sample box and two television cameras are used to inspect the inside of the box. Stopping robot motion after each routine allows time to inspect the radiation detector and television monitor and to make manual corrections if necessary.

In this application, a commercial robot with the desired size and lifting capacity was not available. Consequently, a special three-axis machine was designed using commercial servo motors and drives and computer components. A Rockwell Aim 65 microcomputer was interfaced to three servo drives, a special move pendant, a proximity sensor, end-of-travel limit switches, and the television camera switching circuit. All device driver algorithms, application programs, and diagnostics were written in the Forth language.

Using a unique adaptive control feature, the robot is able to accomodate 2-inch variations in container locations between sample boxes. These variations are due to mounting and fabrication differences. The proximity sensor is mounted to the hand of the robot and it scans over a predefined window to locate the sample container (Figure 8). The sensing algorithm then stores the x, y, and z coordinates of the sample container which are required in applying the lid and in removing and replacing containers. This technique eliminates the need to teach the absolute locations of the container in each of the sample boxes.

Shielded Cell Transfer System

A robotic system to transfer nuclear sources and supplies to four shielded cells in a material testing facility is under development. The process area consists of a long corridor with access doors, one leading to each cell, spaced along one wall. The existing manual process requires operators, wearing protective clothing and assault masks, to open the door behind each cell and to insert or remove supplies. When the cell doors are opened, sources of radiation and potential sources of contamination are exposed.

Figure 6. Sample Aisle Robot

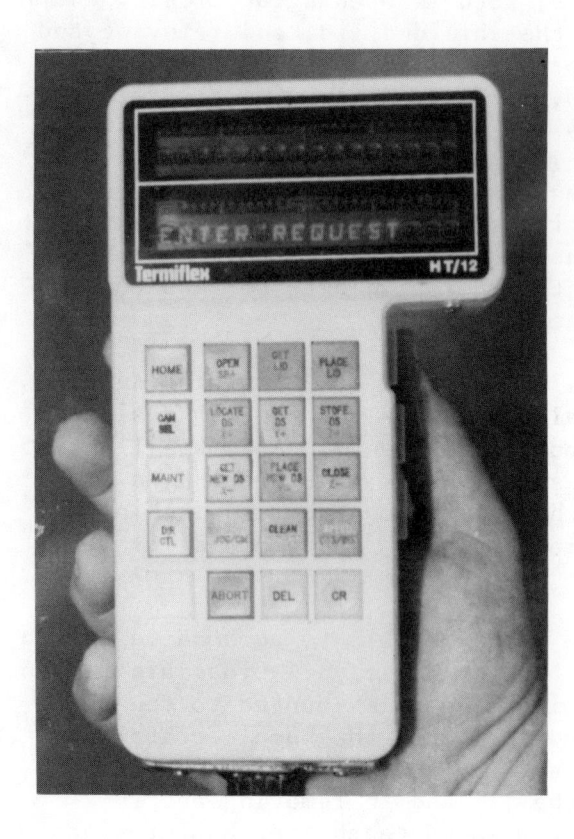

Figure 7. Move Pendant

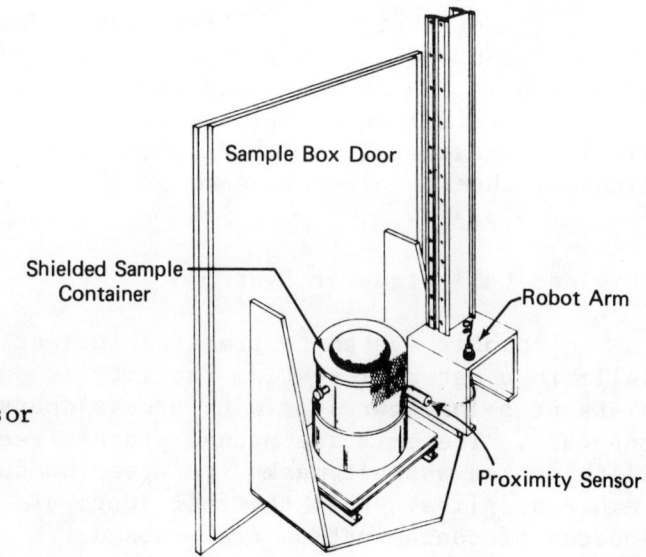

Figure 8. Proximity Sensor

188

The automated process eliminates the need for operators to enter the corridor behind the cells. A glove box separates the corridor from a clean operating area. Materials and supplies are placed on a tray and inserted into the glove box manually. The robot transfers the tray to a desired cell, opens the cell door, and places the materials inside. One end effector is used to perform these functions. It uses a wiper blade to slide the materials off the transfer tray and a hook mechanism to open the cell doors. The robot system is also capable of other operations, including making radiation surveys and contamination smears, and removing and packaging samples from the cells. A variety of end effectors is used to perform these tasks.

The robotic system consists of a GCA 600 industrial robot mounted on a 50-foot ESAB servo welding track (Figure 3). It can travel the length of the access corridor to transfer objects from the designated glove box to the required cell. Different end effectors are adapted to the robot through a quick hand-changing system manufactured by Applied Robotics, Inc.

A six-axis force/moment sensor is mounted to the wrist of the robot (Figure 9). The force sensor is used to verify that objects have been grasped, to detect unexpected collisions, and to detect binding in packaging operations. Work is underway to enable the robot to locate a predefined object by scanning a predefined area. The exact location of the object is pinpointed by approaching it from different directions and storing the location where the object is touched. The sensor will also help the robot to package source samples from a cell into a shielded vessel. A sample is placed in the vessel and sealed with a screw lid. Cross threading and improper placement of the lid generate higher forces than those required for a correct placement. The errors are recognized and corrected by reapplying the lid. An Astek force sensor is being used to develop these techniques.

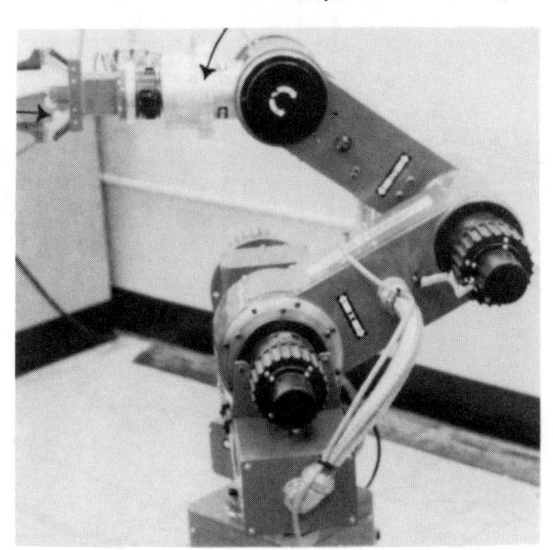

Figure 9. Force Sensor

A supervisory control architecture, consisting of a PDP 11-40 host computer, a Cimroc II robot controller, and the force sensor processor, is used. The application software is menu driven. It asks the operator which tasks he wants to perform and the cell he wants to access. It then performs the operations without relying further on the operator. Television cameras and other sensors for task verification are not required except for general work area inspections.

The robot and the positioning track mount to the floor of the access corridor. A maintenance area where the protective cover can be removed and repairs can be made, will be installed on the end of the robot track. New application programs can be developed in this area for use in the regulated process area.

Telerobot Systems

Du Pont is designing an alpha decontamination and disassembly facility in which to dismantle glove boxes and equipment from obsolete processes at the Savannah River Plant. A prototype of the facility is being fabricated to test various process machines, including a telerobotic system. The specifications for the manipulator were developed by experimenting with both commercial robot devices and power tele-operated manipulators. The final design and fabrication of the telerobot system will be performed by GCA Inc.

The telerobot will be required to perform the following operations: 1) general equipment handling, 2) equipment unpackaging and disassembly, 3) plasma arc cutting, and 4) fastener removal. The many types of equipment entering the facility will need to be processed differently. Consequently, the telerobot will operate in the manual mode the majority of the time. These processes will be performed in a highly contaminated environment. For this reason, a large sealed manipulator is desired. A GCA 6000 power manipulator is being automated for this application. It will be controlled with a GCA Cimroc II controller. The maximum payload of the arm at the tool tip is 300 lb, but the arm can also be used for precise operations. Figure 10 shows the telerobotic systems performing a cutting operation.

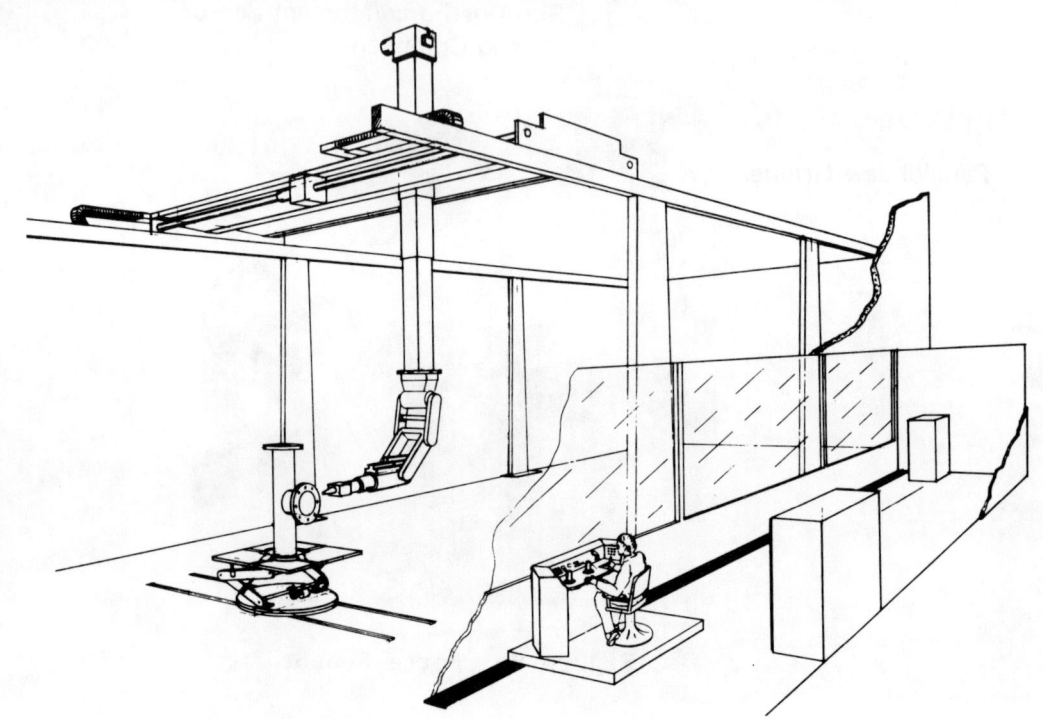

Figure 10. Telerobot System

To enable manual control functions, the system enhances the existing capabilities of the Cimroc controller. Two joysticks are used to move the arm and the gantry axes in tool center point (TCP) and base coordinate motions. One controls the three bridge axes while the other controls the arm joints. Individual joint motions are made using both of the joysticks and three finger switches. A velocity scaling potentiometer controls the velocity of the relative and joint motions.

In addition to manual motions, programmable functions will be used. These motions require careful structuring of tool racks, remote filters, and fixed process machinery. Automatic tool changes, filter removal, and routine equipment maintenance will be preprogrammed while the system is being installed. To verify and check the preprogrammed routines, sensors will be hardwired into the facility.

The automatic mode consists of a macroprogram composed of many application subprograms. The subprograms are called either through a terminal or through special function buttons. Frequently called programs such as tool changes and routine moves to access machines may be called with dedicated function buttons, thus eliminating the use of the terminal for regular operations. The function buttons are used as input signals to the robot controller. A single push button is used to switch control from the manual to the automatic mode.

Several new capabilities have been added to the Cimroc II to enable unique operations in the SRP-ADD facility. One, referred to as the semiautomatic mode, allows the telerobot to make plasma torch cutting operations. It uses both the teach and execute modes of the controller. An operator first teaches a series of points along a cutting surface via a push button switch; the points do not have to be named individually through a terminal. Pressing other switches moves the robot through these points at a speed selected by the velocity scaling potentiometer. When the desired speed and path are determined, the cutting operation is initiated. The torch is attached to the arm of the manipulator through an automatic standoff controller interfaced through the wrist extend joint of the arm. This automatically compensates for material warpage and maintains the tool tip at the correct cutting distance.

The arm will frequently be stressed and will be susceptible to collisions. To avoid damage to the arm and gantry positioner, slip clutches are attached to each joint or axis of the system. These joints and axes can slip without overcoming the inertia of the drive motors and the gear train. Resolvers are mounted directly to the arm joints and through rack and pinions on the gantry axes so that slippage will not alter the coordinate frame of the system.

Modular Light Duty Telerobot

A second telerobot system is being developed by Teleoperator Systems, Inc. to perform general remote handling tasks at the Savannah River Plant. These tasks include: 1) mobile deployment on remote vehicles, 2) use on special stands for fixed operations, and 3) use with overhead bridge crane systems. The design objective for the device is to produce a low cost, modular arm with which to perform manual and automatic tasks. The arm has a 54-inch reach capability and a lifting capacity of 50 lb. When lowered from an overhead crane, the manipulator, including its base, will fit through a 30-inch-dia opening. The telerobot will be modular and will use interchangeable end effectors. It will be able to handle a variety of small power tools. Joint motors are powered from servo amplifiers which run off a direct current 24-volt bus. This allows the arm to be operated on remote vehicles using

12-volt batteries. The arm consists of seven modules which are easily detachable and replaceable through glove ports. A design criteria states that these modules be inexpensive, typically less than $5 thousand each except for the base module. No module except the base will weigh more than 25 lb.

The arm's kinematic arrangement consists of a waist rotate joint followed by three parallel joints: shoulder pivot, elbow pivot, and wrist pivot, and a wrist yaw. Separate wrists modules including one with wrist pivot and roll capabilities and one with wrist pivot, yaw, and roll capabilities can be interchanged. All manipulator joints are protected by slip clutches in case of overload.

In the initial design, the arm is controlled with a single joystick set up to allow simultaneous motions of the parallel joints. Other motions are made with finger switches on the joystick. The robotic functions arm performed using an ICC 3200 robot control system which has several advantages for manipulator control. One advantage is that it derives velocity compensation from the same resolvers used for detecting joint position. Another is that the maximum joint torques can be programmed. The controller will maintain this torque value without shutting off. This feature can be used to limit the forces which the arm can exert; this is useful when performing delicate tasks. The arm and controls are configured for future upgrades, including interfacing a host processor necessary for advanced control statagies and writing the trajectory and inverse kinematic control algorithms necessary for resolved motion capabilities.

REFERENCES

1. Martin, H. L., *Joining Teleoperation with Robotics for Remote Manipulation in Hostile Environments*. Conf paper VI, p 9-1(17) RI/SME Robotics 8 Conf, Detroit, MI, June 4-7, 1984.

2. Asano, K., *Control System for a Multi-joint Inspection Robot*. American Nuclear Society Robotics and Remote Handling in Hostile Environments Conf, Gatlinburg, TN, April 23-27, 1984.

3. Gover, J. E., *Hardened Robotics for Utilization with Nuclear Reactors*. Sandia Laboratories Report SAND84-2367. October 1984.

4. Scoville, J. H., *Robotic Abrasive Process System*. Conf paper VI, p 5-52(10). RI/SME Robotics 8 Conf, Detroit, MI, June 4-7, 1984.

BIBLIOGRAPHY

Dyches, G. M. and Burkett, S. D., *Laboratory Robotic Systems at the Savannah River Plant*. Conf 831198-1. Eastern Analytical Symposium, New York, NY, November 1983.

Bowman, R. E. "How Radiation Affects Engineering Materials," *Materials in Design Engineering Manual No. 173*. July 1960.

Gover, J. E., *Basic Radiation Effects in Electronic Technology*. Radiation Effects in Microelectronics Div 2165, Sandia National Laboratories, Albuquerque, NM 87185. February 1984.

Yount, J. A., *Microprocssor Enhanced Real Time Manual Control of an Industrial Robot*. Conf paper, VI, p 9-18(10). RI/SME Robotics 8 Conf, Detroil, MI, June 4-7 1984.

Fisher, J. J., *Shielded Cell Transfer Automation*, Conf. paper p 139 (7). American Nuclear Society Robotics and Remote Handling in Hostile Environments Conf. Gatlinburg, TN, April 23-27, 1984.

CHAPTER 6

MOBILE SYSTEMS

MANIPULATOR VEHICLES OF THE NUCLEAR EMERGENCY BRIGADE IN THE FEDERAL REPUBLIC OF GERMANY

GERHARD WOLFGANG KÖHLER, MANFRED SELIG,
and MANFRED SALASKE *Gesellschaft für Kernforschung mbH.
D 7500 Karlsruhe, Abteilung Reaktorbetrieb und Technik, IT/PB
Federal Republic of Germany*

KEYWORDS: *manipulators, remote equipment, remote vehicle*

ABSTRACT

The Nuclear Emergency Brigade is equipped with a number of manipulator vehicles that allow rescue actions to be carried out in radiation fields under accident conditions.

One heavy vehicle with a power manipulator for use in the open was followed by the medium-sized radio-controlled MF2 vehicle for action in both open country and large rooms. The EMSM I electric master-slave manipulator was developed for vehicles and hot cells. The light MF3 vehicle, with a new variable chassis geometry and to be used indoors, is under development.

Other vehicles, manpower, organization, and applications are mentioned briefly.

INTRODUCTION

The Nuclear Emergency Brigade was established by the Gesellschaft für Kernforschung mbH., Karlsruhe, Federal Republic of Germany (FRG) on behalf of the responsible federal ministry. It was planned to cope with accidents and incidents in nuclear facilities and during the transport of radioactive materials, if the means available to the operators were not sufficient to control the situation. With the existing emergency organizations, the function of the Nuclear Emergency Brigade is to isolate and remove hazards generated by a nuclear incident.

If local radioactive dose rates measured at the site of an incident are very high, then manipulator vehicles are important devices to be used in rescue measures. They can be indispensable in decisive phases of the action.

A Nuclear Emergency Brigade equipped for versatile application would require three sizes of manipulator vehicles that can be employed in many situations to supplement and support each other:

1. heavy-duty vehicles for outdoor use
2. medium-sized vehicles for use in both open country and large rooms
3. light vehicles for use primarily inside buildings.

The Gesellschaft für Kernforschung mbH, with the cooperation of several industrial companies, has designed and built a number of manipulator vehicles for the Nuclear Emergency Brigade (Kerntechnischer Hilfszug), FRG. The brigade, stationed near the Karlsruhe Nuclear Research Center, is equipped with one heavy-duty vehicle (manipulator crane vehicle) that has been in operation for 7 years. This was followed in 1974 by the medium-sized MF2 vehicle. The light MF3 vehicle is still under development.

MANIPULATOR CRANE VEHICLE (FIG. 1)

The chassis used is that of a three-axle truck carrying a two-section boom with hydraulic drives. It is 15 m long and has a 1000 kp[a] load capacity. The boom can carry a heavy-duty power manipulator of the SM5-B type (Fig. 2). The manipulator was designed by the Gesellschaft für Kernforschung and manufactured by Eisenwerk Weserhütte AG, D4970, Bad Oeynhausen, FRG.

The manipulator is driven by electric motors, has a load carrying capacity of 300 kp in any position, and has six different movements. The

---

[a]kp (kilopond) or kilogram of force is the equivalent of the gravitational force on a kilogram of mass = 9.81 N.

197

Fig. 1. Manipulator crane vehicle with SM5-B power manipulator gripping a 200-litre waste drum.

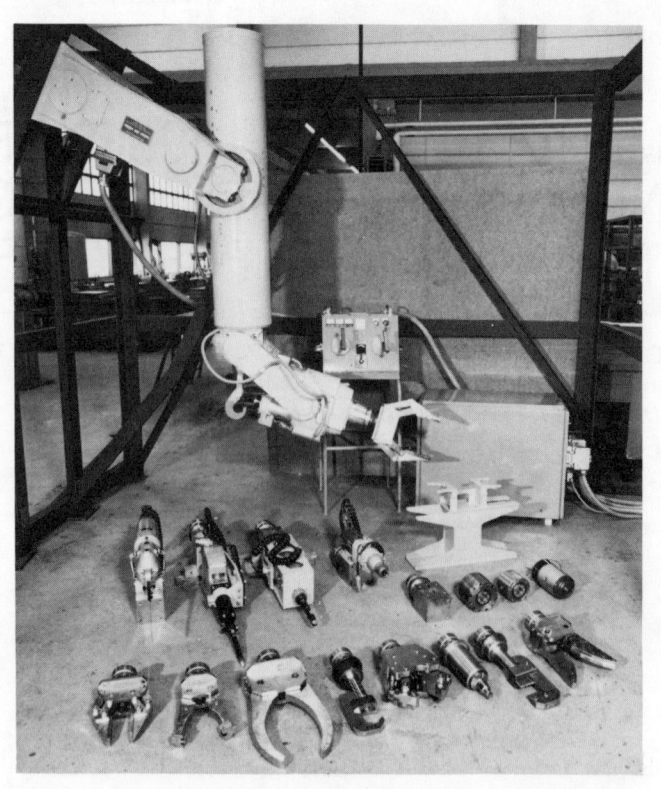

Fig. 2. An SM5-B power manipulator complete with accessories for the manipulator crane vehicle. Top left: mechanical arm with parallel-jaw tongs with 300-mm-wide opening and hook. In the background: an operating console on the left and a switchboard cabinet with a power supply on the right. Front row (from left to right): 140-mm parallel-jaw tongs, multipurpose tongs, pipe tongs, grip hook, parallel screw wrench, tool carrier for socket wrench, cutting hook, and plate shears. Second row (from left to right): tenon saw, hammer, drilling tool, impact screw driver, carrier for electric tools, and, as spares, two gearing units and one electric motor. A device for the remote exchange of the 140-mm parallel-jaw tongs and the grip hook can be seen behind the spares.

parallel-jaw tongs normally used can be exchanged remotely against a grip hook and can be replaced manually by many special tongs and tools that also can be used with the manipulator of the MF2 vehicle. The manipulator crane vehicle has a total weight of 33 tons. The operating station of the manipulator is behind the driver's cabin. It is shielded on its front side by 40 mm of steel. Operations are viewed either directly through shielding windows or by using two television cameras, attached by arms to the left and right sides of the front section of the boom. The manipulator can be operated both from the console on the vehicle and via cable from a distance of up to 50 m. This vehicle can be used only for handling medium activities.

THE MF2 MANIPULATOR VEHICLE SYSTEM AND POSSIBLE APPLICATIONS

The remote radio-controlled cross-country medium-sized MF2 manipulator vehicle, the first device of its type in Europe, was developed within the framework of "Project Nuclear Safety" with the support of the Ministry for Research and Technology. The MF2 was the concept of the Gesellschaft für Kernforschung mbH and was developed by Dr. Ing. W. Ständer of the Polytechnic Institute from the specifications and design drawings.

The MF2 is to be used where radioactive material or radioactively contaminated objects must be treated or removed even under conditions of difficult terrain, where high radiation levels prohibit manual action.

Manipulator Vehicle

Main Specifications. The MF2 is a cross-country tracked vehicle driven by electric motors supplied from lead storage batteries and radio-controlled from television pictures. The working device is a heavy-power manipulator of SM5-D type mounted on a swivel turret (Fig. 3). The main data of the vehicle are:

Total weight	3.4 ton
Length of chassis	3.20 m
Width overall	1.55 m
Height to upper edge of swivel tilting head	2.05 m
Loading height	1.53 m

Fig. 3. View of the medium-sized MF2 manipulator vehicle.

Equipment. The extensive equipment makes not only remote-controlled operation possible, but it also makes it possible to obtain information on damage. The three-jointed multipurpose manipulator has a total of six different movements, each driven by an electric motor. In each position of the arm, objects up to 200 kg in weight can be manipulated. The carrying capacity at the load hook of the elbow joint is 400 kp (Fig. 4).

The large parallel-jaw tong that is normally used has an opening width of 300 mm, and can handle round objects. For better gripping, it has triangular longitudinal and transverse cutouts in the center of the jaws. It can be exchanged for a whole series of different special tongs and tools, resulting in a wide range of working possibilities. The hand of the manipulator has a mechanism for rapidly changing the tools.

The following tools are available:

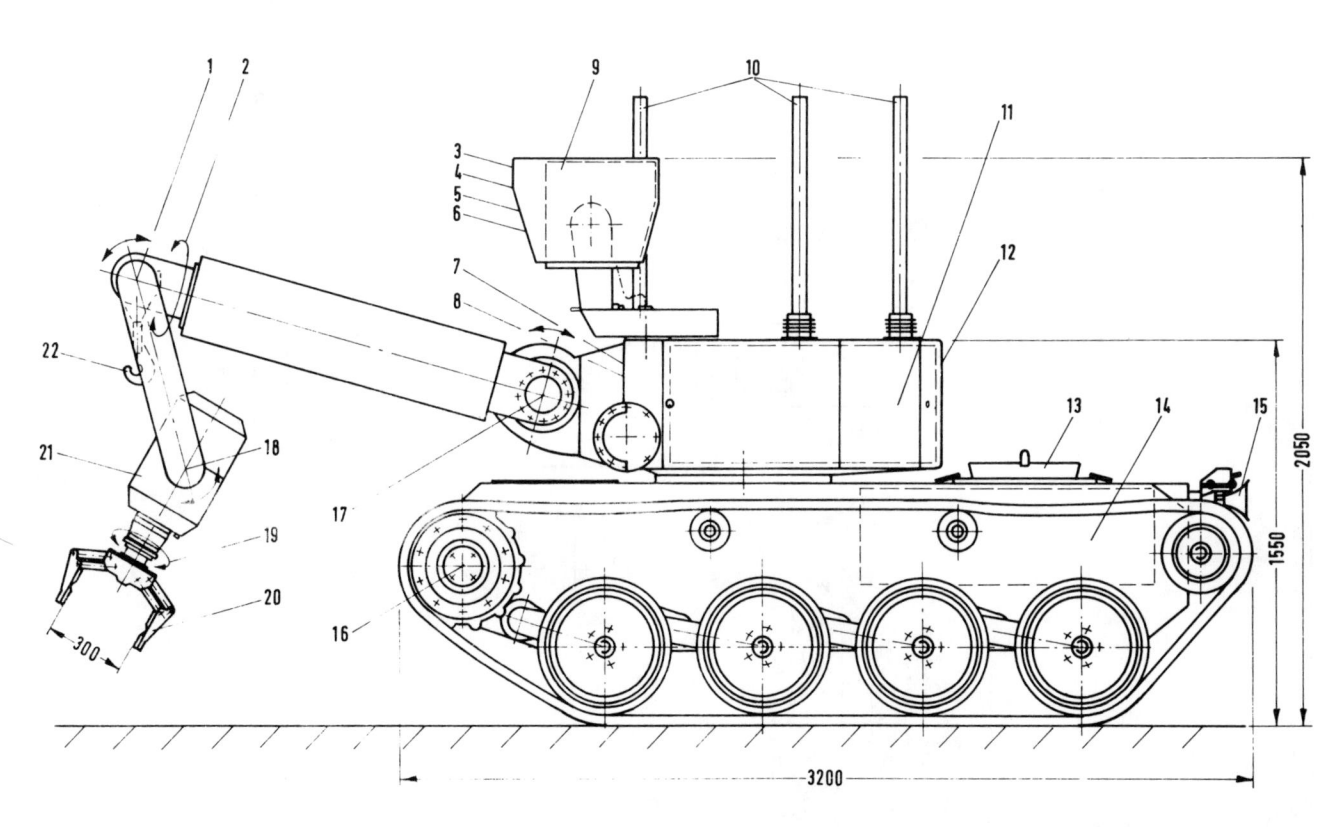

Fig. 4. The MF2 manipulator vehicle—details: 1. elbow joint; 2. arm rotation; 3. mono television camera; 4. searchlights; 5. stereo microphones; 6. stereo television cameras; 7. gamma dose-rate meter; 8. temperature sensor; 9. swivel-tilting head; 10. aerials; 11. transmitters and receivers; 12. swivel turret; 13. sample collecting vessel; 14. lead batteries; 15. automatic trailer coupling; 16. driving wheel with drive motor and gear at rear; 17. shoulder joint; 18. hand joint; 19. hand rotation; 20. large parallel-jaw tongs; 21. SM5-D power manipulator; 22. load hook.

1. Seven special tongs: grip hook, small parallel-jaw tongs, multipurpose tongs with six different functions, pipe tongs, tongs for inside grab, grab for 200-litre drums of radioactive waste, and pin-up jaws for large parallel-jaw tongs to permit gripping bulky objects.

2. Four mechanical tools: parallel-jaw screw wrench, tool carrier for socket wrench, cutting hook, and plate shears.

3. Four electric tools: impact screw driver, drilling tool, tenon saw, and hammer.

Basically, any electric tool operating on 110-V d.c., of the correct size, and provided with an adaptor can be used by the manipulator.

Commercial electric tools are provided with a holder adapted for use on the manipulator and mounted on a standard carrier (Fig. 5) that can be inserted into the hand like the tongs. The carrier has springs to be used for automatic feed and to serve as an elastic link for applying the tool. The spring force can be altered with the grab drive.

A tool carrier for carrying the parallel-screw tongs and for depositing the large parallel-jaw tongs is located at the rear of the vehicle. The television cameras are mounted in a swivel tilting head installed on the turret of the vehicle. It can be moved laterally 1 m to the left. This head contains the following apparatus:

1. one pair of stereo television cameras for observing the operations

2. one mono television camera with zoom lens for orienting at greater distance

3. four searchlights, two distance and two wide beams, that make night work possible

4. two stereo microphones for noise transmission.

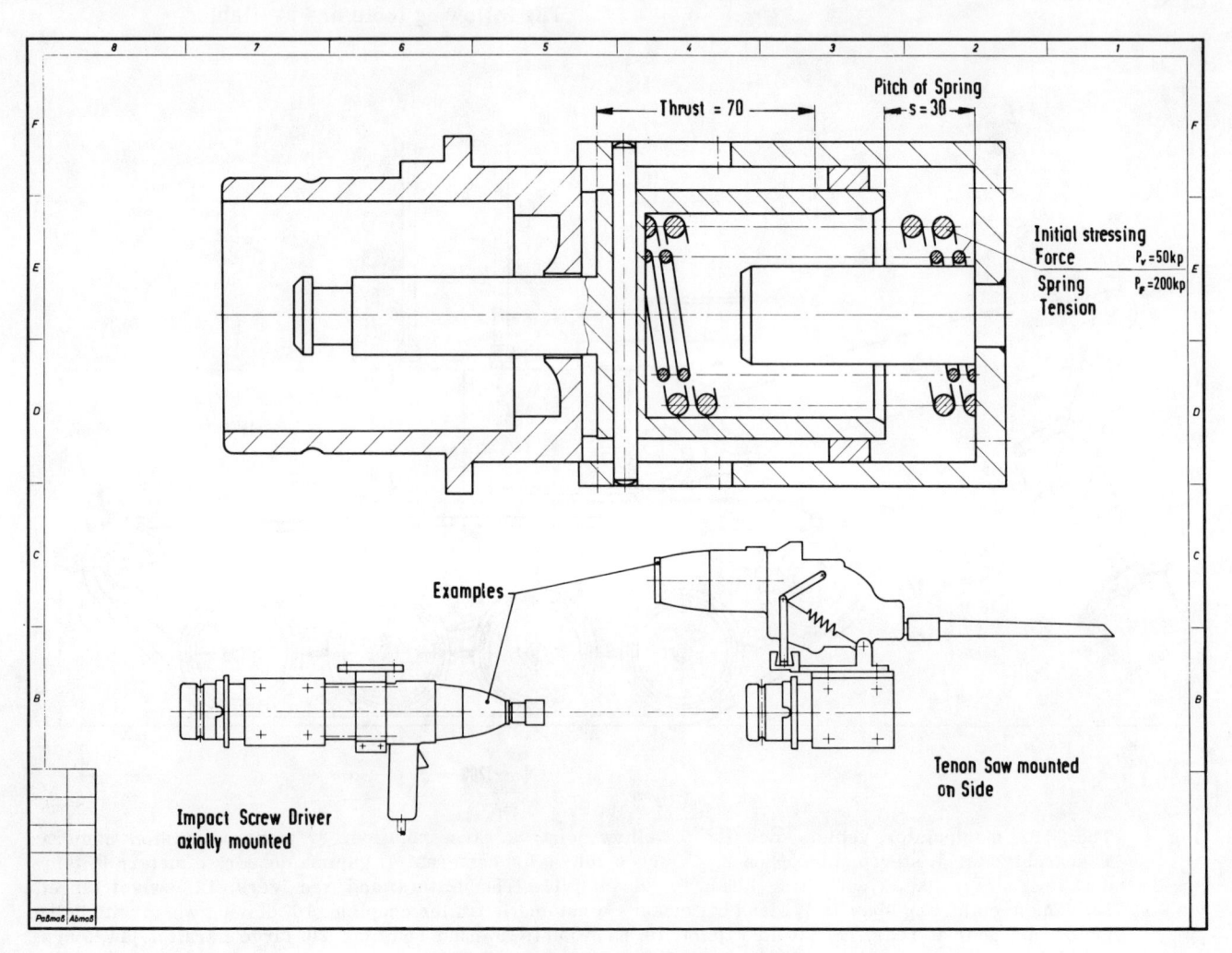

Fig. 5. Carrier for electric tools.

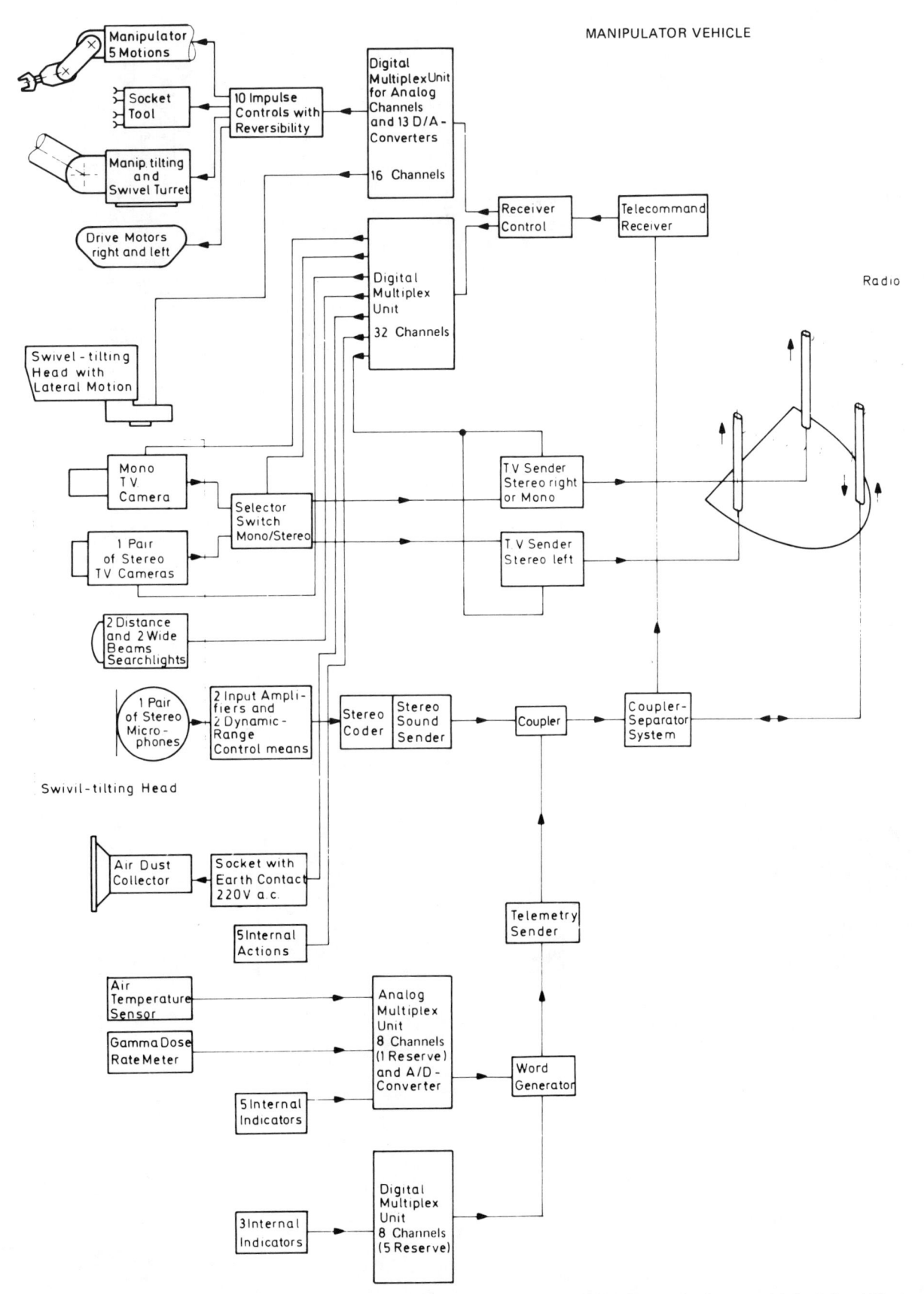

Fig. 6. The MF2 manipulator vehicle (simplified

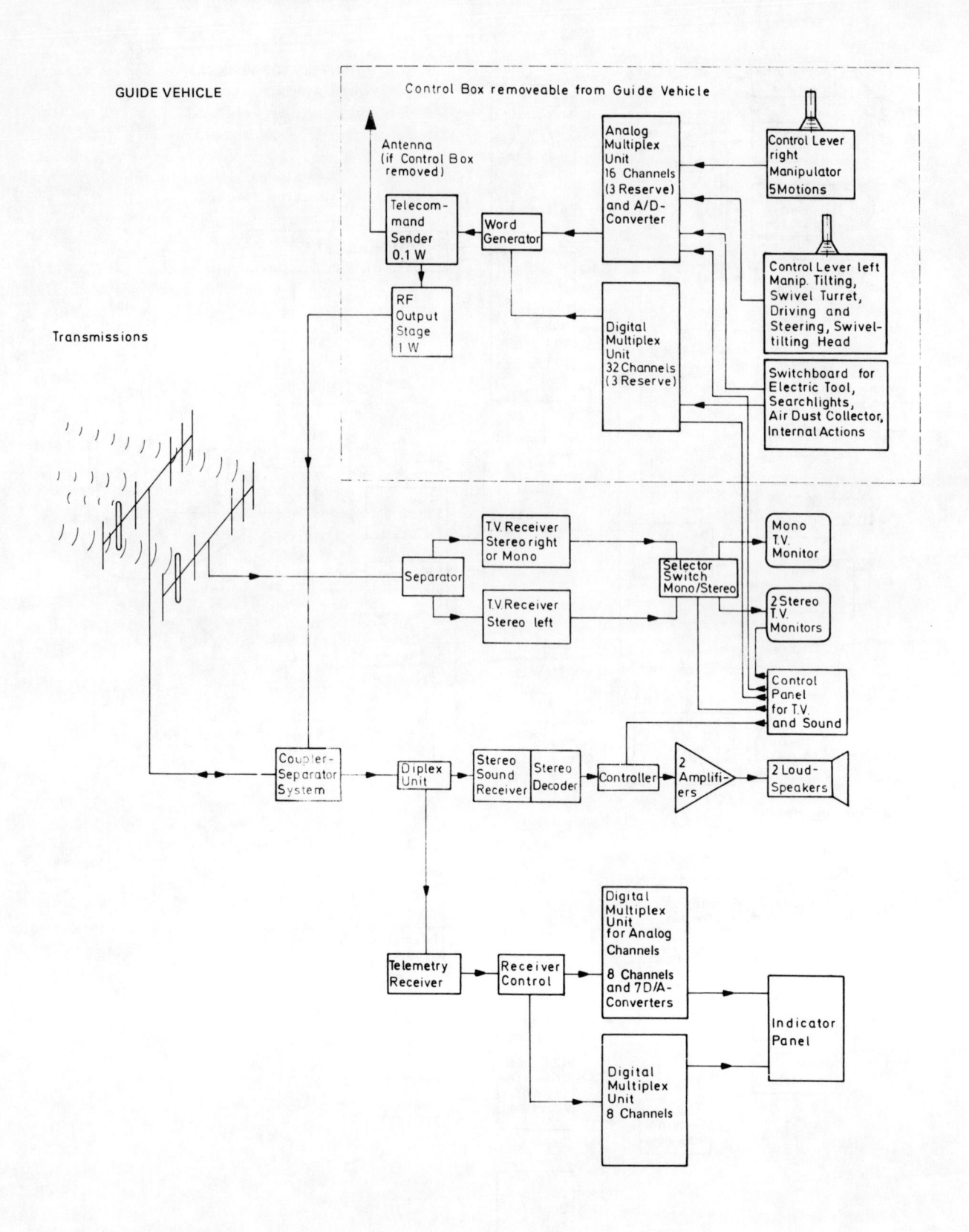

block diagram) for command and information.

The equipment is supplemented by:

1. a gamma dose-rate meter and a temperature sensor in the front of the turret

2. an air-dust collector at the rear of the turret, energy supply via 220-V a.c. converter

3. a sample-collecting vessel on the rear part of the vehicle

4. automatic trailer coupling at the rear.

When desired, one or two small cameras (for stereo photographs) or a 16-mm film camera can be mounted on the swivel tilting head. The lifting eyes, by which the vehicle can be towed away or suspended from a crane, are located at the front and rear. In place of the air-dust collector, any other device powered by 220-V a.c. could be carried, if desired.

Remote Radio Control (See Fig. 6). Primary data communication between the manipulator vehicle and the guide vehicle is by a single half-duplex[b] radio link using 1-W transmitters in the 400-MHz band. All analog data are multiplexed and converted to digital, then combined with multiplexed direct-digital data before encoding for transmission by pulse-code modulation (PCM) with a 300-Hz band width. Command data to the manipulator vehicle can utilize up to 16 analog and 32 digital channels, while 8 analog and 8 digital channels are provided for sending back telemetry data to the guide vehicle. Additional transmitters in the manipulator vehicle (operating in the same 400-MHz frequency range) include a 1-W FM stereo transmitter for directional sound transmission, and two 10-W television transmitters of 5-MHz band width. In addition, the control box can be removed from the guide vehicle (see Fig. 7) and still control the remote vehicle by a self-contained 0.1-W command transmitter, but with no return reception capability.

The range of the transmitters and, consequently, the radius of use of the manipulator vehicle is at least 1000 m.

Vehicle. The rollers of the chassis are suspended on torsion bars. The suspension system can be blocked, especially during action. The speeds of the electric motors in the chassis and the manipulator can be changed continuously by pulse control. The batteries are located in the rear of the vehicle, and the communications systems are installed in the turret.

---

[b]Half-duplex = alternate, one way at a time, independent transmission on a single channel.

Auxiliary Vehicles

The already mentioned guide vehicle, from which the manipulator vehicle is controlled, and a transporter for loading the MF2, are part of the complete system (Fig. 8). The control desk (Fig. 9) is at the rear of the guide vehicle with all controls, indicators, television monitors, and receiver and transmitter installations.

To shorten the required work time, vehicle control and manipulator operations are carried out with only two levers, one with five and the other with six possibilities of motion. The control box with the two operating levers can be taken from the control desk and can be carried on a belt by an operator (Fig. 7). Thus, it also is possible to control the manipulator vehicle independently of the guide vehicle.

The grip force exerted by the manipulator is measured and indicated on the control desk. In addition, the operator can obtain an impression of the forces exerted by a tone that increases in pitch with the increasing grip force.

The electric supply of the guide vehicle is obtained by an auxiliary power system. The guide vehicle contains a charging installation for the batteries of the manipulator vehicle. It also carries a set of reserve batteries and the special tools for the manipulator.

The manipulator vehicle can descend on its own power from the transporter over two self-contained ramps. The complete system can be transported by aircraft. Over longer distances, the medium transport aircraft "Transall" C-160 is available. Over shorter distances, the medium transport helicopter of the CH-53 DG type can be used. In the latter case, the guide vehicle and transporter can be carried only as exterior loads.

Possibilities of Application

The preceding description of the equipment discerns the capabilities of the MF2. It is intended for use in the open or in large rooms where ionizing radiation makes the access of personnel impossible, or greatly limited.

The manipulator, with its auxiliaries, is designed for heavy, but not complicated, work. The use of the remote-controlled vehicles could be necessary to eliminate or remove items causing trouble or other consequences of an accident. The MF2 probably would be used more frequently for maintenance, repair, modernizing, and demolition of nuclear plants. Application outdoors could be necessary in transport accidents involving radioactive material and heavy contamination; e.g., in the countryside.

Remote-controlled operation with vehicles in large rooms has to be considered in nuclear

Fig. 7. Operator carrying the control box on a belt; the MF2 manipulator vehicle.

Fig. 8. The MF2 System—guide vehicle, manipulator vehicle, and transporter.

reactor buildings, reprocessing plants, large hot cells, and deep storage of radioactive wastes.

For nuclear reactors, a natural center of interest, the vehicle might be used on the reactor platform, in pump rooms, in the intermediate storage of radioactive wastes, for changing filters containing radioactive substances, and in the turbine building of boiling-water reactors, providing that there is adequate accessibility.

The MF2 could be used for the following tasks:

1. salvaging and packaging radioactive materials
2. erecting brick shielding walls

Fig. 9. The control desk for the MF2 manipulator vehicle. From left to right: mono television monitor, levers for vehicle and manipulator, and stereo television monitors.

3. suspending tackle in crane hooks
4. transporting objects
5. using tools, including electric tools, for dismantling, mounting, separating, etc.
6. safety tasks in plants; e.g., actuation of operating controls of various types
7. disconnecting and connecting electric connectors and quick-disconnect couplings in fluid lines
8. exploring and reporting damage by observation, listening, measuring, and photographing
9. salvaging objects; e.g., stationary measuring instruments, fragments
10. depositing measuring instruments, television cameras, and illuminating lamps
11. applying markings at locations of high radiation levels
12. clearing operations; e.g., removing obstacles such as fences, barricades, railings, light walls, etc.

In addition to high radiation fields where only remote-controlled execution of operations could be considered, more frequently a transitional stage exists. Here, short time access is permitted with the possibility of combining manual and remote-controlled operations; e.g., carrying out preparations manually and the complicated work steps and activities of longer duration with the MF2.

In principle, a vehicle like the MF2 also could be equipped with tools other than the power manipulator SM5-D; e.g., with a pair of heavy electric master-slave manipulators or with lifting gear with a carrying capacity of 500 kp. Such a combination could have many applications and would permit the execution of complicated operations. The scanning frequency of the channels would have to be raised considerably, and the number of analog channels, especially those for transmitting measured values, would have to be increased.

THE ELECTRIC MASTER-SLAVE MANIPULATOR (EMSM I)

The EMSM I is the first electric master-slave manipulator developed in the FGR. The apparatus belongs to the class of 200 N (45 lb) carrying capacity. Mainly, it is intended for nuclear purposes, especially for use in large hot cells and for medium and heavy manipulator vehicles. The most obvious innovations, compared with previous foreign electric master-slave manipulators, are two additional movements, the electric dead-weight compensation and a load hook at the elbow pivot.

The EMSM I manipulator was developed by the Gesellschaft für Kernforschung in cooperation with Blocher-Motor KG.

Fundamentals

In electric master-slave manipulators, the characteristics of mechanical elements transmitting power between the master (control) arm and the slave (working) arm are reproduced by an electrical system. Movements carried out manually and forces and torques exerted are transferred from the master arm to the slave arm and a genuine feel is transmitted to the hand of the operator corresponding to the forces and torques exerted by the slave arm or affecting it in the opposite direction.

At present, electrical master-slave manipulators are remotely controlled apparatus with the most efficient kind of control installation. The forced parallel operation of both arms effects an essential advantage over manipulators with simple (unilateral) position control and also over those with additional control circuits for power feedback.

Short Description of the System

The new remotely operated apparatus essentially consists of three components connected by a multicore cable, a slave arm, a master arm, and a combined switch and control cabinet (central unit) on the control side. For trial purposes, it was provided with a series of provisional accessories; a carriage of adjustable height for each of the slave and master arms, a stereo television installation with a swivel-tilting head, an assortment of tools, and others (see Fig. 10).

The slave and master arms are essentially of the same design, except for the tongs and handle, and are provided with nine different movements each (see Fig. 11). All nine movements have a bilateral position control with force reflection. They are driven by electric motors through amplifiers. The master arm has electric motors for the force reflection. All drive units of one arm are installed in a common housing above the shoulder pivot from which the power transmission passes to the individual elements of the arm.

The dead-weight compensation is done electrically, apart from the additional springs on the shoulder pivot and not, as usually done, by counterweights.

Arms

The increase in mobility of the arm by an additional elbow pivot and the possibility of rotating the forearm considerably increases the working capability. It is possible to grip deeply into smaller horizontal openings, such as locks or containers, reach around objects, and more easily operate levers. In positioning the tongs, the movements of the arms are distributed to several elements of the arm and their drive units so that their individual speeds remain lower, the apparatus can be moved more easily, and the speed for positioning the tongs increases.

The possibility that an arm may take any of several positions at a certain position of the handle and tongs generally causes no trouble, as

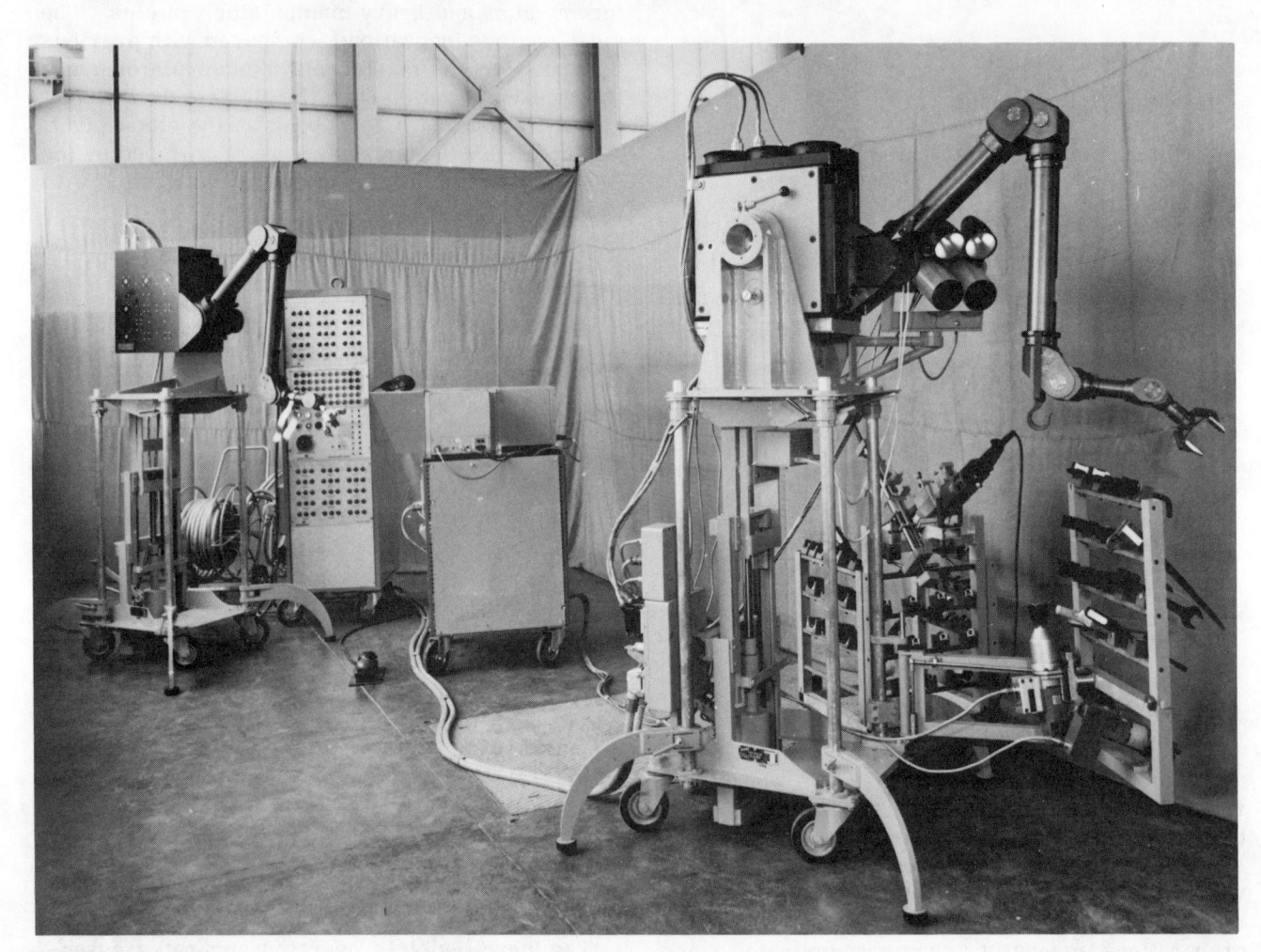

Fig. 10. The EMSM I with accessories. From left to right: master arm on a carriage, cable drum, combined switch and control cabinet, stereo television monitors, slave arm on a carriage with stereo television cameras and tools.

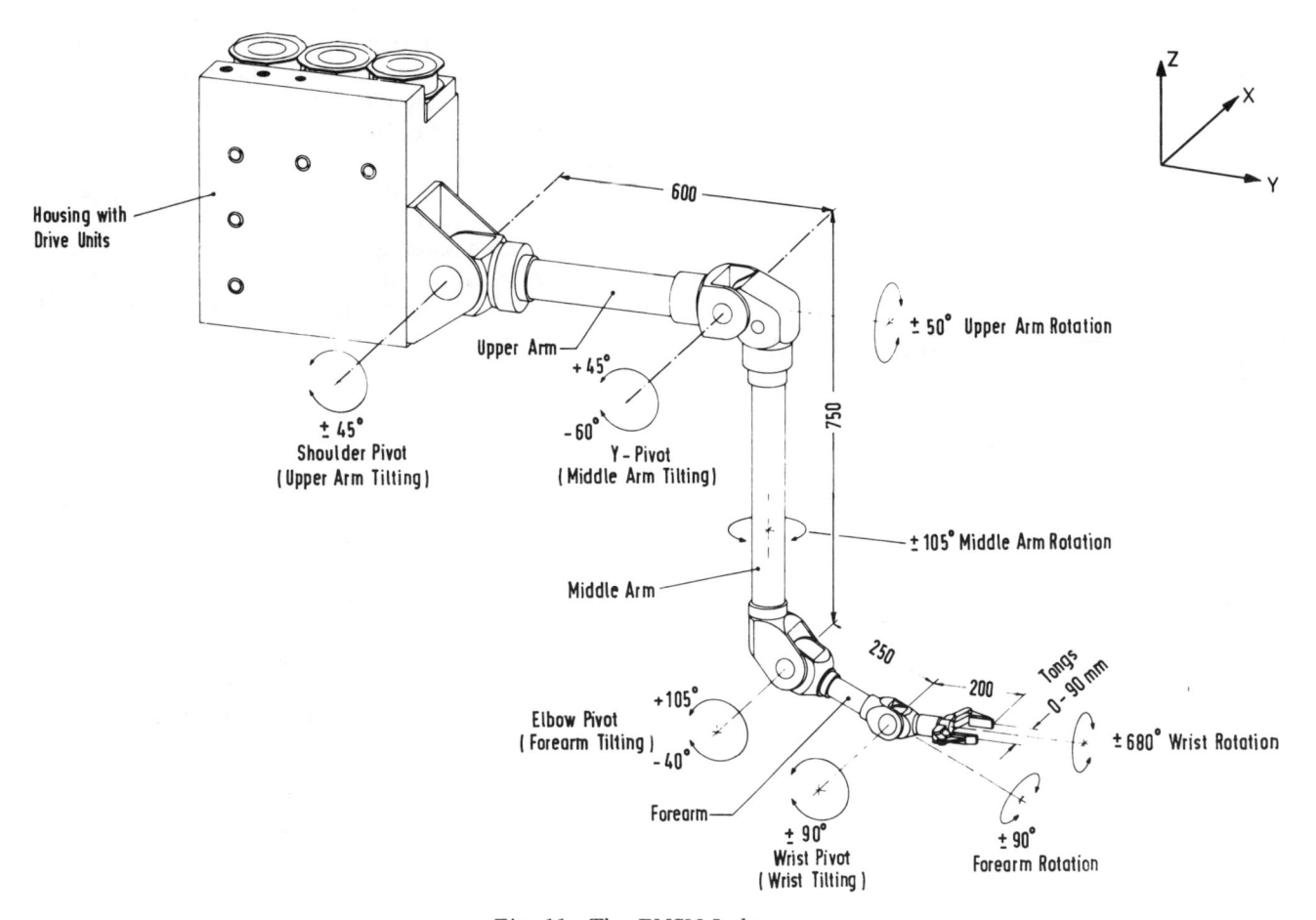

Fig. 11. The EMSM I slave arm.

experiments have shown. So that it is not necessary to use the second hand for wide lateral movements of the arm, a clamp is provided on the forearm of the apparatus as an auxiliary guide, which selectively can be tilted downward, touching the forearm of the operator on both sides. It is open at the bottom so that the operator can extract his arm without difficulty after releasing the handle. The power transmission for tilting and rotating the upper arm, and also for tilting and rotating the middle arm, is carried out primarily by gear wheels and shafts. Those for the other movements are carried out mainly by steel cables, for reasons of weight and space (see Fig. 12).

Each pivot and rotating movement of the following arm element has a common pair of drive units for mutual support. With both units rotating in the same direction, the pivot is moved. When moving to opposite directions, rotation is carried out. The wire ropes can be quickly retensioned manually or remotely manipulated by means of devices installed.

The parallel-jaw tongs and the jaws can be replaced by remote manipulation by using a simple device. The drive units can be very quickly replaced. An additional load hook, on which heavier objects can be hung, is provided at the elbow pivot.

Drive Units

The drive units for the slave and master arms consist of either two or three motors each, a gear mechanism, a position transmitter, and a brake on each motor. Deviating from this, one motor is installed on the master arm for wrist pivoting and rotating movements, and a brake is installed on only one motor for gripping (see Fig. 13).

Two very similar types of three-phase induction motors with a maximum output of 100 and 80 W, respectively, are used for all movements of the slave and master arm. To construct a manipulator of useful dimensions, inertia, and acceleration values, the motors must have very small dimensions in relation to their performance and output moment. A very high radiation resistance is aimed at for all organic materials that cannot be avoided.

Fig. 12. The EMSM I slave arm.

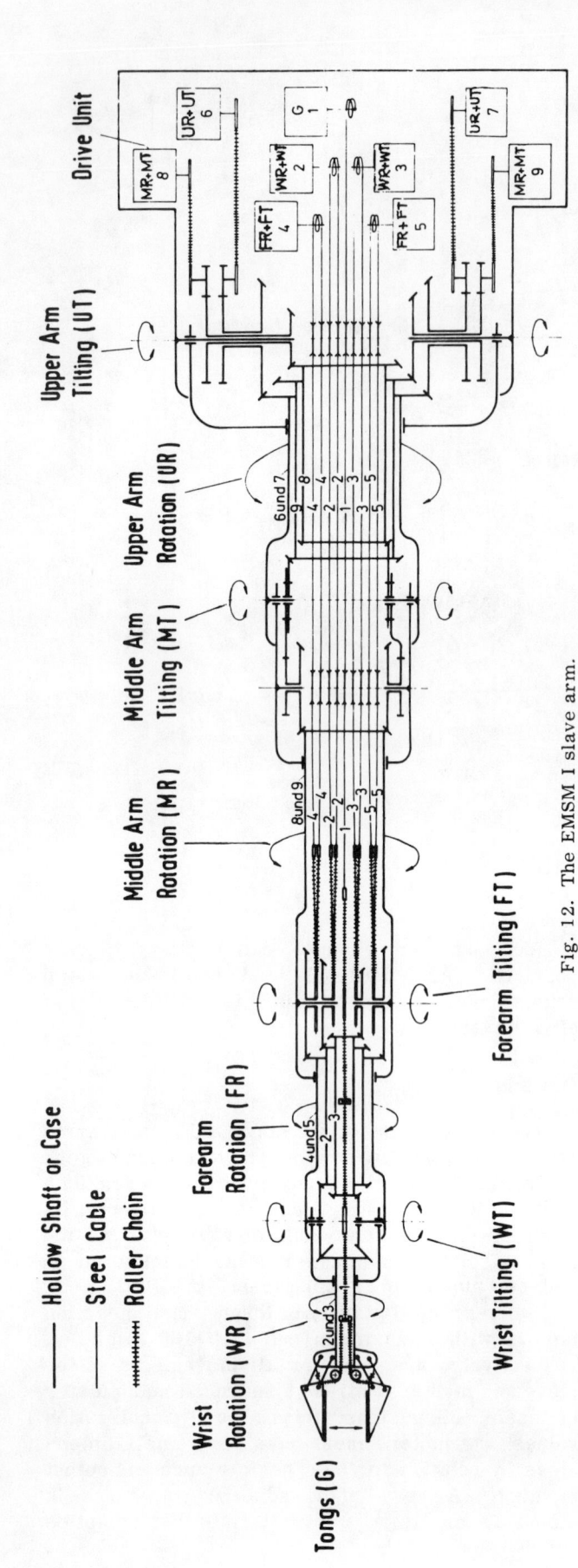

After extensive irradiation tests, an insulation material on a polyimide base was selected. This has resistance against gamma radiation up to 5×10^8 R and permits operation at a temperature of 220°C.

The gears contain two or three step-spur wheels (see Fig. 14); one type of each is used for the slave and master arms. To obtain easy movement, a gear mechanism with a ratio of only 1:3 is used for the units of the control side compared with that of the working side. This is possible because power amplification is used in the middle and upper load ranges and only a fraction of the force exerted by the slave arm has to be exerted by the drive units of the master arm.

In addition to the motors used for the movements, separate motors are installed for the electric dead-weight compensation to produce the forces during working and for the transmission of feel. The motors for drive, as well as those for dead-weight compensation, are followed by separate gear mechanisms of the same gear ratio. The drive wheel of the dead-weight compensation line is coupled to that of the drive through delay springs.

Electric Power Transmission and Force Reflection

The manner of operation for one movement is as follows (see Fig. 15). By actuating the master arm, the position transmitter of the control side is moved. Its signal is passed to a controller. When moving the master arm by a small amount, the corresponding signal from the working side differs from that of the control side proportional to the displacement angle and, if necessary, according to the polarity corresponding to the direction of movement. The controller produces a difference position signal and derives a speed signal from it. The resulting signal of the controller is passed to the amplifier for the slave arm drive. The motors for the slave arm drive are correspondingly modulated and adjust the slave arm in the direction of the movement of the master arm with a delay unobservable by the operator until the same position is reached.

The amount of current at the amplifier output for the drive of the slave arm is passed with the direction of the current to the feedback unit. The resulting signal is, if necessary, weakened at a desired ratio (power amplification on the slave arm) before passing to the amplifier for the drive of the master arm. The motor for the drive on the control side counteracts correspondingly to the movement imposed when exerting a force, so that the operator obtains a feel for the size of the load.

On thermal overload of one or several motors,

209

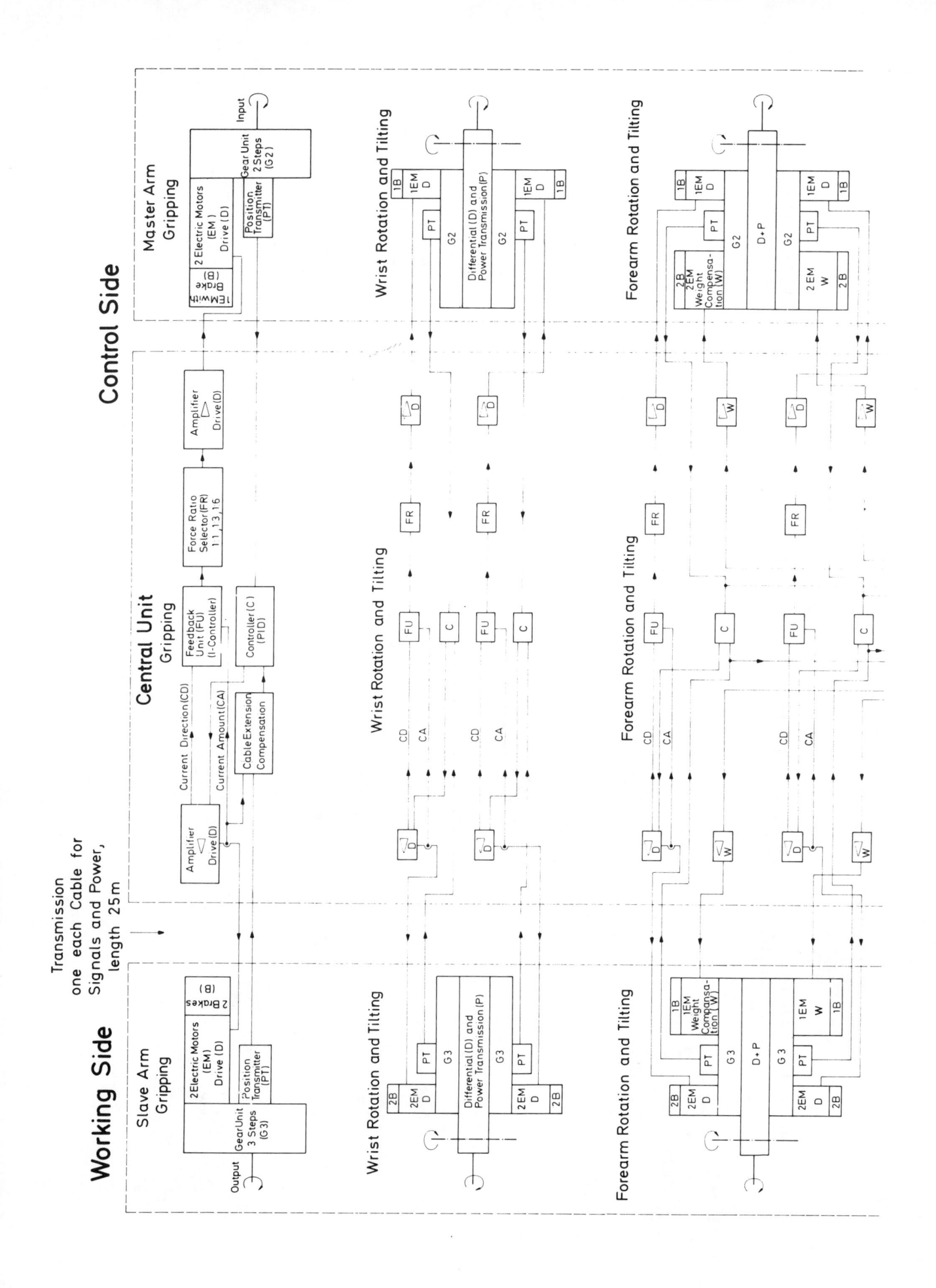

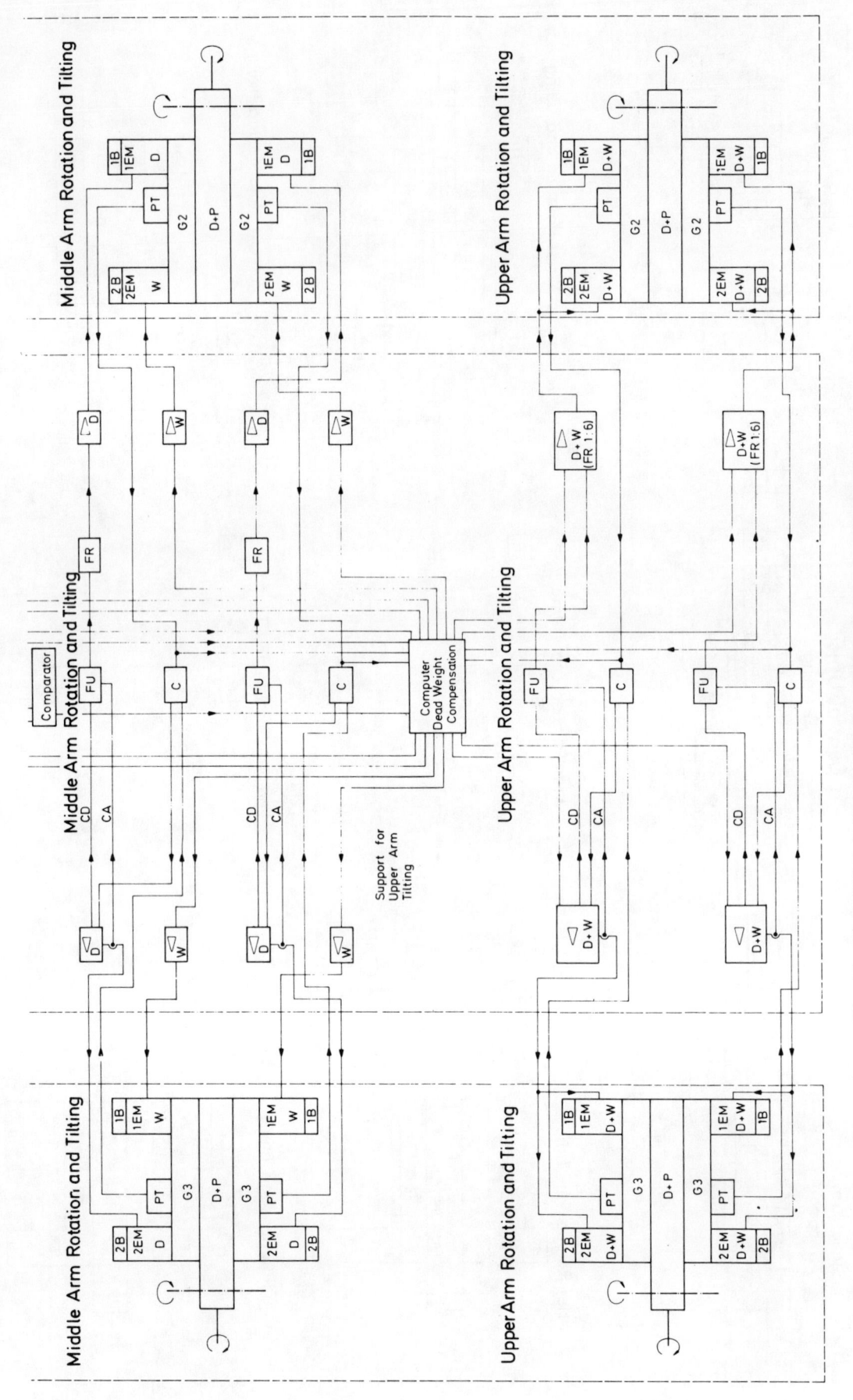

Fig. 13. The EMSM I—block diagram without monitoring and power supply.

a monitoring unit switches off the current to the controller, motors, and brakes of both arms through a time relay or a second relay. It also may act as an overload cutout if it exceeds a maximum permissible deviation of position of one or several elements of the slave arm compared with the corresponding position of the master arm. The brakes are locked in the currentless state.

The overall block diagram (Fig. 13) gives an impression of the extent of the electromechanical system and shows the differences and links existing at the individual movements.

Dead Weight Compensation

The dead weights are compensated at six movements (forearm rotation and tilting, middle arm rotation and tilting, and upper arm rotation and tilting). Separate electric motors are installed for weight compensation. Their manner of operation for one movement is, according to Fig. 15, as follows. To move the master arm a very small degree, the operator only has to start the corresponding gear mechanism and the motor of the drive line while the gear mechanism and motors for weight compensation, coupled to the drive line through delay springs, are still at rest. The signal of the position transmitter of the master arm is passed to the controller and to a central computer for dead-weight compensation. The computer supplies compensating signals corresponding to the movement of the arm element at the upper arm, taking into account the position of the lower elements (see Fig. 13), to separate amplifiers for the weight compensation on the controlling and working side. The motors for weight compensation start with a very small delay compared to the drive motor on the control side. They change their supporting output moment so that it corresponds exactly to the moment produced by the dead weight of the arm. The arm stops in any position without braking.

The electric weight compensation is necessary in practice for a manipulator with nine possibilities of movement, if the disadvantages of counterweights at the lower arm elements, or additional power transmission, are not to be accepted. It is possible to design an arm that can be protected against contamination by a closely fitting nonobstructing plastic booting. However, it must not be overlooked that the expenditure for electric weight compensation is high.

Technical Data

Force ratio (FR) master-to-slave (force reflection) is selectively 1:1, or 1:3, or 1:6 (force amplification). Objects up to 25 kp are lifted with tongs at FR 1:6. Objects up to 60 kp can be lifted on a load hook at FR 1:6. Tongs without a load are moveable in the X-Y plane at FR 1:6 at speeds up to 1.4 m/s.

The forces that can be exerted with the tongs in various directions and at different positions, the force exerted on the handle to start the apparatus without load, and the forces required at the tongs to start the system moving in the opposite direction (force reflection) are given in Table I for when the apparatus is completely switched on. The accuracy of positioning the tongs by viewing with the naked eye from a distance of 2 m is -0, +2 mm.

Tools

To provide the manipulator with flexibility, it was provided with numerous different commercial mechanical and electrical tools (see Fig. 16) that can be set into the tongs. All tools, except the socket wrenches, were provided with adaptors to ensure a better grip with the tongs.

Applications

Since the prototype proved fairly reliable during an extensive trial, commercial apparatus was manufactured without intermediate steps. Regarding more complicated work, we recommend that two slave or master arms be combined into one unit.

For use in hot cells, it seems suitable to have an "overhead installation" per unit, with four additional switch-operated movements. These are: a hoist with multiple telescoping tubes, a rotating mechanism, a crane trolley, and a traveling bridge (see Fig. 17). The drive housings of the slave arms also can be tilted up or down by a switch to increase the working range (Z-indexing motion). The slave trolley also could be provided with a 1000-kp electric lift.

Finally, manipulators can be provided at nuclear reactors as a means of reducing the dose to personnel during inspections, maintenance, and repairs. This assumes that the plant has been suitably designed so that the rooms are accessible to the vehicles.

THE LIGHT MF3 MANIPULATOR VEHICLE

The light MF3 manipulator vehicle presently is being developed by the Gesellschaft für Kernforschung mbH., Karlsruhe, FRG. Primarily, it will be used indoors; however, it also can be used in open country. For this vehicle, a new variable chassis geometry was designed. The chassis and equipment are separate designs. Accordingly,

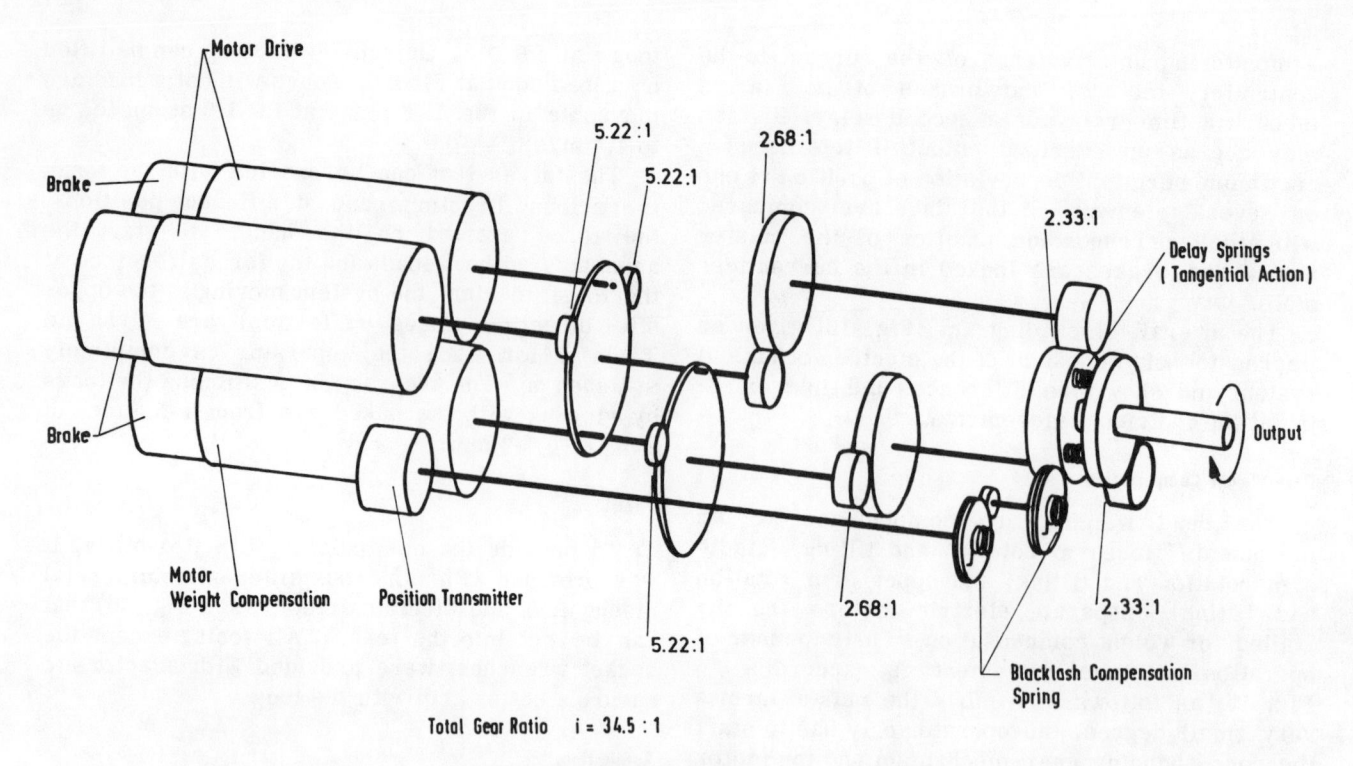

Fig. 14. The EMSM I slave servodrive unit.

various kinds of manipulators and handling devices can be employed.

The chassis has been completed already. The equipment to be used in action has been planned but still must be manufactured. The chassis was produced by Thyssen-Henschel AG.

Design

Basic Principles. The MF3 manipulator vehicle is remotely controlled by television pictures. Picture transmission, control, return signals, and power supply are routed through a trailing cable. The standard working equipment will be two lightweight EMSM II electrical master-slave manipulators (MF3-E version). The equipment of the vehicle includes the two slave arms, a pair of stereoscopic television cameras with zoom lenses, two searchlights, two stereo microphones, an electric socket, a gamma dose-rate meter, and a temperature sensor (Fig. 18).

The complete system includes the operating station with two master arms for control of the slave arms, central units for the manipulators, one foot-control box for the manipulators (on-off), television cameras, one control and indication box, one control box for the vehicle, and two television monitors. In addition, there is a switchbox with a power supply and cable drum. The vehicle and manipulators can be operated by one man.

To make the cable that is used for transmission between the switchbox and the vehicle as thin as possible, a time-division multiplex system will be installed.

The Chassis. The stairs are the chief obstacles in the way of the vehicles. To travel on stairs of a normal gradient, a considerable minimum length is required. However, when there are stair landings and where it is necessary to maneuver in a small amount of space, which is encountered frequently, a short vehicle is needed. Also, the necessary small dimension should be combined with the relatively long range of the manipulators.

These requirements are met by the MF3 vehicle as a result of a special kind of design—a variable geometry of the chassis. The bottom shell of the vehicle is equipped with four oblong chassis members with tracked driving systems. These members can be moved independently around the transverse axes, upward or downward within wide angles relative to the bottom shell. A number of the advantageous features of this design (see Fig. 19)[c] follow:

1. The length can be reduced greatly relative to the normal driving position (Fig. 18), which

---

[c] The figures show the MF3-A version equipped with a light-weight electric master-slave manipulator with four pivots, but this difference is practically insignificant.

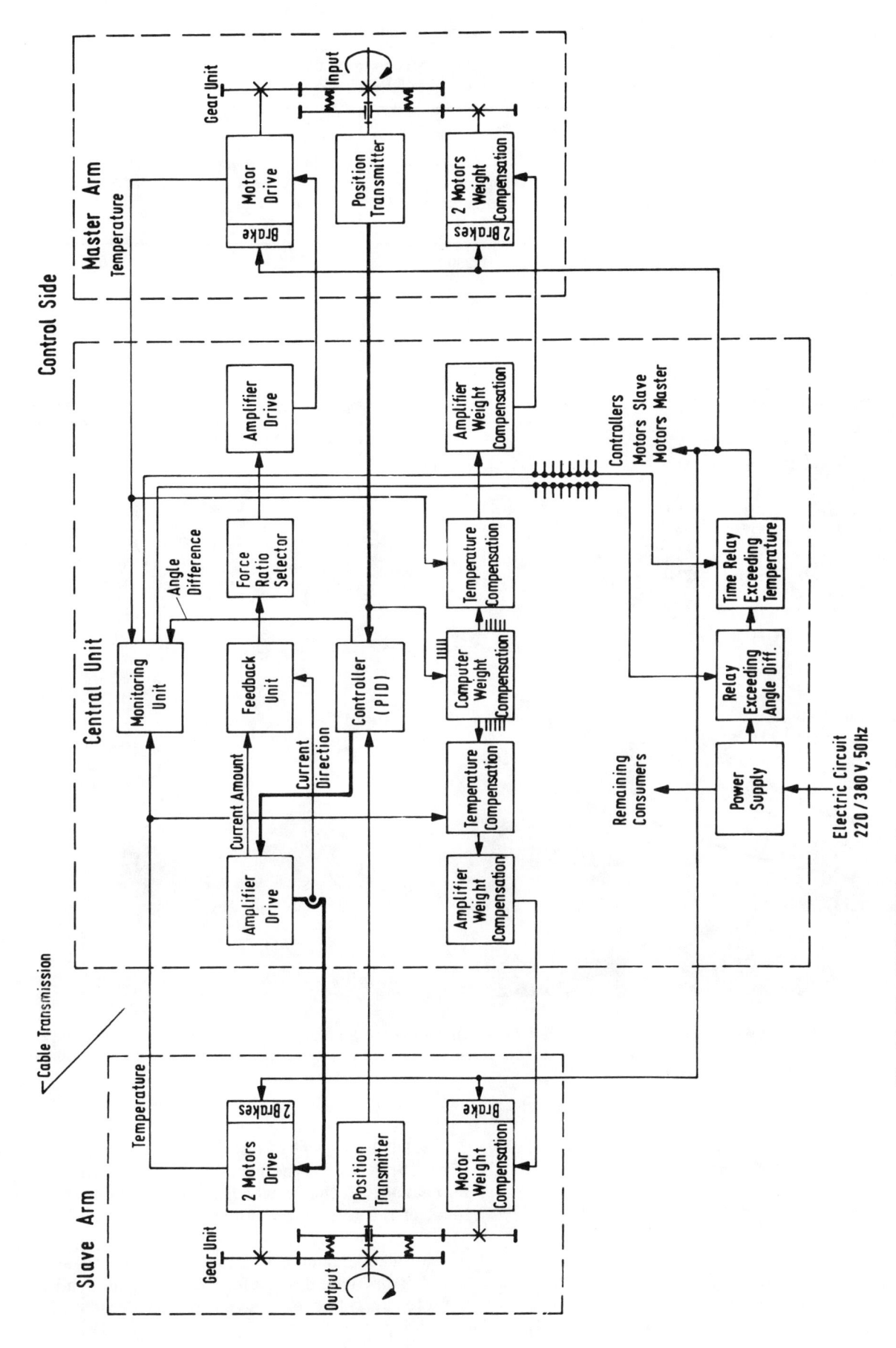

Fig. 15. The EMSM I–block diagram for one motion of elbow or Y-pivot.

TABLE I

Exertable Forces and Idle-Running Forces to be Overcome
at the Handle and Threshold of Feedback

Motions	Exertable Forces (FR 1:6) Tongs in Normal Position (Fig. 11) N	Tongs in Force Direction N	Idle-Running Forces (FR 1:6) N	Threshold of Feedback (FR 1:1) N
X-direction	100	250	2.7	5
Y-direction	250	250	1.25	2.5
Z-direction	80	250	2.5	7.5

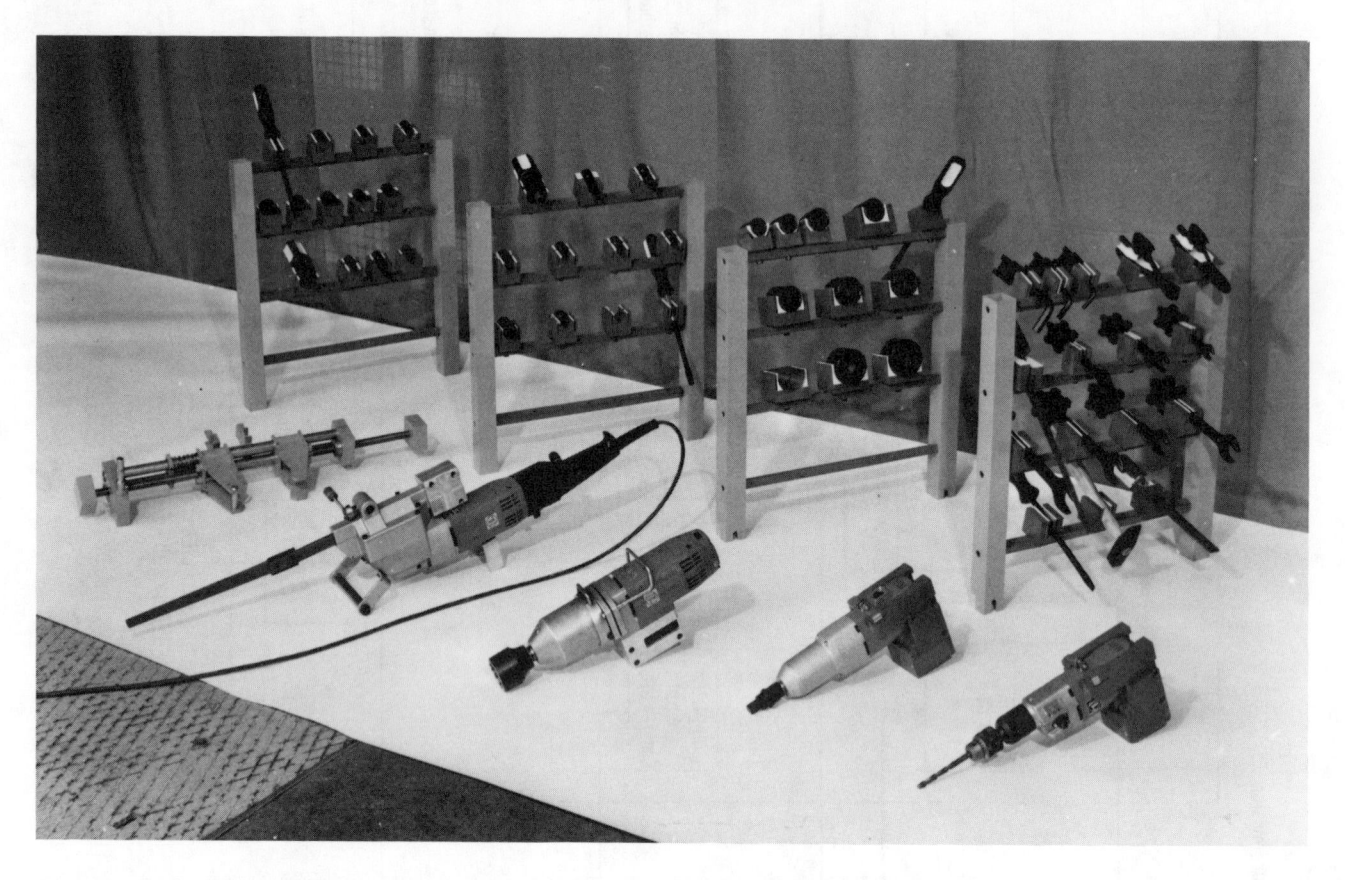

Fig. 16. Tools for the EMSM I. Front row (from left to right): device to clamp a tenon saw on a pipe using only one manipulator arm, tenon saw, two impact screw drivers, and drilling tool. Second row: adapted mechanical tools and socket wrenches.

makes a very small radius for the turning circle (Figs. 19.2, 19.3, 19.6, and 19.8).[d]

[d] If two EMSM II manipulators are installed, the chassis members in Fig. 19.6 can be inclined only as far as +70 deg in front, and +80 deg in the rear. In Fig. 19.2, it also is only up to +80 deg in the rear. In Fig. 19.8, the bottom shell can be inclined only as far as +70 deg, the rear chassis member to +150 deg.

2. The height can be increased greatly (Fig. 19.3) so that the range of the manipulators is increased in the upward direction and the eye level of the television cameras is raised.

3. The ground clearance can be increased. If the bottom shell should touch the ground, the vehicle can get free without any outside assistance.

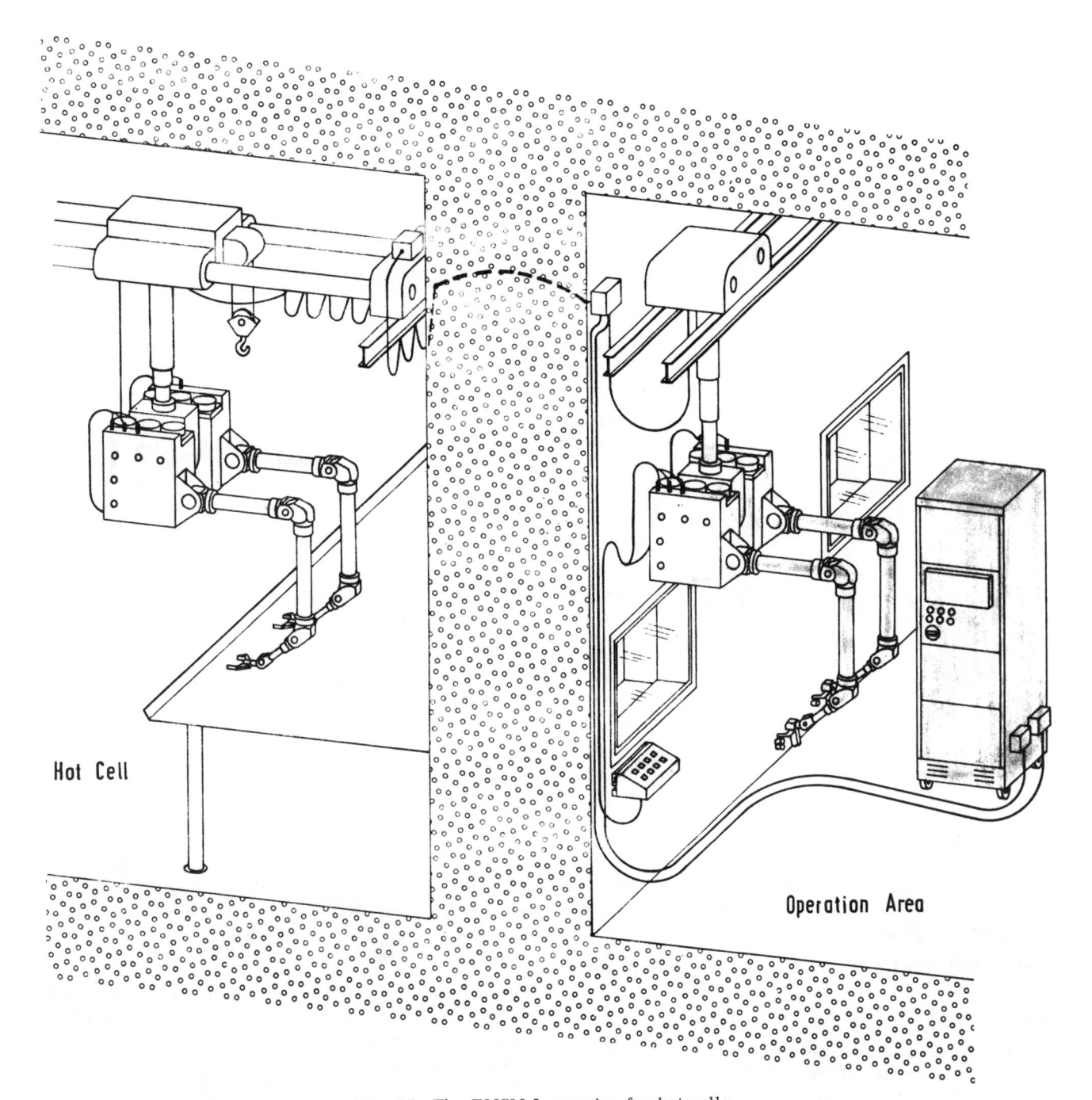

Fig. 17. The EMSM I—version for hot cells.

4. The chassis can be adapted to the contours of the terrain; e.g., hills and valleys.

5. The steering ratio can be reduced by decreasing the length, similar to Fig. 19.5.

6. The climbing height almost can be doubled as against Fig. 18 (Fig. 19.4) without any further adjustment of the chassis members.

7. The range of the manipulator can be increased below the standing level of the vehicle by inclining the front bottom shell downward (Fig. 19.7).

8. Longitudinal or transverse inclination of the bottom shell can be balanced out over broad ranges (Figs. 19.9 and 19.10).

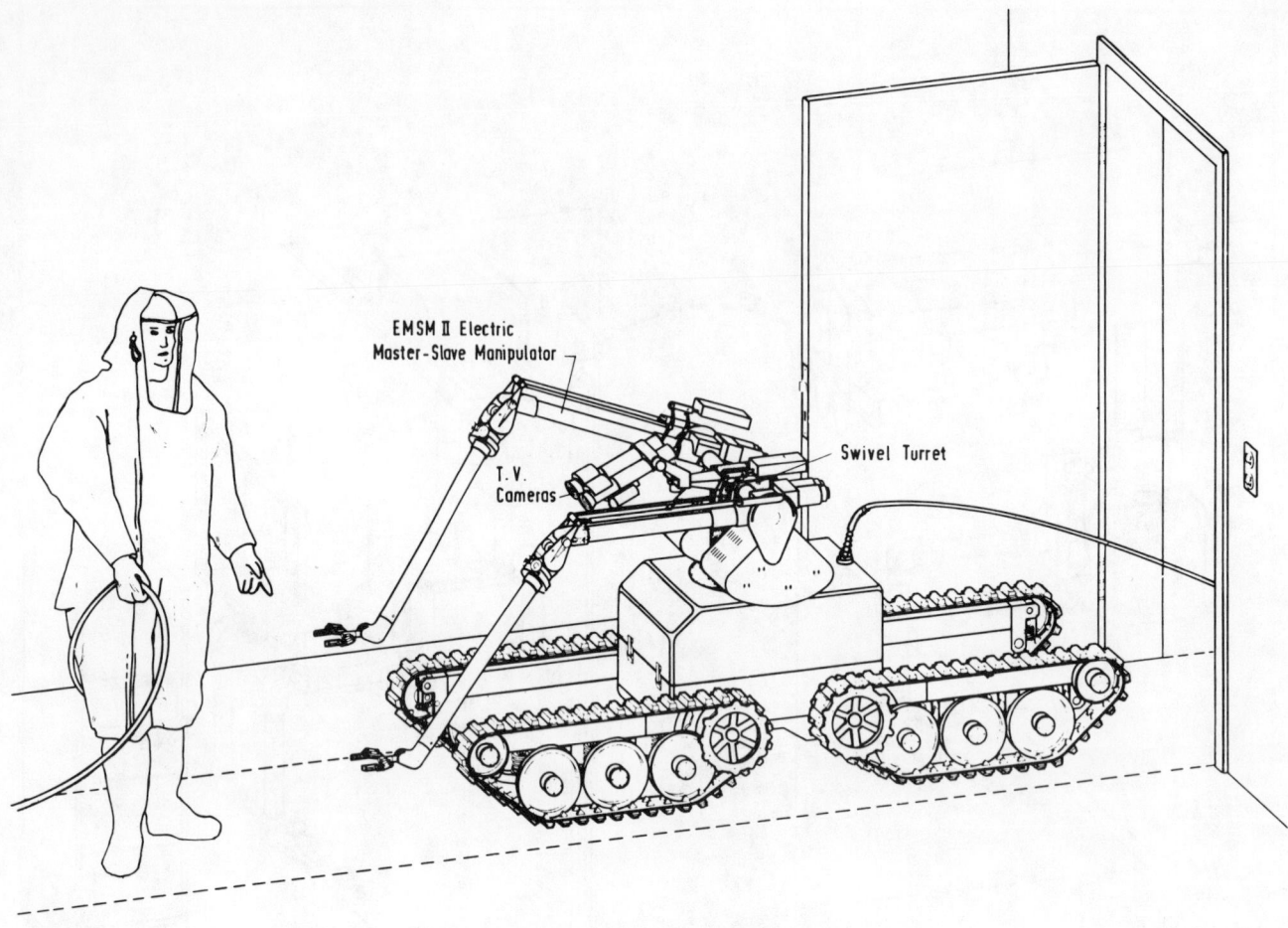

Fig. 18. The MF3-E manipulator vehicle.

9. Two adjustments of the chassis members and two travel movements allow climbing two steps that are very high, relative to the dimensions of the vehicle (Fig. 19.5).

10. Stairs with a gradient up to 45 deg and a step height up to 24.5 cm are passable without any inclination of the chassis members.

11. Because of its four tracks and high driving, and because of its wheels and the slopes of the tracks in the middle, the vehicle will not fall from a level surface to a stair when driving downward (Fig. 19.1).

12. When losing a track, or one track each per side, the vehicle still can be moved with certain limitations.

Several features of this new design also are advantageous for vehicles used for other than nuclear purposes.

The MF3 vehicle is moved and the chassis members are inclined by electric motors. The rollers are individually spring suspended. The tracks have a positive driving system. The bottom shell has a hook attached in the center. One lug each is attached to the left and right sides, front and rear, for pulling a trailer, towing, or for hoisting the vehicle.

Equipment. The planned light-weight EMSM II electrical master-slave manipulators, whose dimensions and weight are tailored to the MF3 vehicle, use the same well-proven principle as the EMSM I type; i.e., the master and slave arms are driven by electric motors via amplifier-connected bilateral-action interconnected position-control circuits. For dead-weight compensation, counterweights will be used. If loads exceed 4 kp, a servoforce amplification is used. The jaws of the tongs can be replaced by the other slave arm without any auxiliary device. A heavy-duty SM5-H power manipulator with 100 kp load capacity would be advantageous (MF3-S version) for completion. This unit would allow the use of powerful mechanical and electrical tools in the rooms.

For test purposes, the existing electric EMSM I master-slave manipulator was mounted on the

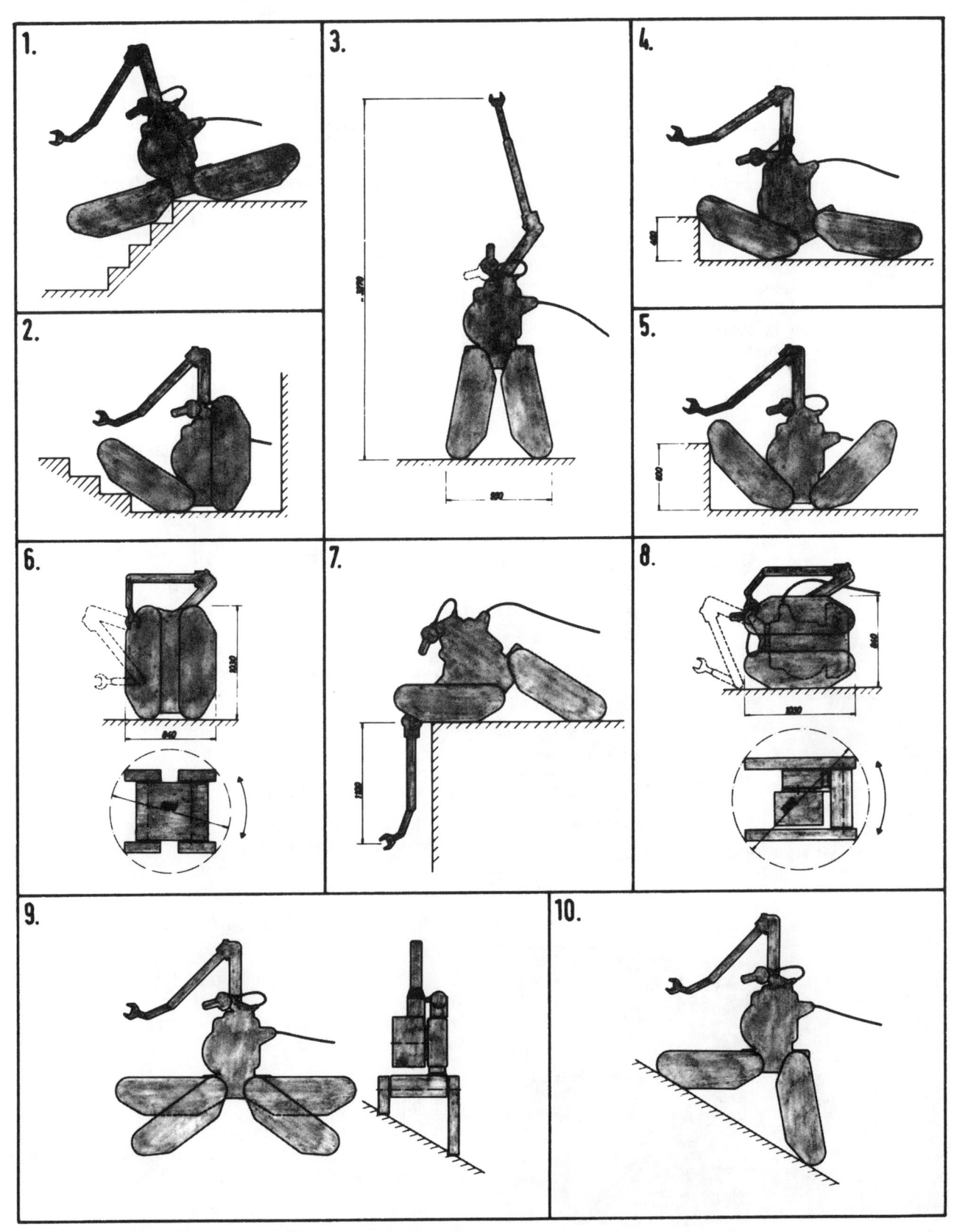

Fig. 19. The variable chassis geometry of the MF3 vehicle: 1. When driving downward, no dumping on a stair; 2. situation at a stair landing; 3. increasing the range in the upward direction; 4. climbing high steps; 5. climbing very high steps with adjustment of the chassis members; 6. shortest length; 7. increasing the range below the standing level; 8. lowest height; 9. transverse inclination balanced out; 10. longitudinal inclination balanced out.

217

MF3 vehicle chassis (Fig. 20). Equipped with stereo television cameras and one gamma dose-rate meter, the MF3 can be used as a reconnaissance vehicle (Fig. 21).

The MF3-E Vehicle and Manipulator (See Table II)

Other Vehicles. Besides the manipulator vehicles, the following heavy-duty mobile equipment is available to the Nuclear Emergency Brigade:

1. One tractor shovel (payloader)—this radio-controlled fourwheel vehicle of ~12-ton weight normally is equipped with a shovel, is driven by hydraulic cylinders, and is designed for earth work. Operation is observed by direct viewing or binoculars. For test purposes, this vehicle was equipped with the SM5-B power manipulator of the manipulator crane vehicle instead of with the shovel.

Fig. 20. Light MF3 vehicle with EMSM I slave arm.

Fig. 21. Light MF3 vehicle with stereo television cameras, two searchlights, and one gamma dose-rate meter.

TABLE II

A. Vehicle Data	
Weight	500 kg
Length (max/min)	2.26/0.95 m
Width	0.71 m
Height (normal/max)[a]	1.05/1.74 m
Working devices	two light-weight electric master-slave manipulators
Viewing equipment	two stereo television cameras with zoom lenses
Length of cable (max range of action)	100 m
B. Manipulator Data	
Type	EMSM II
Load capacity (normal/max)	12/24 kg
Number of movements	6 + tongs
Length of slave arm (shoulder pivot to tip of tongs)	1.77 m
Force ratio, master:slave	1:1, 1:2, or 1:6
Weight slave arm with drive units	45 kg

[a] Up to the upper edge of the television cameras in the basic position.

2. One truck with flatbed semitrailer to transport the tractor shovel over longer distances.

3. One self-propelled shielding wall with a mechanical master-slave manipulator, shielding lead either 50 or 150 mm thick, demountable components, three wheels, and an electric motor.

4. One small fork-lift truck to mount the shielding wall.

5. One Syntelman manipulator system with two arms, a carriage with four wheels, a cable transmission, exoskeleton control, and stereo television cameras.

6. One small truck to transport the Syntelman system.

7. One small radio-controlled vehicle with one gamma dose-rate meter, four wheels, and a gasoline motor.

8. One truck with a container to transport the decontamination equipment.

9. Two trucks with a container for personnel decontamination.

10. One truck with a container and equipped with a whole-body counter for assays for the incorporation of radioactive material. (One container with decontamination equipment, for decontamination of personnel, and one with a body counter that can be connected with tents to a station.

11. One truck with a container for evaluating personnel glass dosimeters and radioactive samples.

12. One truck as a changing room for the emergency crew and as the command center.

13. One vehicle for sampling (Unimog S type) and one trailer with an inflatable boat.

14. One semitrailer to transport radioactive materials (for two 200-litre drums, shielded with 13-cm lead and 2-cm steel).

15. One truck to transport various tools and devices.

16. One workshop semitrailer.

17. One vehicle equipped with an elevated working platform (the length of the boom is 18 m).

18. One trailer with a telescopic light tower.

19. Two combicars for radiation measurement and personnel transportation.

20. One truck semitrailer with a cover.

21. Two trucks with a cover.

CONCLUSION

The Nuclear Emergency Brigade presently has an operating staff of five men who are primarily responsible for the maintenance and upkeep of equipment. Should operations become necessary, ~120 men, who are employed at other jobs at the Gesellschaft für Kernforschung, can be mobilized.

In addition to an automobile park, a multitude of equipment, tools, and work clothes are stored in the depot of the Nuclear Emergency Brigade. The Nuclear Emergency Brigade frequently assists in operations, maintenance, and repair work performed in nuclear-power stations and on the nuclear-propulsion ship, the *Otto Hahn*. The objective of all this is to train personnel and to test equipment.

Two operations have required the help of the Nuclear Emergency Brigade. The manipulator-crane vehicle was used to recover a 12,000-Ci ^{60}Co radiation source that could not be introduced into its shielding. Also, last year the MF2 manipulator was used to replace two active filters in the Biblis-A nuclear-power station, since the device provided for this purpose malfunctioned. Fortunately, it has not been necessary to operate the Nuclear Emergency Brigade in an accident situation.

We plan to set up a separate company to operate the Nuclear Emergency Brigade. This will be funded by the operators of nuclear-power stations. The duties of the company probably will be extended to include assistance in repair operations.

Teleoperators in the Nuclear Industry

J.A. Constant and K.J. Hill,
Atomic Energy Research Establishment, U.K.

Summary

This paper describes some of the work being carried out by the Robotics Section at Harwell.

Teleoperators are remotely controlled machines capable of carrying out operations, using some form of information link, at the command of an operator.

The paper describes ROMAN, a teleoperator designed and built at Harwell to meet a specific requirement of the Reactor Development Laboratories at Windscale. This unit is to operate in a radioactive environment in order to reduce the radiation dose rates received by personnel who carry out the tasks at present.

PART 1

REMOTELY OPERATED MOBILE MANIPULATOR (ROMAN)

1. Introduction

ROMAN was designed and built to operate in the maintenance area behind a line of irradiated materials handling cells in the Reactor Development Laboratory of the AEA Windscale. Operations inside the boxes are normally carried out from the front face by means of long handled tongs through the lead shield wall with viewing through lead glass windows. Access to the boxes is through removeable back panels, in turn shielded by a 200 mm thick steel and lead block wall which are removed by means of a fork lift truck when access is required.

To effect a box entry to carry out maintenance, remove unserviceable equipment or install a new piece of equipment, it is necessary to reduce the levels of radiation by removing the majority of active materials to an adjacent cell. Even so the residual radiation will limit personnel access to a few minutes per day, and a signidicant amount of effort and time may have to be expended in decontamination to achieve an acceptable radiation level.

To keep personnel radiation dose levels to a minimum it was decided that a remotely operated mobile manipulator should be designed and manufactured to remove and replace the back panels of the boxes, remove and replace equipment, and carry out adjustments and minor maintenance within the cells. It was to be capable of lifting loads up to 35 Kg at a radius of 2 meters and have the ability to place the load into the bottom of a container one meter deep.

The specification was established by the RDL Windscale staff in close conjunction with the Robotics Section staff of Harwell's Engineering Division. The completed machine is shown in Figure 1. The layout of the area in which ROMAN is used is shown in Figure 2.

2. Design Considerations

To achieve the mobility required to maneuver the mobile platform in confined spaces and the ability to negotiate floor gulleys and obstacles up to 100 mm high under control while carrying its full load, it was decided to use a pair of simple notched tracks - 75 mm wide rather than to use wheels. A tracked vehicle can be steered more easily in a confined space and can achieve neutral turns which is not possible with a wheeled vehicle. Tracks are cheaper and easier to replace than tires, the frequency of changing being determined by contamination rather than wear.

As a general policy it was decided to make the operation and the controls of the vehicle as simple as possible consistent with good reliability. This would ease maintenance which is of particular importance in the radioactive environment in which the vehicle is to work.

Consideration was given to the different methods of providing power. Electrical mains power was chosen rather than batteries because of weight and charging problems, in either case a control cable is required. This decision means that the combined power and control cable is of moderate size. As there is no regular pattern of tasks to be performed a pre-programmable control system was not considered appropriate. It is not possible to control the teleoperator from any one position as the inside of the cell is only visible from the cell face, it was therefore decided to control from either a master control panel positioned at the end of the maintenance corridor behind the cells, or from a slave control panel located at the working face of the particular cell in which the manipulator is being used.

Various types of motors were considered for driving each of the individual movements required. The space available for motor drives at the manipulator gripper was limited due to the restriction placed by the size of the entry holes into the cells.

3. Design

The design may best be described by considering the main elements of the basic system in turn.

3.1 Mobile Platform

The mobile platform has a variable speed from 0-60 m/min both forward and reverse, is capable of negotiating rough floors, floor gulleys, ducts, rails etc, surmounting obstacles up to 100 mm high and climbing a $15°$ slope.

It was decided that the method of constructing the mobile platform should be to design a single gear box using one on each side. The two gear boxes are joined together with two channel sections back-to-back to carry the manipulator and a light steel plate carries the driving motors and electrical equipment. This makes it a simple matter to vary the length and the width of the platform for other future applications.

The gear box is designed to be driven by an electric motor and the gear ratio was worked out to give a vehicle top speed of 60 m/min at the motors maximum running speed of 2500 r.p.m. The overall reduction of 47 to 1 is achieved by the use of a two stage reduction spur gear train and a chain sprocket drive to the wheels. The gear box is oil lubricated.

After studying the relative cost of either fabricating a gear box or using light alloy castings, it was established that the costs were not significantly different for manufacturing one pair and the use of castings would show positive savings if any further platforms were constructed in the future, thus it was decided to use LM6 aluminum alloy castings.

The mobile platform is driven by two 750 W motors. One motor fitted in the front of the platform drives the right hand track and the second motor at the rear drives the left hand track. Steering is achieved by controlling the speed and direction of rotation of each motor. Electromagnetic brakes are fitted to stop the drive to the tracks when the power to the motors is removed. These brakes are operated through a chain drive rather than directly on to the rear of the motor to minimize the width of the vehicle, and the braking is progressive. The brakes operate on the 'fail safe' principle so that in the event of a complete loss of power the tracks are locked.

Standard timing belts are used for the tracks with a 20 mm thick section of rubber bonded onto the back to prevent the track running off the driving wheels.

In the event of a power failure or the occurrence of a fault which would stop the platform with the brakes on, built in jacks are provided which can raise the platform far enough to place standard skates under and so permit the platform to be towed. These jacks are also used to raise the platform when a track change is being carried out.

The decision as to the type of motor to be used was determined by its duty, size, voltage and not least its availability. As there is only a limited requirement for climbing ability in the operation of the platform 200 V dc shunt motors are used. This permits the use of standard thyristor controllers which are readily available commercially.

The controllers compare the armature voltage with a voltage representing the set speed, any difference being amplified to cause the necessary speed adjustment to be made. Current limiting circuits protect the motors under stall conditions. The two controllers are carried on the platform to reduce the number of power and control leads required in the cable link. The power to the drive cards is isolated 15 seconds after the completion of operation. This is a safety precaution to prevent the vehicle being driven in the event of a thyristor in the control card developing a fault.

The relays used to energize the drive and manipulator motors are housed together with the controllers and a low voltage power supply in compartments, protected by simple covers, adjacent to the main drive motors. In these positions the electrical equipment is not in the direct radiation path from any of the radio active cells and is therefore unlikely to receive any significant radiation dose.

A pendulum type switch is fitted so that in the event of the platform approaching an unstable condition the driving motors are stopped and a warning is given to the operator.

Power 'ON' lights and an "Emergency Stop" button are fitted on the vehicle.

A PVC apron, which covers the top of the platform to assist decontamination, can be replaced relatively cheaply.

3.2 Manipulator

The manipulator is capable of handling a load of 35 Kg in any orientation with the arm fully extended and has seven degrees of freedom of movement.

To arrive at the optimum dimensions of the various components of the manipulator to enable it to achieve the required degrees of movements and to ensure that it would be able to reach all of the internal faces of the cell, a very simple cardboard model of the box arrangement and a meccano model of the mobile platform and manipulator were made.

The degrees of movement determined for each action are shown in Figure 3 and are as follows; each movement is independently operated.

Waist	Rotation of $360°$.
Shoulder Action	From $30°$ below horizontal to a vertical position - a total movement of $120°$.
Elbow Action	$135°$ either side of the upper arm - total movement of $270°$.
Fore-arm Rotation	$360°$ about the fore-arm axis.
Wrist Elevate Action	Total elevating action $135°$ from a position in line with the fore-arm downwards.
Wrist Rotation	Continuous in either direction.
Gripper Action	Parallel jaw action from 0-120 mm with a closing force of 18 Kg. Due to the limited amount of access through the back entry, the overall length from the wrist pivot and the end of the gripper is limited to 250 mm.

After studying the relative costs of fabricating the various components compared to using light alloy castings it was established that even on a unit basis there was no significant difference in cost, and positive savings would result by the use of castings if any further manipulators are constructed. Some weight saving is achieved and the finished appearance of castings was considered to be superior to fabrications. Because these components have to carry higher loads than the platform gear boxes, LM6 was considered to be too soft and LM 25 aluminum alloy, heat treated, was used.

From previous work it had been established that the optimum tip speed of the grippers for manipulative work should be 2000 mm per minute. This speed is too slow for approach work so the waist, shoulder, and fore-arm elevate movements are fitted with a high speed drive to give a tip speed of 10,000 mm/min which is operated by means of a separate switch.

The loads on the rotating waist are carried by a journal mounted vertically in two bearings. The upper one is a deep groove ball bearing carrying the top radial and thrust loads and the lower one a phosphor bronze bearing immersed in oil. The drive is from a 400 watt dc motor through a 70 to 1 standard oil filled gear box into a 58 to 1 spiroid gear and pinion which is oil immersed.

The loads of the shoulder pivot are carried by a journal mounted on a pair of deep groove ball bearings. The pivot action is driven by a 200 watt dc motor through a proprietary oil lubricated gear box into a 58 to 1 spiroid gear and pinion running in oil. The spiroid gear is attached to the journal.

The loads on the elbow pivot are carried by a journal mounted on a pair of deep groove ball bearings. The pivot action is driven by a 200 watt dc motor into a gear box specially designed to fit inside the arm thus keeping the arm as compact and clean as possible. The gear box which is oil lubricated has a three stage straight spur train with a reduction of 34 to 1 into a 58 to 1 spiroid gear and pinion.

The fore-arm rotate motion is mounted on miniature precision ball bearings and is driven by a 30 watt integral geared d.c. motor with a 15 to 1 gear reduction into a purpose built gear box with 40 to 1 single spur gear into a 58 to 1 spiroid gear and pinion. The spur gears and spiroid gear and pinion are oil lubricated.

The loads on the wrist pivot are carried by a journal mounted on a pair of deep groove ball bearings, grease lubricated. The pivot action is driven by a 30 watt integral geared d.c. motor with a 15 to 1 gear reduction into a purpose built gear box with a 4 to 1, two stage spur train, into a 50 to 1 worm and wheel, grease lubricated.

The wrist rotate is driven by a 25 watt integral geared d.c. motor with an 80 to 1 gear reduction into a purpose built gear box with a 1 to 1 bevel gear into a 12.5 to 1 worm and wheel which are oil lubricated. The loads are carried by a journal mounted on a pair of miniature precision ball bearings.

The gripper open and close movement is driven by a 25 watt d.c. motor into a 2 to 1 bevel gear and then a 3 to 1 single spur gear which drives a lead screw with 2.5 mm pitch square left and right hand threads. The drive motor which is integral with the gripper unit is fitted with slip rings so that continuous rotation of the wrist is possible.

The manipulator is mounted on a simple fixed pedestal which can be readily modified to optimize its height for a particular requirement. The manipulator is of modular construction so that it can be readily dismounted at each joint and removed from the platform, and the wrist terminates in a bayonet fitting so that a range of gripper units can be readily interchanged.

PVC gaiters are provided from the wrist to the top of the platform to protect the manipulator and to assist in decontamination. These can be replaced relatively cheaply.

The cable harness, from terminals in the base of the vehicle, includes plugs and sockets at the breakable joints as well as at the connections to the motors thus allowing for easy replacement in the case of motor failure and maintenance.

Small universal motors are used with gearboxes for the waist, shoulder and elbow movements. A special controller has been designed and manufactured for these motors, it includes a current limiting circuit which protects the motors in the stall condition. Stalling can occur if the operator tries to move the manipulator not realizing, while watching the gripper, that some other part has come up against an obstruction, eg the box entry.

The motors, which drive the fore-arm rotate and wrist movements, are all low voltage permanent magnet motors with a power output of approximately 30 W chosen primarily because of their compactness. These motors are integral geared units and their direction of rotation is easily reversed. The speeds of these motors are not varied but their operation is again current limited so that they will not burn out if stalled. The gripper open and close drive uses a similar motor without an integral gear box.

Electro-magnetic brakes are fitted to the shoulder and waist movements to prevent any over-run or run-back when the power is removed. The small d.c. motors are all dynamically braked, the low efficency of the gear ratio between each motor and drive prevents run-back when the power is removed.

Limit switches are incorporated in all of the manipulator movements with the exception of the gripper 'open' and 'close' action. The gripper action is designed to maintain its grip on an object being held should the power fail.

3.3 Control Consoles

The platform and manipulator are controlled from either one of two control consoles, which are free standing and portable and of such a height to suit standing operators. Each control panel includes control handles for the platform drive and the manipulator movements, two speed switch for some of the manipulator movements, tilt alarm and reset buttons, push buttons and indicator lights for power supply, and control position selector switch and indicator lights. A control console is shown in Figure 4. The master console is normally located in the maintenance area, out of direct radiation paths, where the level of radioactivity is low. This is generally referred to as Control Position 1. Control Position 2 is at the face of the particular cell where the manipulator is being used. This is as shown in Figure 2.

Except for the 'Emergency Stop', controls can only be operated from one control position at any one time. Initially, operation usually takes place at Control Position 1 when the selector switches on both the control panels must be selected to positions CP1. The lamp indicators will show the positions of both switches. The operator at Control Position 1 would normally drive the teleoperator to the cell in which work is to be carried out, remove the back panel of the box and position the manipulator within the cell. Control is transferred to Position 2 by the operator at Control Position 2 turning his selector switch to CP2, this will give an indication on the panel at Control Position 1 that the second operator is ready to take over control. It is only after the master operator has moved his selector switch to position CP2 that the teleoperator may be controlled from Position 2. A similar sequence is employed when the first operator requires to take over control again.

The master control console is fed from a standard 13 A mains supply point at 240 V ac 50 Hz which provides the total power for the platform and manipulator. A current-operated earth-leakage circuit breaker is fitted to provide personnel safety in the case of an earth fault or an operator touching a live point, the unit trips the supply in the event of an out of balance current between live and neutral occurring in excess of 30 mA.

A control cable interconnects the two control consoles and a combined control and power cable links the master control panel and the platform. In this application, rather than use a cable reel, it was considered adequate to pay out the cable to the platform as required by hand. Power is supplied to the platform by energizing the main contactor using the start button on either of the control consoles.

The platform drive is controlled by means of a specially designed joystick which, by moving it forward or backwards, controls the forward or reverse drive and by rotating it in either direction steers the platform in the direction the joystick is rotated. The distance the handle is moved forwards or backwards from its normal position determines the speed and direction of the platform. A differential gear mechanism in the drive handle adjusts the position of the setting of two centre-tapped potentiometers which regulate the thyristor drive units controlling the speed of the left and right hand drive motors. A "dead man's handle" is incorporated so that the platform will stop if the handle is released.

In the event of the platform tilting to a position approaching an unsafe state of equilibrium the pendulum operated switch stops the motor and gives warning to the operator by an intermittent sound alarm and flashing light. There is an over-ride switch fitted to the tilt safety circuit and it is the operator's responsibility to use this to drive the platform in the correct direction to reach a stable condition.

The manipulator control handle, which incorporates a "dead man's handle", is designed to give a good ergonomic analogy of the operation of the manipulator movements. All of the seven movements of the manipulator are controlled from this single handle as shown in Figure 5.

Operation of a separate spring return switch in conjunction with the respective controls on the manipulator handle selects either fast or slow speeds for the waist, shoulder and elbow movements.

Conclusions

Experience has shown that there are two major problem areas in controlling teleoperators at the end of long cables. The first is in handling the trailing cable and the second is the information link between operator and teleoperator.

With ROMAN which is designed to operate in a speicifc area and always in sight of the operator these problems are not serious, but with teleoperators operating out of sight and in unknown situations at distances of up to 100 meters they will be severe.

Considerable progress has been made at Harwell in overcoming the information link problems and further work aimed at achieving cordless control is in progress.

With the increasing radiation levels which will be encountered in the Nuclear industry, the possibility of the lowering of maximum permissible radiation levels for radiation workers, and the growing social resistance to working in hazardous environments, there will be an increasing demand for teleoperators once the technical feasibility and reliability has been proven.

Acknowledgements

The authors are pleased to acknowledge with thanks their gratitude to their colleagues who have contributed to the design, construction and testing of the teleoperators. They are also grateful to Mr. G. W. Peagram, Head of the Project Engineering Group, Harwell, for his help and encouragement throughout the projects.

Figure 1

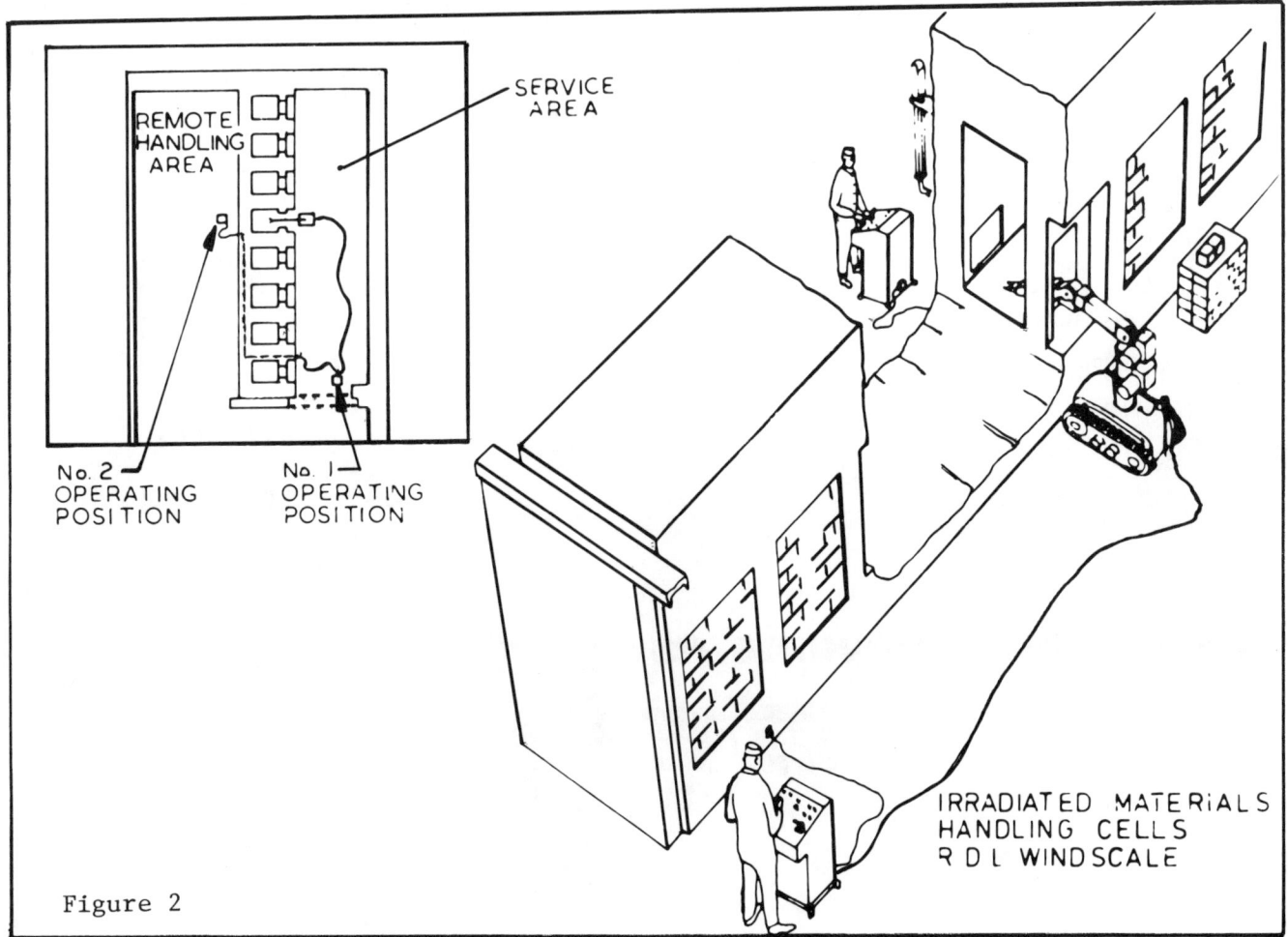

Figure 2

No. 2 OPERATING POSITION

No. 1 OPERATING POSITION

REMOTE HANDLING AREA

SERVICE AREA

IRRADIATED MATERIALS HANDLING CELLS
R D L WINDSCALE

227

Elbow action 270°

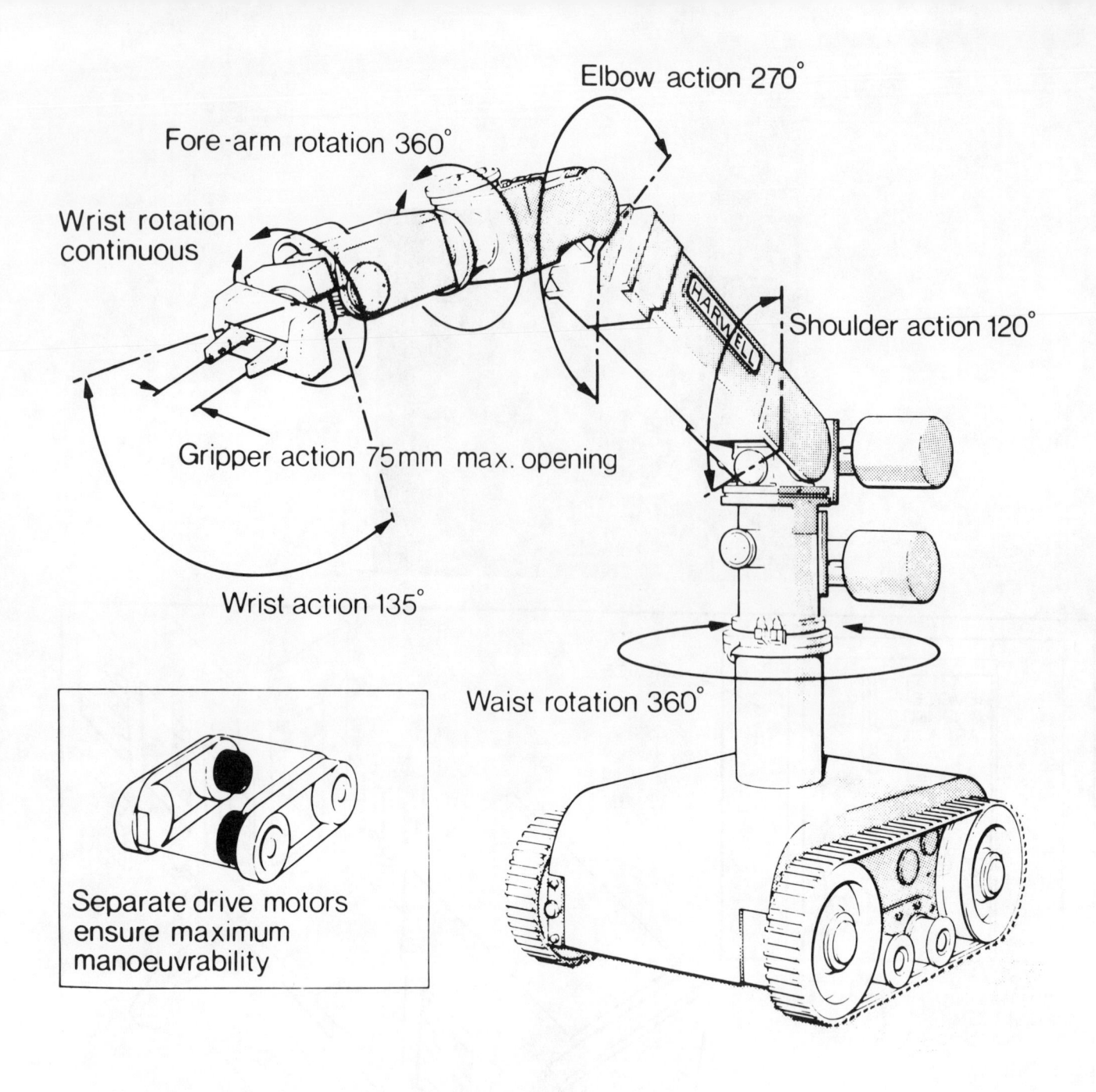

Fore-arm rotation 360°

Wrist rotation continuous

Shoulder action 120°

Gripper action 75mm max. opening

Wrist action 135°

Waist rotation 360°

Separate drive motors ensure maximum manoeuvrability

Manipulator movements

Figure 3

Figure 4

228

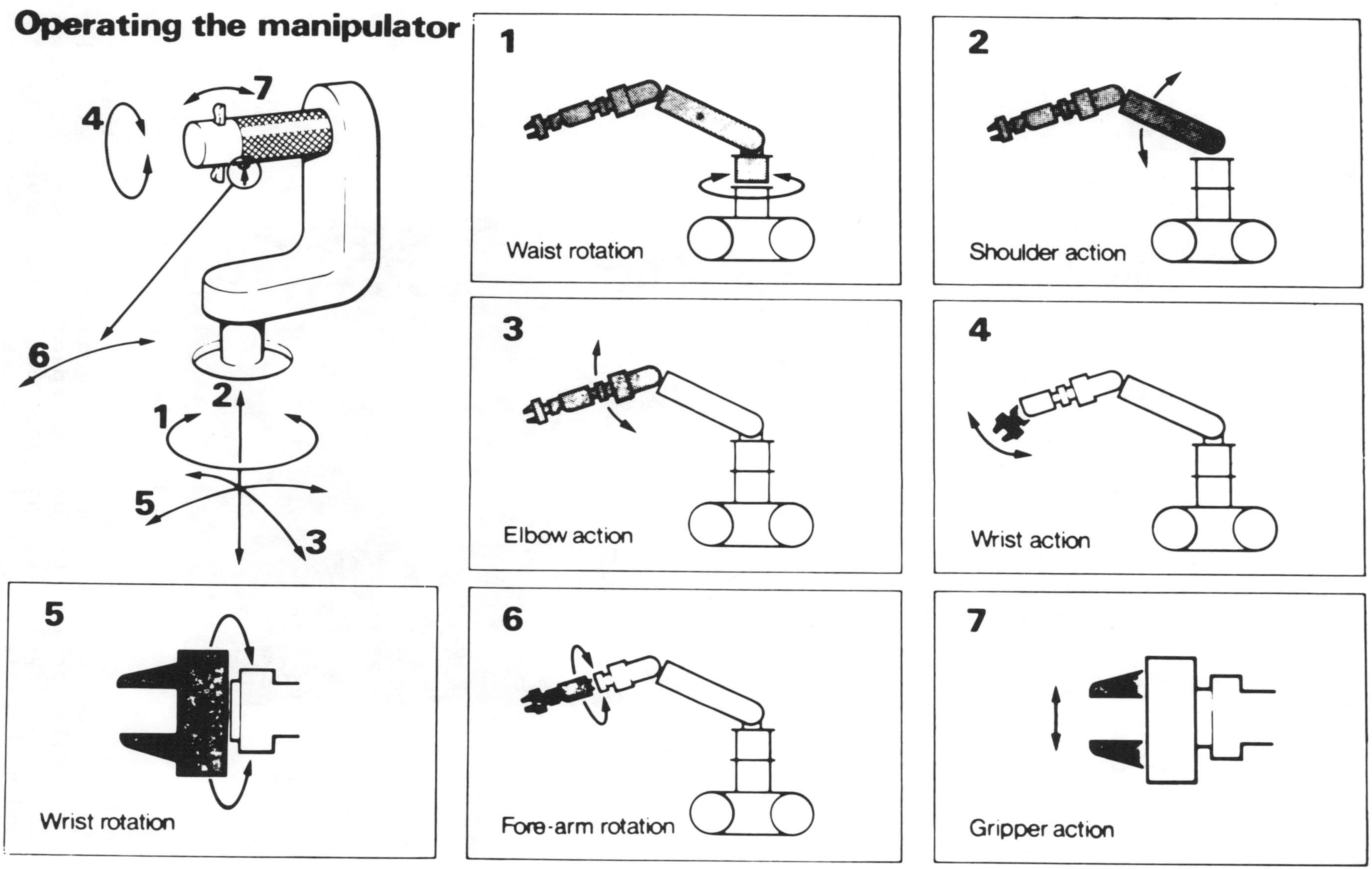

Figure 5

Presented at the RI/SME Robots West Conference, November 1984

Technology for Mobile Robotics in Nonmanufacturing Applications

by Thomas G. Bartholet
Odetics, Incorporated

ODEX I, the first functionoid, was recently introduced as the demonstration unit pictured in Figure 1 in order to exhibit revolutionary and unique capabilities in mobile robotics. As a functionoid, ODEX represents a new class of mobile robotics particularly adaptable to a wide range of functions, even those which cannot be fully planned in advance. ODEX is capable of providing remote access as a man-replacement in highly varied and cluttered surroundings without necessitating facility additions or modifications. As this paper will describe, this new set of robotics technologies for the first time offers agility with high strength, simple operation, and high reliability with ease of maintenance. As a result, product designs resulting from ODEX present new and major opportunities in the areas of operations, maintenance, inspection and surveillance to reduce human exposure to unsafe or unhealthful situations and to reduce costs in these areas.

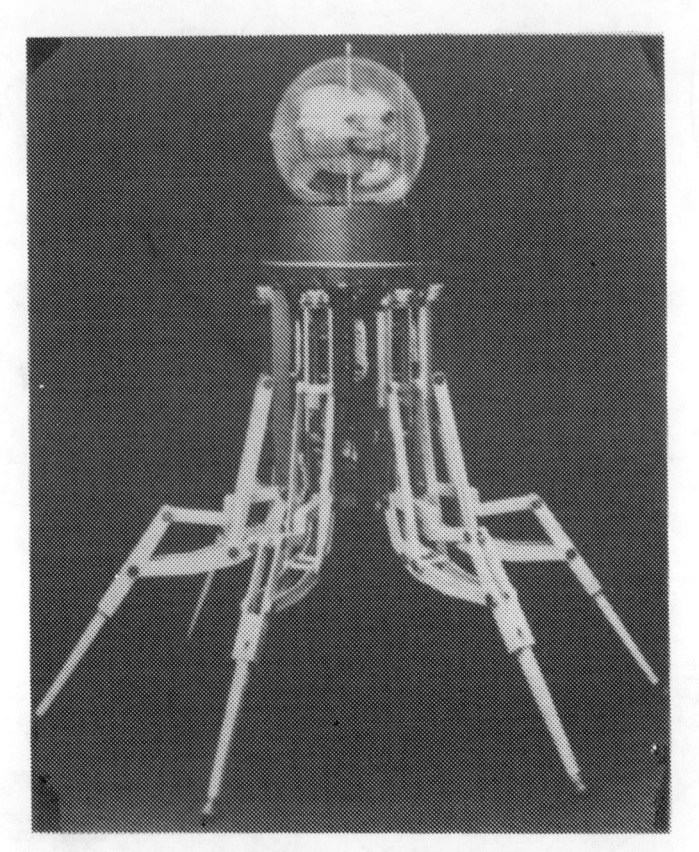

FIGURE 1 - ODEX I

The robotics advance represented by ODEX results from the extension of sophisticated technologies combining mechanical design with microprocessor controls for high performance spaceborne tape recorders. Odetics, Inc. has played a leading edge role in developing high capacity and high reliability tape recorders for the National Aeronautics and Space Administration (NASA), the Department of Defense (DOD) and the majority of the free world's space programs. Such tape recorders employ advanced techniques in servo-control, such as optical feedback methods. Variable tape speeds and positioning are controlled to tight tolerances in order to achieve the necessary performance and reliability criteria. Quality controls to NASA and DOD specifications in design, manufacture and inspection are integral to these activities.

From this technology base as a corporate resource, it was realized that an unexpected advance in mobile robotics was possible. Therefore, a combination of design objectives was established which, if accomplished, would clearly demonstrate the actual realization of such a major advance. These design objectives are listed in Table 1. Though ODEX I has thus been designed as a capability unit, a basis for product development was clearly in mind for applications ranging from military uses and rough terrain exploration to nuclear activities and mining, among others.

TABLE 1

DESIGN OBJECTIVES FOR GENERAL
DEMONSTRATION PURPOSES

- Advanced locomotion for walking
- Ability to traverse rough terrain
- Ability to go where humans go
- Omni-directionality for instantaneous change of direction in tight maneuvering
- Ability to climb and descend
- Ability to change walking profiles as well as height
- High strength-to-weight ratio as a multiple rather than a fraction
- Agility
- Stability as a platform for instrumentation and work packages
- Self-contained power
- Extreme power efficiency
- High controllability
- Ease of operations
- High reliability with ease of maintenance and repair

As the resulting demonstration unit, ODEX I has an impressive set of performance specifications. In this particular design configuration, ODEX can assume a variety of profiles from tall through squat to low profile, and from a compact tuck to narrow to wide widths. These are pictured in Figure 2. Associated performance parameters are listed in Table 2.

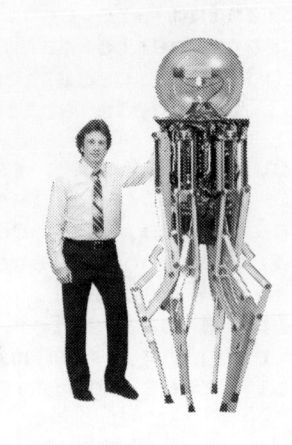

TALL

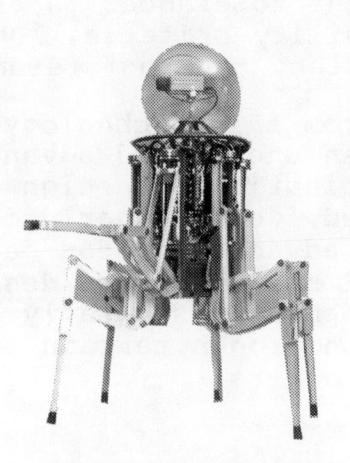

WIDE ARTICULATED

TUCK

NARROW

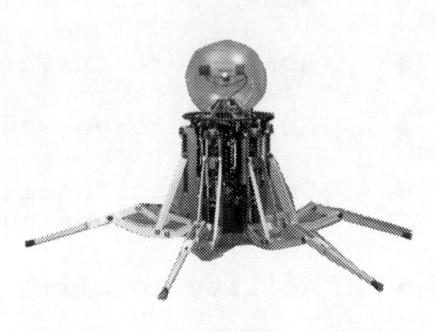

SQUAT

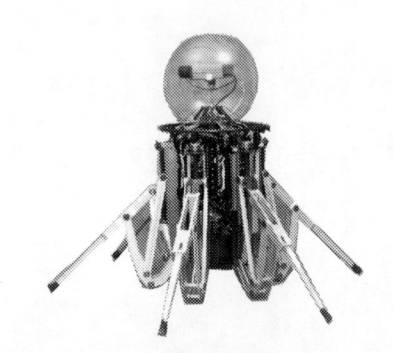

LOW PROFILE

NARROW ARTICULATED

FIGURE 2 - ODEX PROFILES

TABLE 2

PERFORMANCE PARAMETERS FOR
GENERAL DEMONSTRATION PURPOSES

- Height

Squat/low profile	36 inches
Tucked	48 inches
Tall	78 inches

- Width

Narrow	21 inches
Tucked	27 inches
Squat	72 inches
Low profile	105 inches

- Stepping height — 30 inches

- Maximum angle of ascent/descent — 45 degrees from horizontal

- Maximum speed (level ground) — 1.5 mph

- Weight (without payload) — 370 pounds

- Payload capacity

Lift (six legs)	2090 pounds
Strength-to-weight ratio	5.6X
Carry (slow speed)	1680 pounds
Strength-to-weight ratio	4.5X
Carry (normal speed)	960 pounds
Strength-to-weight ratio	2.6X
Single articulator	450 pounds

- Power Consumption

Normal walk	350-450 watts
Standby	2 watts

These particular performance characteristics are important only as a demonstration and should not be viewed as indications of, or even limitations on, performance capabilities for future product configurations designed for selected applications. Nonetheless, two examples of its performance range are shown in Figures 3 and 4, where ODEX is seen stepping into a pickup truck and also lifting the same truck.

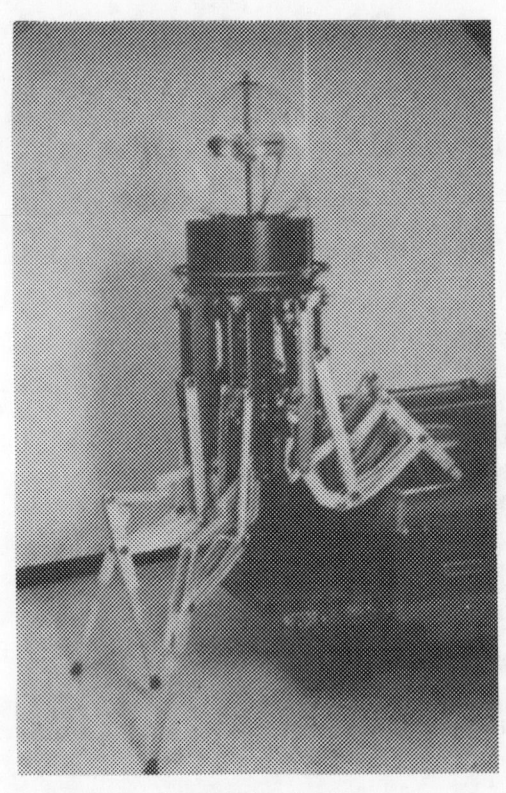

FIGURE 3
ODEX STEPPING INTO THE PICKUP TRUCK

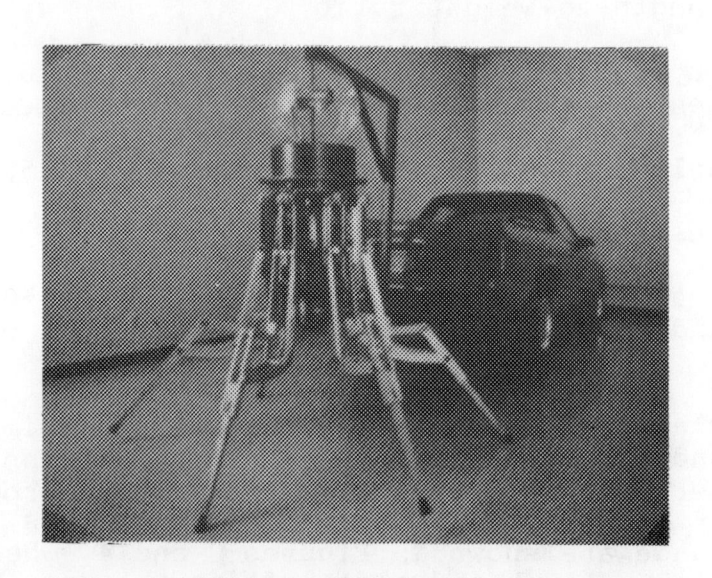

FIGURE 4
ODEX LIFTING THE PICKUP TRUCK

Integration of mechanical and electronic design in ODEX I is the secret to its success. The normal walking mode is based on an alternating tripod gait in order to combine strength, agility, speed and stability. Each of the six legs or articulators are driven in three degrees of freedom by three DC motors. The parallelogram construction of the articulators is one technique used to maintain strength throughout a wide range of positions and reach.

The drive mechanism of ODEX I consists of six identical and interchangeable articulator assemblies. Each assembly consists of one articulator, three drive motors with appropriate gear assemblies and one Level 1 computer. The Level 1 computer contains preprogrammed software to execute all the detailed motions involved in any articulator action needed to change profile and height, to walk, or even to act as a manipulator arm.

The six articulator assemblies with their Level 1 computers are coordinated and controlled by an onboard Level 2 computer. Again, preprogrammed software automatically selects and directs the intertwined motions and positions of the six articulators as a coordinated group. The onboard computing power with its advanced software makes possible the rapid, smooth and precise movement demonstrated by ODEX I.

As a result, the human operator of ODEX has a very simple control box (see Figure 5) to deal only with the highest level of commands. Operating ODEX is as easy as driving a car. Anyone, with a little practice, can do it. The operator's controls consist of:

- A joystick to set ODEX in motion or alter its path in any direction. ODEX references its direction from the location and orientation of the control box so the operator need not be aware of any other artificial frame of reference. Speed is a function of how far the operator pushes the joystick in the desired direction.

- A rotation knob to cause ODEX to rotate on its central, vertical axis. The direction and extent of knob rotation controls the direction and speed of ODEX's rotation.

- Separate knobs to allow the operator to vary the height, stride length, stance width and stepping height of ODEX anywhere within its performance range.

- A thumbwheel selector/indicator to choose the desired walking mode or articulator function. Even a crab walk to straddle obstacles is demonstrated as one possible variation. A step climbing function is available and the operator can take independent control of one or more articulators to use them as arms or manipulators.

The mechanical design and computing power of ODEX permits even further flexibility and versatility. Any of these commands to walk, to rotate, to change height or stride and so on, can be executed simultaneously or in combination. The operator simply activates those

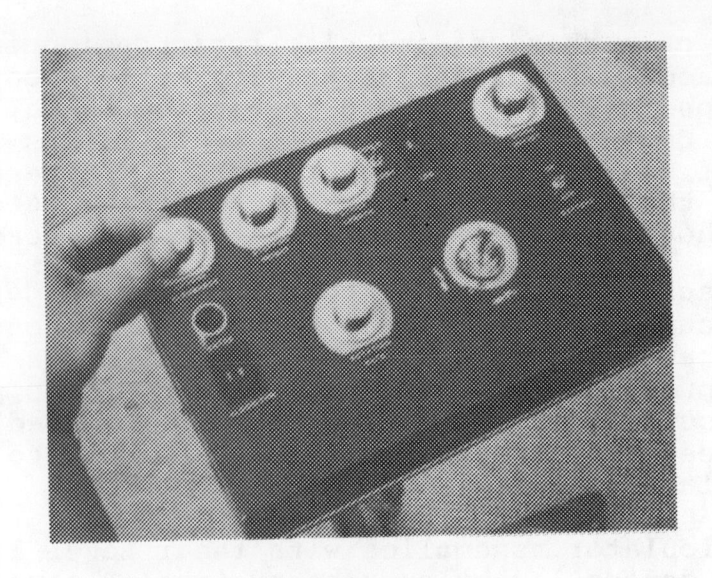

FIGURE 5
ODEX I CONTROL BOX

controls simultaneously or in combination. The onboard computers and software interpret these high level commands and superimpose the necessary, detailed commands to execute the combined motions desired. It's as simple as that.

ODEX I remains totally free to perform all these operations unimpaired by any cables for power or control. At present, all electrical power is provided by an onboard storage battery. Solely for demonstration purposes, we have used a standard 24-volt, 25-ampere hour aircraft battery. Based on this power supply, ODEX I can operate up to one hour without recharging. For longer, more power-hungry operations, alternative power sources could be used including larger batteries, generators or even fuel cells. If necessary, power lines from remote power outlets could be played-out from an onboard drum to minimize drag restrictions on ODEX's mobility.

Commands are sent to and information received from ODEX via high capacity digital radio links. Such digital techniques are readily expandable to cover data requirements for instrumentation or work packages which ODEX may carry. Alternative RF frequencies or wire and optical technologies are available to avoid any problems or concerns with electro-magnetic interference. To deal with transmission obstacles, RF repeater, wire, fiber optic, or direct optical techniques may be used.

There are many benefits resulting from these new robotic technologies combined and embodied in ODEX I. First is unprecedented load-carrying capacity. Prior to ODEX, the field of robotics spoke of strength-to-weight ratios as a small percentage. We now speak of strength-to-weight ratios as multiples and can increase them substantially over those demonstrated by ODEX. For the first time, mobile robotics providing the accessibility of ODEX have become feasible as remote vehicles and operators for practical work packages containing manipulators, tools, sensors, cameras and other instruments.

Additional major advantages of the ODEX technologies are the demonstrated versatility and flexibility. ODEX I represents a man-replacement and then some. It can go where humans go - through doors and narrow passages, on stairs and landings or into other tight spaces. To do so, few if any modifications to existing facilities or current procedures are required, unless desirable changes are made possible by the advent of ODEX. For example, since most nuclear facilities were designed for man access, ODEX-like vehicles can deal with them, however variant their designs and sizes. Also, since ODEX is so versatile and responsive to teleoperated commands, it can be used for tasks which were not anticipated or which cannot be totally planned in advance. Furthermore, an ODEX-style functionoid can be used for multiple tasks to gain full utilization for maximum effectiveness and efficiency. Robots of dedicated design, and therefore limited usefulness, will be superceded by fully adaptable functionoids like ODEX for many applications.

Another major benefit resulting from our technology-base and design-approach is high reliability with easy maintainability. As has been stated, the entire engineering and production base for ODEX is derived from a NASA and DOD environment. As such, quality levels and disciplines are high. Design, manufacturing and testing are fully documented and traceable. Components have been selected and designed for high reliability. Mechanical and electronic systems have been configured to achieve long mean time between failure (MTBF). Modular design with readily available components provides for easy and fast maintenance. Routine, scheduled servicing can assure realization of the design reliability.

Since ODEX I was introduced, several design enhancements have been added. These include:

- A gyroscopic based inertial reference system to sense the vertical direction with respect to gravity;

- Force threshold or contact sensors in the lower leg assemblies to locate terrain surfaces wherever they may be;

- Improved algorithms programmed into the onboard computers to incorporate the added inputs from the inertial reference system and leg sensors and to add more sophisticated walking capabilities for irregular terrain;

- A pan and tilt television camera system with endless 360-degree rotation capability as well as a tilt range from -90 degrees to +30 degrees.

Figure 6 shows the improved ODEX I in walking sequences on stairs and random ramp surfaces. The inertial reference system keeps the

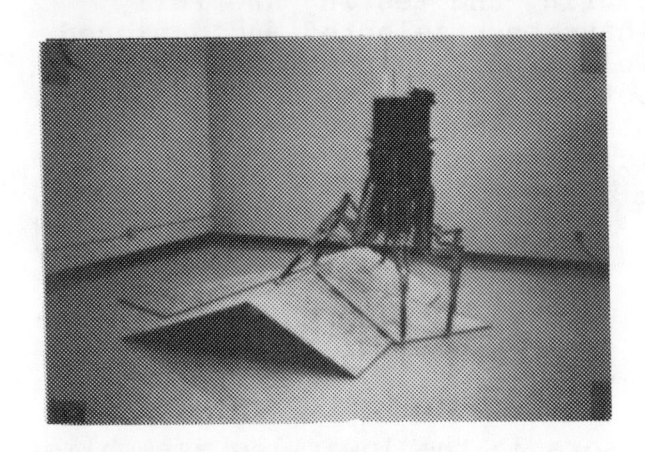

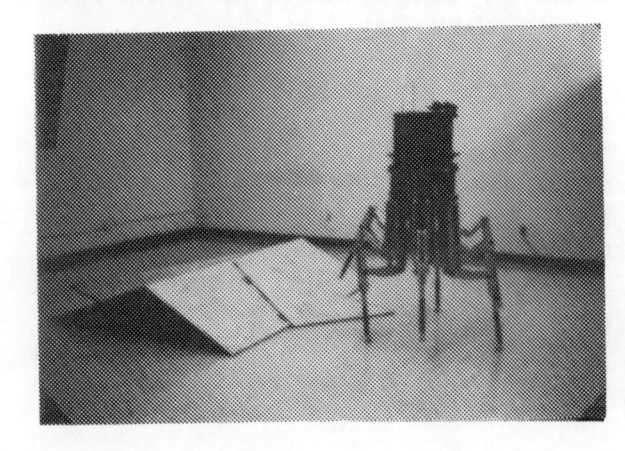

FIGURE 6

functionoid within two degrees of vertical at all times. Stair climbing and descending algorithms use the leg sensors to locate stair edges and, therefore, to find the correct foot placement for the desired direction of travel. ODEX has no prior knowledge of the stair dimensions and, in fact, each stair may have unique dimensions. Versatility is further assured in that the algorithms do not require a precise alignment of ODEX with the direction of the flight of stairs. In addition, ODEX can operate on curving stairs or on flat surfaces in this mode. As a result, transitions onto or off flights of stairs are smooth and stable.

On irregular terrain surfaces as simulated by the random arrangement of ramps, the improved ODEX is in a mode of operation which seeks to place each foot at a particular position to accomplish the desired path of travel. If a surface is contacted permaturely or later than expected due to surface irregularities, the ODEX then automatically decides whether or not this actual foot placement is satisfactory for the desired travel and whether or not it is within the stable envelope of the vehicle. If so, that foot placement is accepted and the robotic system adjusts accordingly in order to continue forward. If not, an alternative foot placement is attempted. If none can be found, the ODEX will stop until the operator can view the situation and choose an alternative direction of travel.

At no time does the remote operator have to concern himself with the actions of the legs. These are automatically and totally controlled by the onboard computers of ODEX. The teleoperator simply uses the camera system to view his surroundings and to drive the body of ODEX in the desired directions under appropriately selected modes of operation.

In the future, added sensors will further speed the operation of ODEX and will expand its versatility. Proximity sensors based on ultrasonics will eliminate the extra motions and time now required. Laser scanning techniques will provide computer-friendly 3-D information on the surroundings to further expand computer capabilities for increasingly autonomous operations. Work packages with manipulators, end effectors and appropriate sensors are being formulated.

As these technologies advance, mobile robotics for non-manufacturing applications will see a wide range of uses. These include:

- Maintenance robotics inside congested facilities such as nuclear power plants.

- Physical security patrolling of sensitive installations.

- A wide range of field logistics and material handling.

- Hazardous situations such as firefighting and explosive ordnance disposal.

- Military battlefield applications.

Presented at the American Nuclear Society 20th Conference on Remote Systems Technology, 1972

VIRGULE VARIABLE-GEOMETRY WHEELED TELEOPERATOR

JEAN VERTUT, JEAN-PIERRE GUILBAUD, GUY DEBRIE,
JEAN-CLAUDE GERMOND, and FRANÇOIS RICHE
Commissariat à l'Energie Atomique,
Saclay, France

KEYWORDS: remote vehicle, manipulator

ABSTRACT

VIRGULE is a French acronym referring to a radio-controlled rescue vehicle capable of operating over rough terrain either indoors or outdoors. The teleoperator is equipped with at least two TV cameras and one pair of MA22 master-slave manipulators, each capable of exerting 12-kg forces in any direction, and of lifting up to 30 kg in certain positions. The four-wheeled vehicle has more flexibility than a track vehicle: It can move in a straight line in any direction (referred to its own axis) and can steer about any center of rotation, controlled by a single control handle with three degrees-of-freedom. When the vehicle is moving, only one arm can be operated. A 24-channel digital communications link is used for control, and another one for feedback. These links may be radio or coaxial cable. Batteries provide short-term autonomous operation; power mains (indoors) or a gasoline-engine-driven generator (outdoors) are used for long operation.

HISTORY

The Virgule project was initiated five years ago at CEA to provide equipment suitable for emergency operations related to possible shipping accidents and accidents in nuclear facilities, either outdoors or indoors.

As a first step, a PAR I vehicle was ordered. It is now installed in a small truck, designated GMTT,[1] operational day and night to face any emergency. The truck works both as control center and as transport for the PAR unit. Several source recoveries in hot laboratories and irradiation facilities have been successfully executed. The most demanding operation we have done was in a chemistry laboratory, which consisted of dismantling an accidentally explosive circuit, and included cutting pipes under cooling by liquid nitrogen.

Concurrently, the second step, started four years ago, was the construction of a prototype test vehicle[2] (see Fig. 1). On satisfactory test operation, the third step, initiated two years ago, included the development of a new, improved vehicle with variable geometry, now completed, and a new compact servo master-slave manipulator,[3] Model MA22, to be completed in the Spring of 1973. The complete unit is to be assembled in 1974.

GENERAL CONCEPT OF VIRGULE I TELEOPERATOR

This work finds its place among other past or present research[4-8] on teleoperators for various

Fig. 1. First test vehicle completed in 1969. Dimensions: 1.4 m square × 0.5 m high (56 × 56 × 20 in. high).

environments. The general concept must lead to an operational unit, and also to serve as a test unit for system studies and for experiments concerning efficiency of teleoperators (e.g., time to reach a work zone and do a specific task). Efficiency has been the primary factor in selecting the following basic system components:

Manipulators

Manipulators[3] are bilateral, master-slave, force reflecting units. This family is known to approach the best time efficiency compared to others, due to both perfect combination of different motions and to force reflection.

Vehicle

Because of viewing system limitations, the drive system must be very flexible: An original analog computer is employed to provide individual, coordinated directional- and speed-control for each of the four wheels to allow vehicle travel in a circular path about any possible center of rotation, and also straight-line travel in any direction relative to vehicle axis.

Viewing

The main TV camera is placed between the two manipulators. Pan-tilt head control will be tested and compared with an automatic control servoed to tong position, by which one tong will be always visible on the display screen. It facilitates work, and for exploration the camera looks where the manipulator's finger is pointing and also gives a very good feeling of the environment. A secondary camera, directly oriented by the wheels, will be tested for vehicle driving.

Communication

For a short distance, up to 300 m (1000 ft), two miniature coaxial cables will be suitable. A container will deliver coiled cable as required by the moving vehicle. After operation, cable could be coiled down or disposed as waste. For long distance, and especially outdoors, a two-way radio link will be used. Modular construction allows choice, depending on environmental conditions, to get the best reliability and acceptable weight.

GENERAL DESCRIPTION

The vehicle has a fixed rear axle narrow enough to pass through standard doors (80 cm; 32 in.); the front axle is retractable to the same width (with limited stability) but is normally extended to its maximum width (120 cm; 48 in.) for maximum stability. The manipulators are mounted on a pivoting, articulated support, by which they can be closely retracted to provide maximum stability when the vehicle is moving over uneven terrain, or when the front axle is narrowed. When the front axle is extended (as in normal operation), the articulated support can position the manipulators to cover an extended

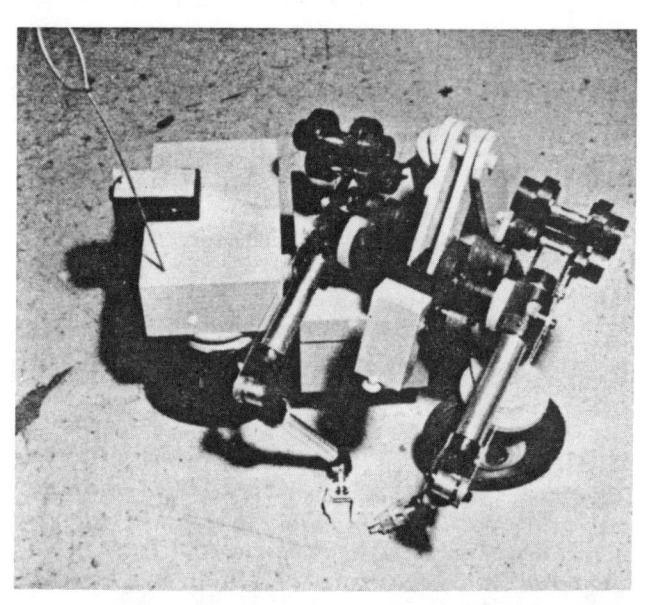

Fig. 2. Model of VIRGULE I, side view. Shown with support turret at right rotation limit and manipulators working at ground level.

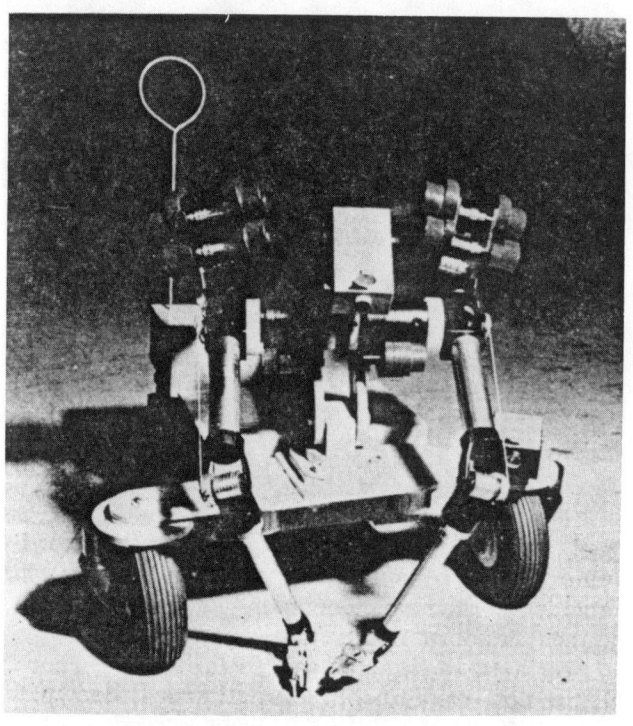

Fig. 3. Model of VIRGULE I, front view. Maximum width: 1.4 m (56 in.).

work volume up to a height of 2.5m (8 ft), and also to reach the ground at both sides and in front of the vehicle (Figs. 2-4).

Since the front wheels are independently retractable, the vehicle can be operated in an asymmetric configuration with only one wheel retracted.

On the central section of the vehicle heavy components as batteries (30 V, 40 A-h) are located. Power driving circuits for manipulators and vehicle are on top of them.

The tops of the equipment racks on center and rear are nice working surfaces for the manipulator proper, and also to fit tools and ancillary equipment.

The structure of the vehicle is rotatable around an axial tube (see Fig. 5) and weight repartition corresponds to this triangle stability disposition.

Fig. 5. Vehicle chassis with front wheels in normal (extended) operating position. Shown with right front wheel on 22-cm (9-in.) step to illustrate rough-terrain capability.

Provision is made to improve rear-wheel stability reaction by a spring load when the rear axle pivots.

When an extra gasoline-engine-driven power supply is to be carried, it will fit at the extreme rear. Nevertheless, it appears better to use another small vehicle to carry it, with the interesting feature of carrying an additional TV camera to look at the main vehicle. They could then plug in when they arrive at the work area.

VEHICLE DESCRIPTION AND DATA

The vehicle now completed is illustrated in Figs. 5-8. The total weight is 160 kg (350 lb).

The articulated disposition of the axles around the main tube is very effective in crossing bad obstacles (see Fig. 5).

The wheel unit (see Fig. 6) includes separate built-in motors for propulsion and for steering, using harmonic drives. Each wheel assembly weighs 28 kg (60 lb). Unlimited steering is possible by use of brushes, and orientation is given by a potentiometer. A shaft projects at top to fit any accessory supposed to follow the steering motion (TV camera or lights).

Front Axle Variable Geometry

The front axle is fitted with two pivoting arms (Fig. 6). An electromagnet (not shown in the picture) releases a spring-loaded lock pin.

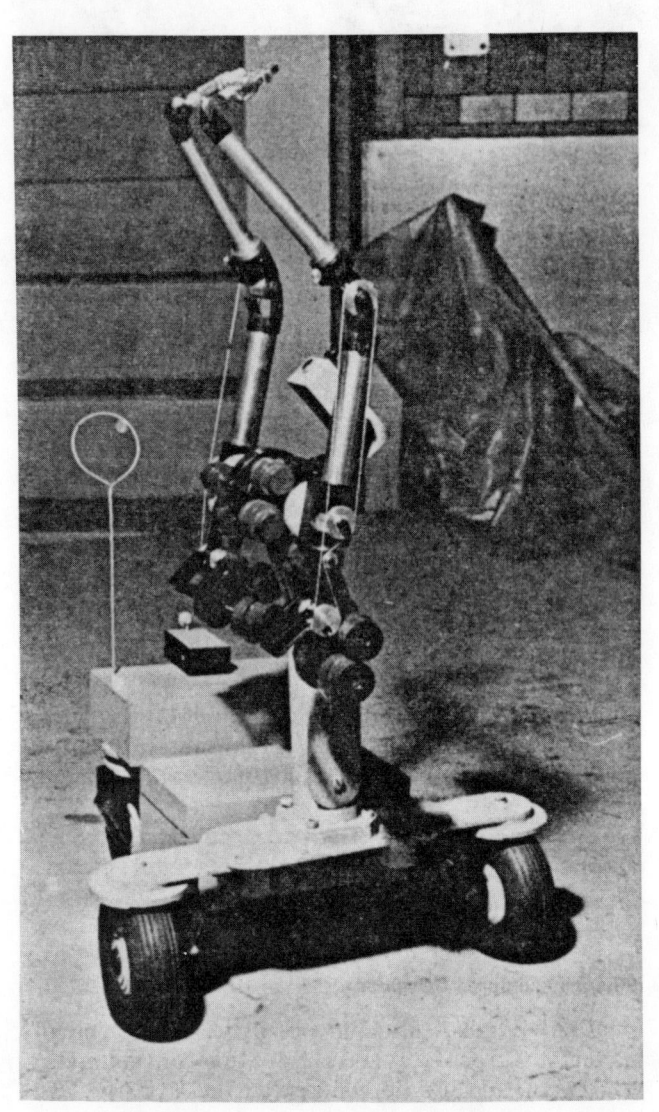

Fig. 4. Model of VIRGULE I. Shown with manipulators indexed for maximum vertical reach: 2.5 m (100 in.).

242

(a) (b)

Fig. 6. Retraction of right front wheel. (a) Normal, extended position, (b) Wheel driving arm into retracted position.

In Fig. 6a, the wheel is in normal position. To retract, the wheel is first steered perpendicular to the arm, and the lock release is activated; by just operating the wheel propulsion (as seen in Fig. 6b), the arm is pivoted and the lock will engage automatically. Then the wheel has to be again steered to the right direction. All must be done with the vehicle stopped. Figure 7 shows use of this to pass through a standard door.

Wheel Servo Drive and Obstacle Crossing

Steering and propulsion are controlled by servo drives—details on steering computation are given later. Effect of very low pressure tires and speed servo are additive for providing the extremely good obstacle crossing as demonstrated in Fig. 8.

The vehicle can climb steps 22 cm (9 in.) high—greater than the wheel radius. This is due to the good cooperation of all resting wheels when one meets the obstacle.

As visible in the figures, the regular small airplane tires could be improved for adherence.

STEERING CONTROL OF THE VEHICLE

Steering and Speed Computer

The original steering computer is a purely mechanical one as an application of geometry. The principle is demonstrated in Fig. 9. The basis is that all steering pivots must be located on one circle. Then wheel axes AO, BO, etc., must converge to O as the center of rotation of the ve-

243

Fig. 7. Vehicle with both front wheels in retracted position, passing through standard door.

hicle. From O, we are considering the diameter O M N. The center O can move along this diameter from infinity (vehicle goes straight) to any position. When O is outside the circle, it has a corresponding point P inside the circle called a conjugate point, satisfying the relationship: (PM/PN) = (OM/ON). Then the pencil AO, AM, AP, AN is harmonic, and AM and AN are the bisectors of angle OAP. So we mechanically converge small rods to P for each wheel, the rods pivoting with potentiometer shafts, to control steering of corresponding wheels with opposite angles. Then the wheel axes are converging to O as the rods are converging to P. When we get O inside the circle, we switch the servo loop to the same angle and we directly control O instead of P. So we have a continuous steering from infinity to the vehicle center about pivot points along diameter MN. This diameter also can be rotated, so the final computer enables the pivot point O to move into all possible positions, giving the possibility of going straight in any direction (see Fig. 10) and to steer about any point up to the vehicle center.

The computer also gives rotational speed for each wheel.

Of course the mechanical analog computer used in the control is a reduced-scale model of the vehicle geometry.

Control Handle

The computer is located at the control and the different potentiometer signals are transmitted to the vehicle. On the vehicle, servo drives are fed with these signals to control speed and steering of the wheels. A single handle with three degrees-of-freedom is used to control the whole vehicle (see Fig. 11).

The control handle is suspended below a small horizontal bracket attached to the top of a vertical post which in turn is movable in a fore-and-aft direction (Motion I), and rotatable about its vertical axis (Motion II). The handle can also rotate about the axis of the horizontal support bracket up to 90 deg to the right or left of its normal vertical position (Motion III).

Motion I controls the speed of the vehicle, forward and reverse (stopped at center), in a straight-line movement along the direction defined by the axis of the horizontal bracket, Motion II (see Fig. 10). Vehicle movement along a circular path is controlled by Motion III; the axis of rotation is determined by the intersection of a line perpendicular to the handle (in a vertical plane) with a horizontal base plane below the control. A combination of Motions II and III can place this axis of rotation at any desired point; at both (left and right) 90-deg extremes of Motion III, the axis of rotation is at the vehicle center. When moving in a circular path, the direction and velocity of the vehicle are still controlled by Motion I of the control.

MULTIPLEX COMMUNICATION SYSTEM

Communication Requirements

Control board to vehicle:

Manipulator position control	7 × 2 analog channels
Vehicle propulsion and steering	4 × 2 analog channels
Pan-tilt control	2 analog channels
Auxiliaries	20 to 50 on-off channels

Vehicle back to control board:

Manipulator feedback	7 × 2 analog channels
TV	2 or 3 video channels

The digital multiplexing unit fits into a standard $17\frac{1}{2}$-in. rack.

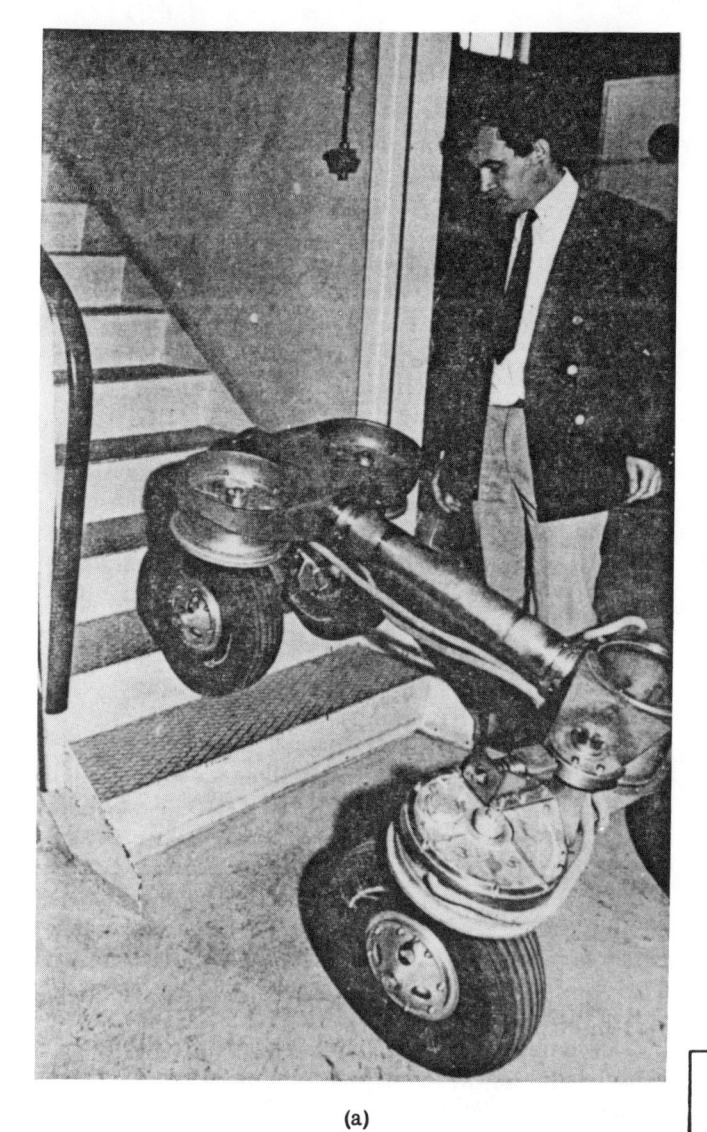

Fig. 8. Vehicle on staircase. (a) Rear axle on edge of step; extended front axle on level floor. (b) Climbing step with front wheels retracted.

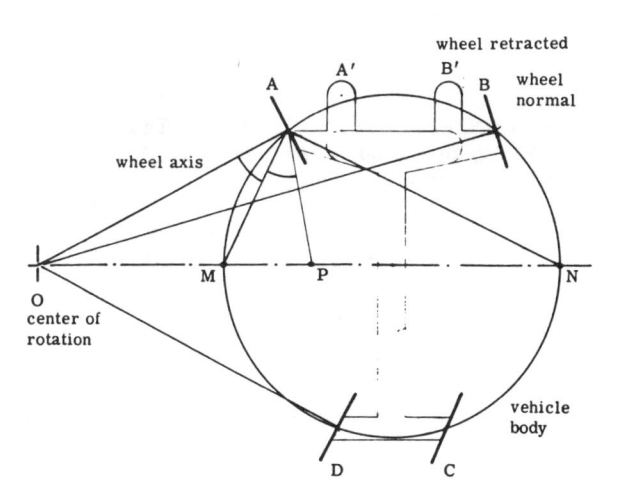

Fig. 9. Diagram illustrating vehicle steering geometry.

Fig. 10. Vehicle with wheels aligned for travel in straight line at 45 deg to vehicle axis. Control handle would be vertical and horizontal support bracket rotated to left 45-deg position.

245

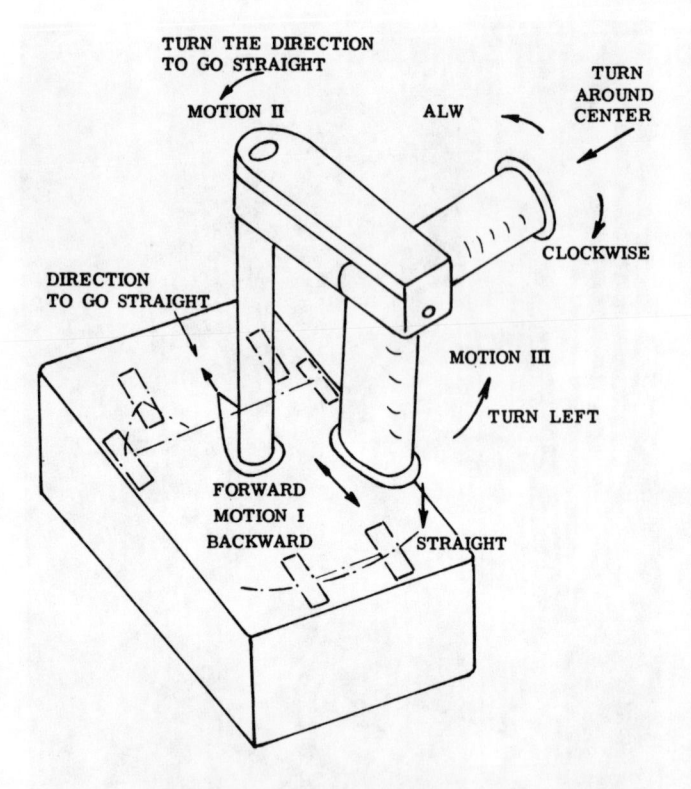

Fig. 11. Integrated control handle with three degrees-of-freedom.

Motion I: Selects forward or reverse, and controls speed.

Motion II: Establishes direction of straight-line movement (relative to vehicle axis).

Motion III: Establishes radius of curvature of circular path, variable from infinity (for straight-line travel) to zero (for rotation about vehicle centerline), and also sense of rotation (clockwise, as shown, or counterclockwise).

We have 16 continuous analog channels, and any 8 can be switched to get 24.

This means that we operate both manipulators, plus one TV, and can switch any of them to drive the vehicle. Each 24 analog channels have 4 on-off signals available for auxiliaries.

On the way back, only 16 channels for manipulator feedback are available, with corresponding on-off signals for indicators on the control board.

Other Special Features

A special check is effected to reject erroneous messages; if a message is rejected, the last message is kept in memory and there is no risk of the manipulator moving accidentally. Loss of communication stops any function. In particular, when one manipulator is switched off, all position references are still in memory and brakes are applied.

This unit can be operated either via coaxial cable or radio link. Up to now, experiments have been performed only with coax, one for control-to-vehicle, the other for feedback, and two additional for TV.

CONCLUSION

This teleoperator appears to be the first one including bilateral-servo, master-slave manipulators, and this development and test program will bring out much interesting data for all-purpose teleoperators for space or undersea application, especially in the exploration of teleoperator efficiency in different situations.

ACKNOWLEDGMENTS

We would like to especially acknowledge the following for their help. Messieurs Lantieul and Corfa from Societe SERI, Messieurs Bailleux, Bonneville, Fanton, and Picot from Ecole Superieure d'Ingenieurs d'Electronique et d'Electro Technique, and all the people in "Section for Equipments in Hostile Environment" at Saclay.

REFERENCES

1. J. VERTUT et al., "Rescue Vehicles for Radioactive Environment," *Proc. Intern. Conf. Peaceful Uses At. Energy, 4th, Geneva*, No. 617 (1971).

2. Colloquium on Developments in Teleoperators, University of Denver (1969).

3. J. VERTUT, C. FLATAU, et al., "MA 22—A Compact Bilateral Servo Master-Slave Manipulator," *Proc. 20th Conf. Remote Sys. Technol.*, 296 (1972).

4. E. G. JOHNSEN and W. R. CORLISS, "Teleoperators and Human Augmentation," NASA SP-5047, NASA Office of Technology Utilization (1967).

5. G. B. HOMER, "Mobile Manipulator Systems," *Proc. 14th Conf. Remote Sys. Technol.*, 129 (1966).

6. C. HUNT and F. C. LINN, "The Beetle—A Mobile Shielded Cab with Manipulators," *Proc. 10th Conf. Hot Labs. Equip.*, 167 (1962).

7. J. W. CLARK, "The MOBOT II Remote Handling System," *Proc. 9th Conf. Hot Labs. Equip.*, 111 (1961).

8. H. KLEINWÄCHTER, "Technical Description of SYNTELMANN," Report Q 320, Lorrach, Federal Republic of Germany (1970).

Presented at the 1983 International Conference on Advanced Robotics.
Copyright © Japan Industrial Robot Association

WALKING ROBOT FOR UNDERWATER CONSTRUCTION

Yoshitane Ishino, Technical Research Center, Komatsu, Ltd.
Toshihisa Naruse, Technical Research Center, Komatsu, Ltd.
Toshiyuki Sawano, Technical Research Center, Komatsu, Ltd.
Norikazu Honma, Electro Technical Research Center, Komatsu, Ltd.

2597, Shinomiya Hiratsuka-shi, Kanagawa Pref., 254 JAPAN

ABSTRACT

The "ReCUS" System, an underwater survey robot developed by Komatsu, Ltd., will be introduced as a practical application of robotics for critical works. "ReCUS" is an 8-legged walking robot which shows great mobility and stability on uneven grounds. Due to this feature, it is applicable to a wide diversity of underwater works.

1. Introduction

Robotizing underwater operations in ocean development has made progress especially in the offshore oil and gas resource development field, where many submersible robots have been working.

Of the submersible robots, approximately 90% are of tethered and untethered free swimming type, more than 90% of which are reportedly engaged in observation and video/photographic documentation. The tethered and untethered free swimming type robot is advantageous to inspect or make a survey of structures while making effective use of its three-dimensional maneuverability. On the other hand, it is difficult for the robot of this type to put out a significant force.

As compared with the free swimming type, the bottom travelling type robot is poorly manueverable. However, a very large working reaction is available from the robot. Therefore, the bottom traveling type is suited to the execution of an underwater civil engineering work. Conventionally, crawler type machines, e.g. underwater bulldozer, has been employed as the bottom travelling types. Its travelling capacity, however, is limited to a relatively hard and flat bottom of the sea.

Introduced herein is "ReCUS" (Remotely Controlled Underwater Surveyor), a seabed robot developed by Komatsu, Ltd., as an 8-legged walking type robot for underwater construction. This robot can even move on grounds that are too uneven of crawler machines. Besides, it is highly stable against a working reaction. And applying the "ReCUS" system to various underwater works will be discussed hereunder.

2. 8-Legged Walking Type Underwater Survey Robot "ReCUS"

2-1. Configuration

Given in Photos 1 and 2 are the remotely controlled underwater surveyor "ReCUS" which Komatsu, Ltd. has developed to examine the lay of rock bed for the foundation work of the underwater bridge piers in response to the request made by Honshu-Shikoku Bridge Authority. This system is equipped with TV and still cameras and ultrasound transceivers in the body. Remote operations by an operator on the mother ship permit the robot to take TV/photographic pictures and measure topographical features of the sea bottom while walking and moving with 8 telescopic legs. (See Fig. 1 and Fig. 2.)

Photo-1 "ReCUS" the underwater surveyor

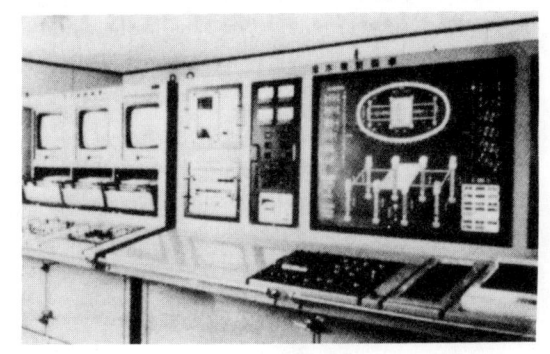

Photo-2 Control panel

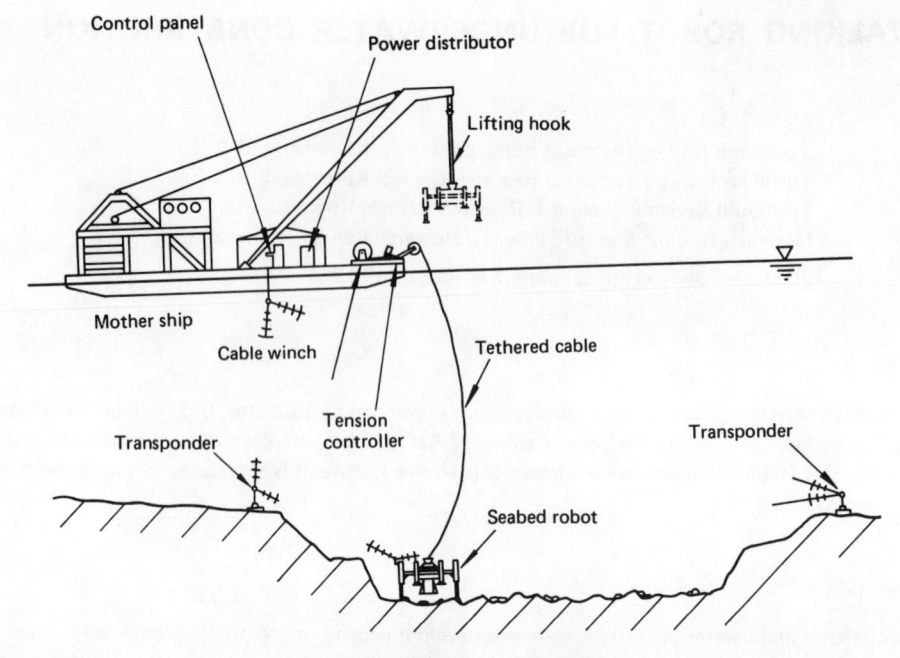

Fig.-1 Operation of ReCUS

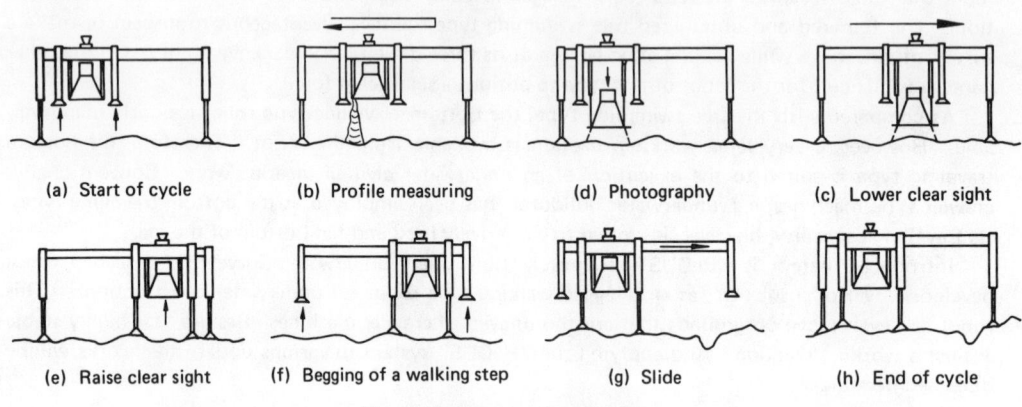

Fig.-2 Sequence of Inspection
(a) Start of cycle
(b) Profile measuring
(d) Photography
(c) Lower clear sight
(e) Raise clear sight
(f) Begging of a walking step
(g) Slide
(h) End of cycle

The reason why the 8-legged walking type robot has been employed is because high manueverability and stability on uneven ground can be expected, not with standing the fast tide undersea. The work site in which the robot was used is 70 meters deep with the maximum tidal current of 6 knots. Besides, the rock which has been blasted has an unevenness of as high as 2 meters. In such working situations it is difficult to apply another system, e.g. crawler travelling type robot.

In 1976, Komatsu started to study the 8-legged walking type robot mechanisms and their control methods by the use of a small machine weighing 1.6 tons. This study allowed us to have a footing for the practical use of the system. After that, we accomplished the 8-legged walking type underwater surveyor for the first time in the world in March, 1979. Table 1 shows the specification of the present robot system.

2-2. Construction

As shown in Fig. 3, the robot body has eight legs which can be independently extended and retracted with hydraulic cylinders. With both inner frame and outer frame made to slide to an fro, the robot is designed to be capable of walking and moving. At the same time, a turntable is provided at the center so that the robot can change its direction by swinging its frames. (See Fig. 4.)

Fig.-3 ReCUS Body

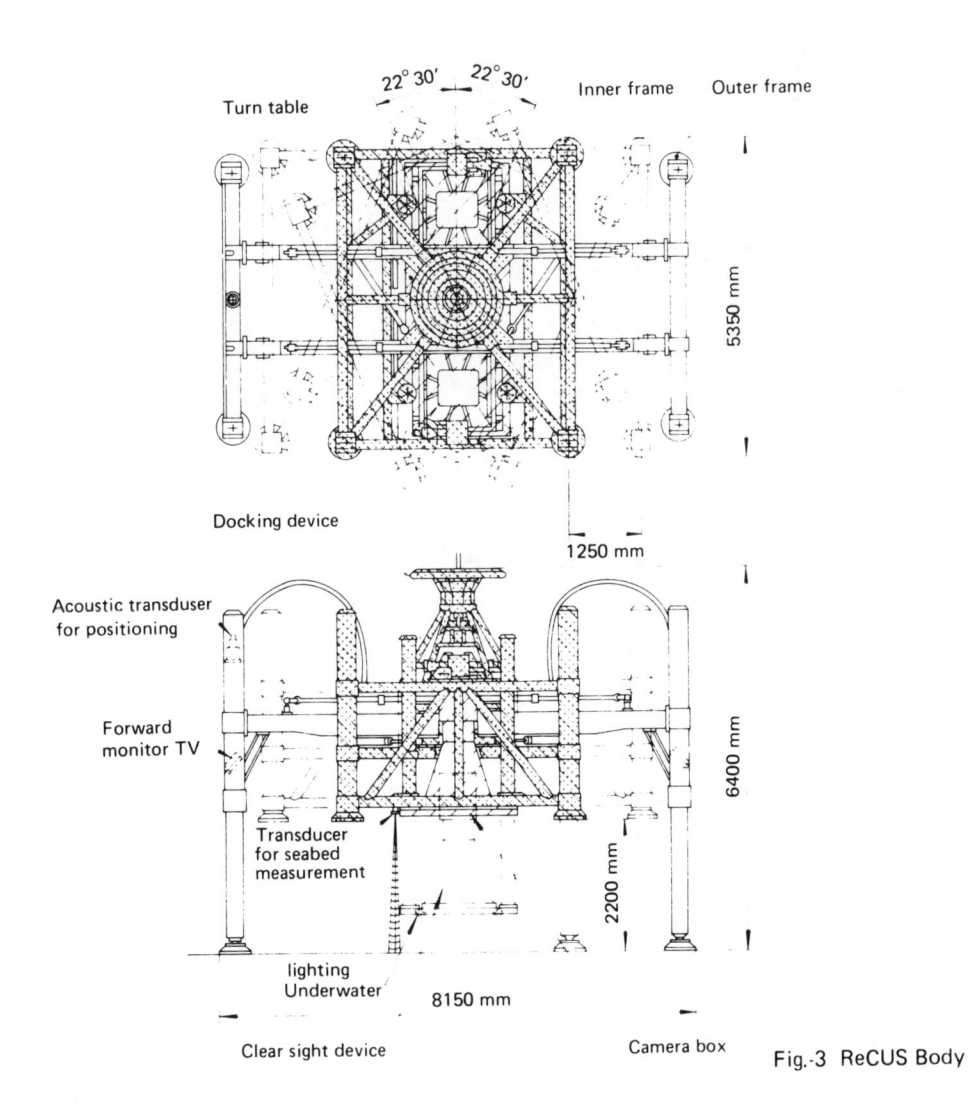

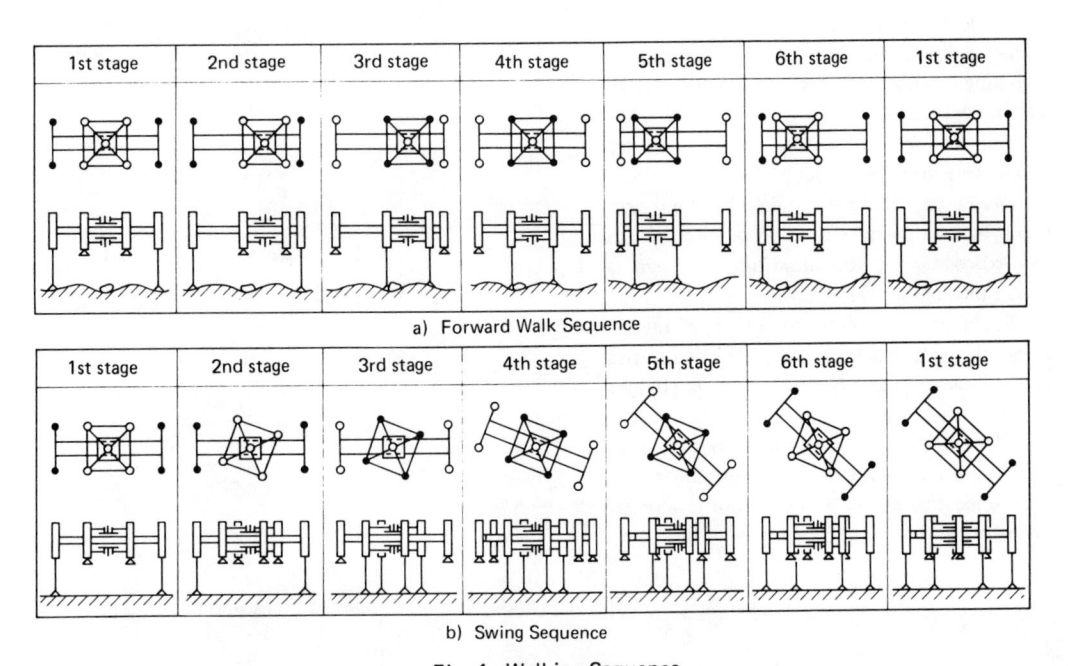

a) Forward Walk Sequence

b) Swing Sequence

Fig.-4 Walking Sequence

In the lower part at the center of the body, moreover, three sets each of TV and still cameras as the main components of the robot system, are provided, each with a clear sight device. The clear sight device is filled with clear water as shown in Fig. 5, so the robot can effectively take video/photographic pictures of an object in a turbid situation under the water. A docking device with a bell mouth shaped guide is provided at the center in the upper part of the body frame. This device has the function of automatically coupling the robot body with a lifting hook on a crane on the mother ship, which process used to be done manually by divers.

Walking of the robot is controlled by the use of a single board computer located in a control panel on the mother ship. It is mainly used for the walking sequence control in straight walks and directional changes and for the horizontal posture control. The walking sequence control covers forward/backward move instructions and directional change angle instructions to put the robot into automatic walking operation. The horizontal posture control, moreover, automatically regulates the landed legs' length so as to keep the robot body posture normally within ±1 degree by the use of a gyro.

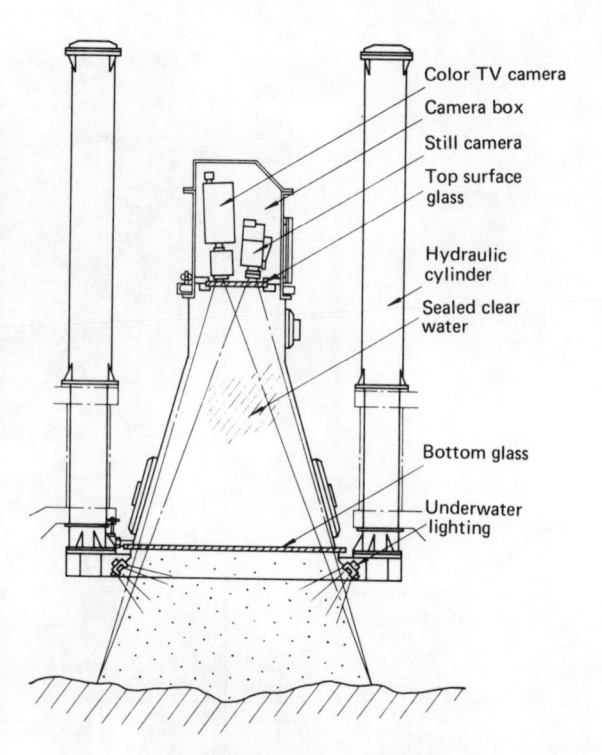

Fig.-5 Clear Sight Device

2-3. Performance

Photo 3 is the mosaic picture of the seabed where the bridge pier is to be founded, obtained by the robot. In other words, the picture was obtained by connecting by turns those photos taken by the robot while it was walking and moving on a 1 thru 2 m high uneven bottom of the sea after blasted with an explosive.

Conventional proximate photography by divers has only allowed photography in a narrow range. The photo taken by the robot, however, has permitted us to precisely grasp the exposed rock bed situations, etc. over a wide range on an overall basis.

Fig. 6 is a measurement graph of underwater topographical features. The figure has allowed us to precisely read an uneven height of the sea bottom.

For working efficiency, the robot requires a survey time of approximately 6 minutes per step while the survey area per hour is approximately 120 m².

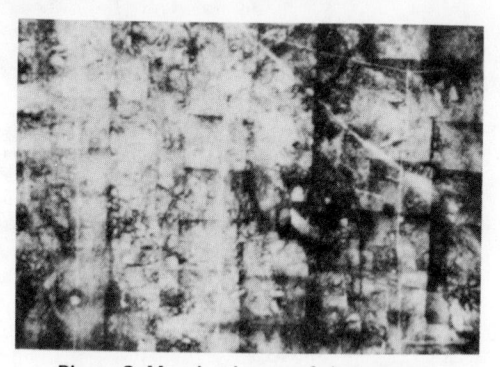

Photo-3 Mosaic picture of the seabed taken by "ReCUS"

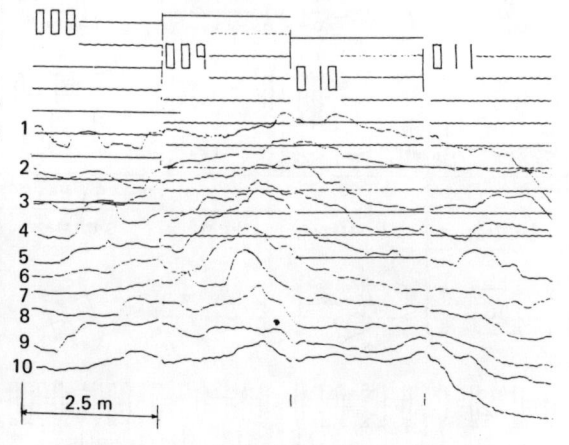

Fig.-6 Profile of Seabed Measured by ReCUS

Table-1 Specifications of "ReCUS"

• Maximum operating depth	70 m
• Tidal current operating restrictions	3 knots (in operation), 6 knots (stand by)
• Operatable range	Within 100 meters radius from the support vessel
• Allowable height difference of the ground	Within 2 m
• Underwater base	Gravel or rock bed
• Travel speeds	78 to 242 m/h
• Gradeability	20 degrees (in level position)
• Body weight	29 tons in air, 22 tons in water
• Machine dimensions	Length: 8.15 m, width: 5.35 m, height: 6.4 m
• Walking stroke	2.5 meters/step
• Leg cylinder stroke	2.2 m
• Swinging angle	±22.5°
• TV/still camera photographing angle (with 3 cameras)	1.1/1.5 x 4.5/4.7 m (500 mm ahead the clear sight glass)
• TV camera	3 color TV cameras, 2/3", power lens: f = 9.5 to 95 mm
• Still camera	3 cameras, motor-driven Leica size, film capacity: 800 pcs., Lens: f = 24 mm
• Monitor	18 inch color monitor x 3
• VTR	3/4 inch U-matic
• Underwater lighting	500 W halogen lamp, 40 lamps
• Topographical measuring system	400 kHz ultrasound pulse echo scan, 10 sections
• Position measuring system	Ultrasound triangulation constructed by the robot and 2 transponders, 60 kHz, measuring range: 20 to 300 m in distance, Recorder: XY plotters
• Signal transmitter	Binary code, 6100 bound, base band transmission
• Computer	PDP 11/03 memory 56 K bytes
• Drive system	Electrohydraulic drive
• Submersible motor	3 phase induction motor, 37 KW/440 V
• Tethered cable	200 meters long, diameter 62.2 mm

3. Application to Various Underwater Operations

3-1. Rubble Leveling Robot

The sector in which mechanization has most lagged behind among the work execution techniques in the harbor/port construction may be deemed as leveling the rubble mound for a breakwater footing. The request for mechanization of the rubble leveling work which is currently being executed manually by divers is very keen among the concerned to increase the working efficiency and to secure safety in operations at a large water depth.

Therefore, various methods have been studied and developed to mechanize rubble leveling. Even at present, however, it may be safely said that a definitive method superior to the divers' work has not been still found out to level the rubble for a break water footing.

A conventional method, e.g. the method of tracting a soil-moving plate with a working vessel or underwater bulldozer, however, involves difficulties in correctly controlling the soil-moving plate, because the vehicle body rolls and pitches under the influence of waves and uneven rubble mounds. As a result, it was impossible to accurately level a rubble mound.

An application of the 8-legged walking type underwater robot to the rubble leveling work, however, seems to be very promising, considering that the robot is highly maneuverable and stable on uneven ground and has the function of allowing the body to be adjusted very precisely to reference leveling plane. Thus, the robot may be considered as a solution to the problems mentioned above.

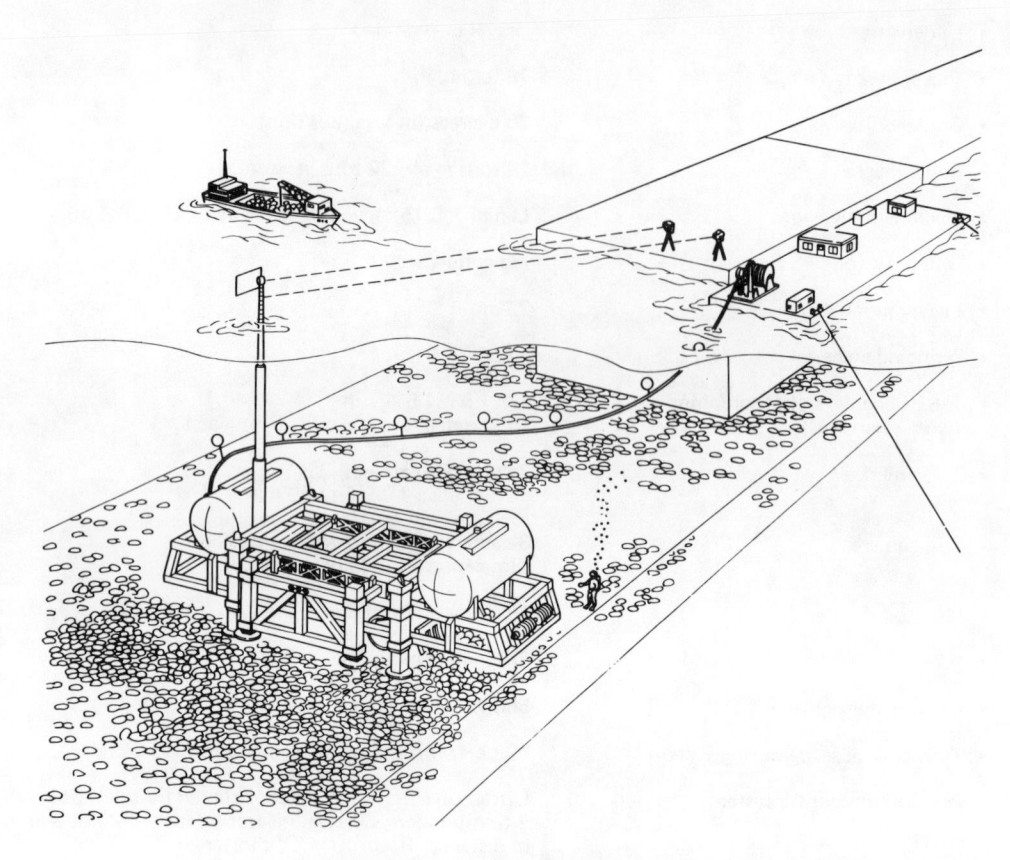

Fig.-7 A Scheme of Rubble Leveling Robot System

Given in Fig. 7 is a concept of the rubble leveling robot system. In other words, the body is solidly and flatly supported in place with four legs. With working bogie horizontally fed on the main frame as rails, a rake is used to push out the riprap forward. After that, the surface is finished by means of roller. This work is executed by turns while the robot is walking and moving so that the rubble mound can be entirely accomplished.

3-2. Application to Other Underwater Operations

To make the most of its features, the 8-legged walking type robot is considered applicable to the following underwater operations:
(1) An underwater excavator equipped with rippers or rotary cutters.
(2) An underwater drilling machine equipped with drills.
(3) A pipeline trench digging machine provided with dredgers and cutter pumps.
(4) A handling robot with manipulators for maintenance of underwater structures, etc.

4. Conclusion

As an example of robotics for critical works practically applied, the 8-legged walking type underwater survey robot "ReCUS" has been introduced.

This 8-legged walking type underwater robot has the following features:

(1) Since it can clear an obstacle, the robot may maneuver at a sharply uneven location where a crawler system, e.g. an underwater bulldozer, has difficulties in travelling.
(2) With the body supported by means of four legs, the robot can control its posture horizontally so that it may work stably against a tidal current resistance or working reaction.
(3) The robot capable of travelling forward well and moving at a constant pitch permits a work to be executed regularly.
(4) The robot is designed so that all sorts of working equipment can be attached to the body chassis. Therefore, it is highly applicable to a variety of heavy operations.

To meet the increasing demand for operation in the depth, developments of underwater robots are making rapid progress, thanks to recent advance in electronics. From now on, we will increase the applicability of this robot system so that it can work at a far deeper location and carry out a far more sophisticated operations.

Finally, we would like to express our highest appreciation for the kindest guidance given by the concerned in Honshu-Shikoku Bridge Authority in our developing the "ReCUS."

REFERENCES

1) "Remotely Operated Vehicles" U.S. Dept. of Commerce National Oceanic and Atmospheric Administration.
2) Yoshinori Kudo, "Remote Controlled Underwater Surveyor" Construction Machinery and Equipment, Vol. 16, No. 5.

APPENDIX

Reprinted with permission from Argonne National Laboratory, Copyright © 1951

Philosophy and Development of Manipulators*

R. C. Goertz
Argonne National Laboratory

ABSTRACT

Since radioactive materials are harmful to the body, it is necessary to shield scientists and workers from these materials. Other than for alpha emitters, the shielding generally consists of walls suitably thick and heavy to attenuate the radiation to safe levels. It is then necessary to carry on experiments and operations behind this wall (generally an enclosure). Manipulators are needed to handle the radioactive materials and apparatus. This paper will concentrate on general purpose manipulators--those that are capable of a large variety of manipulations.

The basic requirements for such a manipulator are examined from the point of view of the manipulations to be performed and from the point of view of the operator, including the cybernetics of such a system. These basic requirements for a good general purpose manipulator are believed to be:

(a) The manipulator must have a minimum of seven degrees of freedom arranged so that an object can be gripped, placed at any point in a specified volume, and be oriented at any given angle.

(b) Controls must be provided for each of the seven motions so that the manipulator can be controlled from the opposite side of the wall. It is preferred that these controls be arranged so that the desired response occurs from a natural movement of the controls.

(c) Means must be provided for the operator to clearly see through the shielding wall. Large thick windows are suitable for macroscopic work.

(d) Means should be provided to indicate the loads being applied to each of the seven degrees of freedom. It is believed that the best way to do this is to reflect all or part of the load back to the control handles.

The viewing system is generally not part of the manipulator as such, but should be considered as a link in the manipulation circuit.

An example of a manipulative system embodying all of these basic requirements is the master-slave manipulator Mod 4 mounted in a shielded enclosure (cave) with a large thick window.

With the advent of reactors there have been introduced many new problems of handling radioactive materials by some means of remote control. Since many radioactive isotopes are produced and since products, including these isotopes, either in their pure form or as impurities, must be tested and handled, new methods for shielding personnel from the dangerous radiations must be provided and handling devices must be designed and constructed. These handling devices must quite often work through shielding barriers or by some remote control means and the variety of such handling devices is very large. Since this is true and since this paper is limited in time and

* Proper illustrations were not available upon publication and have therefore been deleted from the text with permission of Argonne National Laboratory.

scope, it will be primarily concentrated on a more specific type of handling device commonly called a manipulator. Being even more specific, the remarks will be still further confined to the discussion of the basic requirements for a "general purpose manipulator." A broad definition of this general purpose manipulator is that it performs the functions that would normally be performed directly with the hands of the scientists and operators if the materials were not radioactive. It must be pointed out that even the best manipulators at present do not even approach the dexterity and manipulative ability of the human hand. Therefore, to be more specific about the definition, a general purpose manipulator can be classified as a manipulator for performing operations on radioactive materials such as moving them from place to place, turing and orienting them at different angles, placing them in various kinds of apparatus and manipulating some of the apparatus used on the radioactive materials.

In order to design good manipulators, several things have to be taken into consideration. In general, the first of these is the function or work that must be performed and the second is how the manipulator is to be controlled and operated.

Function of the General Purpose Manipulator

In this paper the general purpose manipulator will be defined as a manipulator that does a wide variety of functions. These basic functions can be enumerated as gripping an object, moving it from place to place within a confined volume and orienting it at any given angle. The size and weight of the object to be carried and forces that the manipulator can exert are not important to this discussion but will be assumed to be in the vicinity of 0 to 20 pounds--something that the hand could do unaided on non-radioactive materials.

It is necessary for the manipulator to carry out and perform these functions while it is on one side of a shielding wall and the operator is on the other side of the wall or at some considerable distance. It will be assumed that the manipulator is within a shielded enclosure and the operator is just outside this enclosure.

To perform the above functions, the manipulator must have three degrees of freedom of displacement, three degrees of freedom of rotation, and one degree of freedom for controlling or closing of the grippers or tongs. The minimum number of independent motions is therefore seven. The manipulator can, of course, have more than seven degrees of freedom, and the number of degrees of freedom can be increased without limit as long as suitable linkages and controls are provided. In some cases it is desirable to have more than seven degrees of freedom in order to essentially change the shape of the manipulator. In this case the degrees of freedom are not independent but could be reduced to seven in the vicinity of any specific position and orientation. On the other hand, there can be additional degrees of freedom which are independent. These might be in the form of having individual fingers instead of simple grippers or tongs. The human arm and hand can certainly be classified as a general purpose manipulator; it has somewhere between twenty and fifty independent degrees of freedom.

The Human Being as an Operator

Since the general purpose manipulator is a manually controlled device, it is necessary to include all the links required in the manipulative circuit. These can be enumerated somewhat as follows: (a) the operator wishes to execute some particular function, (b) he then moves the controls in such a manner that the manipulator will perform this function, (c) during this time he will observe the action of the manipulator and the materials and apparatus being manipulated or worked on, and when he is satisfied that the desired operation is complete, he will move the control handle back to its natural position and/or stop the manipulator from further action. There is one thing that has been left out of this circuit that is very important to many manipulations--a means of informing the operator as to the loads and forces he is applying on the object or apparatus. Examining the entire reactions and motions through the complete closed loop, it is found that the operator executes a command by moving his hand, gets a signal back through his eyes, and preferably gets another signal (feel) back through his hand. Certainly this is true when the human being uses his own hand as a manipulator.

From this brief discussion of the steps and circuits that are involved in executing a manipulation, several things become quite obvious. The first of these is that the operator must be able to control the manipulator, the second is that he must be able to see the material and apparatus, as well as the business end of the manipulator, and the third is that he must be able to sense the forces that are being exerted. Each of these important steps will be discussed separately later.

It might be well to point out that some of these important links to be incorporated into the design of a good manipulative system were reached in stages that spread over several years. Some of the early manipulative systems did not include all of these features and many of the present systems still lack some of them. A Remote Control Group was started at Argonne in 1947 at which time there existed some rather crude and simple manipulators that worked over a shielding wall. Mirrors and periscopes were used for viewing. It was evident that this viewing system was wholly inadequate if good manipulation were to be achieved. It also became apparent that a good control system must be devised or long training periods would be required for good operation. An electrical manipulator was completed that year similar to the rectilinear electric manipulator, and it was found that the manipulator performed all of the desired functions but had such a high rigidity that it would tend to damage materials and apparatus with which it worked. It became evident that some means of measuring forces in all of the seven degrees of freedom was required if smooth and delicate manipulations were to be performed even when working with ordinary materials and apparatus. The design of the master-slave manipulator and of the large thick windows were started about that time.

In order to design a good general purpose manipulator, it is necessary to study the human being, first, using his own hands as manipulators and second, using his hands to control mechanical or electromechanical manipulators. The dexterity and controllability of the human hand as a manipulator is amazingly good, and it is important to understand why it is good and under what circumstances it is good. First of all, it is important to understand that the human mind controls all of the useful motions of the hand and that it does so through the nervous system by commanding the muscles to contract or relax and that there are two other bits of information that the mind uses--sight and feel. The mind initiates the command for the muscles to drive the hand to the desired position and uses some sort of computing mechanism to automatically position the hand to approximately the desired position. This can be illustrated by simply observing a pencil on the table, then closing the eyes and picking up the pencil and placing it in some new position. Under these conditions it will be noticed that the hand did not grip the pencil precisely on dead center nor did it place the pencil in exactly the spot that was expected. If this same function is repeated with the eyes open, the accuracy of picking up the pencil and placing it in the desired position is much improved. This characteristic has been explained on the basis that when the eye continuously views the action, the brain recomputes the necessary further motion required to accomplish the desired result. The computing mechanism in the brain has been quite highly developed. If one studies small children and the training they receive through their lifetime, it will be noted that the degree of controllability depends upon experience and learning. For this reason it is important that the controls of the manipulator include the use of all this stored up information in the brain.

Another part to the over-all dexterity and controllability of the hand is the sense of feel and the fact that the arm and fingers will comply or yield when forces are exerted upon them greater than the forces that they initiate. In other words, taking the arm as a manipulator, it is a system that positions by direction from the brain, feels the load by a nerve system transmitting this load to the brain and yields to overloads. These characteristics are very important and should be considered in the design of manipulators.

Although the number of functions that are involved in simply picking up a pencil and placing it in a new position are quite complicated and large in number, the total process takes place at rather high speed, and the large number of functions and information is therefore not a burden but of tremendous value in performing the desired function. Of course, if these things did not take place simultaneously, awkwardness would result. Similarly, in the design of a manipulator, if many of the features of the human hand and arm are included, they must respond simultaneously and at rather high speeds or the system will be awkward.

Control of a General Purpose Manipulator

Since the general purpose manipulator being discussed here is one that performs its functions within a shielded enclosure, it is necessary that the controls for each of the motions be brought out so that the operator may activate any or all of these motions as he desires. This might also be included as part of the definition of a general purpose manipulator since it means that the manipulator responds to the commands of the operator only at the time the operator gives the command and does not perform any functions in an automatic fashion. In other words, the manipulator is a manually controlled manipulator having no automatic characteristics other than limit stops, overload indicators, etc. It might be noted at this point that the manipulator has been described as a manipulator which essentially opens and closes a pair of grippers and moves and orients them at the command of the operator. It will be noticed that these functions are essentially mechanical. Of course, the controls for the manipulator are also mechanical since they must be operated by the hand. To sum this up, the input or controls for the general purpose manipulator and the output of the manipulator are mechanical. These mechanical motions and forces can be transformed into different forms of energy for transmitting the information from the control to the manipulator or vice versa. For example, a manipulator that is driven by electric motors can be controlled by switches at the control box.

In the case of the electric manipulator, the operator's hand closes a switch which sets one of the degrees of freedom into motion at a constant velocity. In the case of the master-slave manipulator, the motion of the handle drives the manipulator in exact positional synchronism for all degrees of freedom. This illustrates two different methods of controlling a general purpose manipulator, each having exactly the same degrees of freedom, one being controlled in velocity and the other being controlled in displacement. It is probably not possible to say which of these systems is the most desirable for in different circumstances the emphasis would shift from one method to the other; for example, it has been found that the master-slave manipulator is much easier to control than the electric manipulator as long as the displacements are within arm reach or a little more. On the other hand, if the manipulator were to move over a large volume, it might become extremely awkward for the control handle to move over a similar volume. It is quite possible that a combination of displacement and velocity control would give good controllability.

Another feature of the control of a general purpose manipulator has to do with the arrangement of the control handles or switches and whether or not all degrees of freedom are activated simultaneously or separately. Referring again to the rectilinear electric manipulator, it is seen that there is an individual switch for each individual degree of freedom. On the other hand, the master-slave manipulator has essentially one handle with a pair of movable levers. The handle of this manipulator is so arranged that six of the degrees of freedom are activated simply by moving the handle in the desired direction and rotation. The seventh degree of freedom is activated by closing the thumb and finger on the movable levers. This motion closes the jaws on the slave end of the manipulator. The master-slave manipulator is considered to have the advantage of having all seven degrees of freedom activated at one time and so oriented that the control of these motions is executed by simply moving the hand in the direction and rotation that the operator desires the tongs to move.

Visual Systems to be Incorporated in the Manipulative Link

We live in a world of three dimensions of space, and the human being is very fortunately equipped with two eyes so that he can better observe all three of these dimensions. Since a good manipulative system must not impair the operator, it is necessary for him to see his work in three dimensions as he would normally see it with his unaided or unobstructed vision. Of course it is, at the same time, necessary to provide shielding between the work being done and the operator. One successful means of giving a fairly good three-dimensional view through a shielding barrier is by using large thick windows. The metallurgy cave MOD 2 equipped with master-slave manipulators MOD 4 show a laminated glass window 4-1/2 feet long, 2 feet high and 3 feet thick in the shielding wall of the enclosure (cave).

It is also possible to get three-dimensional viewing by using binocular periscopes or mirrors of sufficient size to include vision for both eyes. With either of these systems there is one important thing to bear in mind, and that is that the objects seen should be oriented in their natural way and not distorted nor inverted, nor reversed. It has been conceived, and properly so, that a reversed image is suitable providing the manipulator also reverses its motion. However, the gravitational vector should always appear to be downward. If this simple rule is not followed, it is necessary for the brain to go through extra computations in order for the operator to perform rather simple manipulations.

Another method of viewing in three dimensions is by means of stereo television. This system will undoubtedly be useful where long distances are involved or where no other optical system has been or can be installed. At present the resolution and depth perception with this system is not comparable to good large windows of the order of 3 feet thick or so.

It should be remembered that there are several methods that the eyes and brain use for judging relative distances. Probably the most pronounced of these for short range is binocular vision, that is, the use of two eyes, one displaced from the other to measure distance by triangulation. The eyes and brain also make use of the apparent diminution of size of objects as they are farther and farther away. Another important prop is the apparent relative velocities of objects when the head and eyes are moved.

The Sense of Feel or Force Reflection

Another very important characteristic that the general purpose manipulator should have is an indication of the loads or forces that it is exerting. This requirement became apparent while using manipulators that did not have it. There are several possible methods for measuring and indicating these forces. One of the simple methods is to put strain gages on each of the seven degrees of freedom and indicating the loads on meters. This system has several disadvantages, however, that became apparent when a closer investigation of the system was made. In the first place, it is very difficult for the eyes to scan seven meters as well as watch the manipulation. For this reason this system was never built. At present it is believed that the best method for getting this information back to the brain is by placing these loads back on the hand as they would be if the hand were doing the work directly.

Since the characteristic of force reflection is considered to be very important for the relatively low powered general purpose manipulator, manipulators have been built to include this characteristic. Such a manipulator is commonly called a master-slave manipulator. This model is all mechanical and has no electric drives whatsoever. The efficiency of each of the seven degrees of freedom of this manipulator is high enough so that the manipulator will work from either end equally well and so that most of the force experienced at the slave tongs is reflected to the master handle. Several models of mechanical master-slave manipulators have been built and are in use.

In order to build an electrically powered manipulator with force reflection, it is necessary to build rather complicated positional and force servomechanisms. Two such experimental force reflecting servomechanisms have been reduced to practice, and it is planned to build electric manipulators with force reflection in the near future. The electric system will undoubtedly cost more money than the all-mechanical manipulators, but it will have several possible advantages. One of these advantages is that it will not be necessary to have mechanical linkages from the control handle to the manipulator proper and therefore, the operator and manipulator can be separated by greater distances. Another advantage is that the force reflected to the control handle can be either smaller or larger than the force or load at the manipulator. Likewise, the relative motion of the control handle and the motion of the manipulator do not have to be one to one. It is hoped that at least in some cases these advantages will outweigh the increased cost.

There are two distinct and basic reasons why a general purpose manipulator should have force reflection to the control handle--one of these was mentioned as the need for informing the operator as to the magnitude of the forces he was exerting. An additional reason which is of equal or greater importance is that the manipulator should yield to extraordinary forces and actually reverse its motion unless the operator forces it to go forward. It is a little difficult to fully explain this requirement, but one example of the need for reversibility or yielding of the manipulator can be stated as follows: suppose the manipulator is being moved in a horizontal direction and runs into an object at a small glancing angle. It is easy to show that unless the manipulator yields, the force perpendicular to the object can get very high even though the force in the direction of motion is relatively low. There are, of course, many other situations, particularly in the angular motions, where large torques can result if any one independent degree of freedom does not yield. With this situation in mind, it also becomes apparent that if the manipulator is to work with delicate materials and apparatus, a minimum of six or seven degrees of freedom should be activated and be reversible during such manipulations.

Other Methods of Handling Radioactive Materials

It must be pointed out that the above remarks are primarily directed to the design of general purpose manipulators and do not necessarily apply to other types of manipulators. In cases where the manipulator is to be used as a manipulative crane primarily for handling heavy objects, it is quite probable that such a device might have less than seven degrees of freedom and have no force reflection.

On the other hand, the general remarks can very possibly be applied in the design of manipulators other than the general purpose variety. Plans at Argonne call for the development of manipulators capable of carrying loads in excess of 100 pounds and for the development of micromanipulators sensitive to very low forces but incorporating all of the features of a good general purpose manipulator as described in this paper.

INDEX

A

Acceleration, 20-21, 57
Accidents, 240
Accuracy, 9, 28, 177
Acquisition assembly, 132-133, 135
    definition, 131
Actuation, 24, 27, 55, 155
Adaptive control algorithms, 184
Agility, 97-98, 230-231
Algorithms, 9, 15, 84, 88, 175, 187, 237
    adaptive control, 184
    climbing, 239
    control, 144
    motion, 175
Amplifiers
    backup drive, 149
    electronic, 155
    servopower, 149
Angle of ascent/descent, 233
Angular velocity, 57
Anti-fouling processes, 113
Applications, 9, 30
    foundry, 161
    manipulator, 158
    non-nuclear, 52
    nuclear, 175-193
    orbital maneuvering vehicle, 122-123
    radiation, 182
    remote, 59
ARAMIS, See: Automation, Robotics, and Machine Intelligence Systems
Arms, 24, 42, 70, 96-97, 103, 145, 154, 192, 198, 206, 211, 260
    design, 81
    electromechanical, 96
    force-reflecting, 59
    manipulator, 124, 142
    mechanical, 150
    mini, 21
    mobility, 206
    pivoting, 242
    robot, 181, 188
    sections, 182
    service, 177
    slave, 42, 56, 60, 62, 207-208, 212-213, 218
    systems, 15
Articulation, 55, 235
Artificial Intelligence, 41
Assembly, 28, 128
    acquisition, 132-133, 135
    cassette, 146
    space, 121
    vehicles, 119
Astrophysics, 127
Authors
    Bartholet, Thomas G., 230-239
    Charles, J., 89-100
    Coiffet, Philippe, 41-53
    Constant, J.A., 220-229
    Crouch, Donald S., 131-141
    Debrie, Guy, 41-53, 240-246
    Edahiro, Kyosuke, 108-115
    Fabert, Anthony, 23-29
    Fisher, John J., 175-193
    Flatau, Carl R., 19-22, 54-58
    Follin, John F., 163-174
    Foote, A.L., 23-29
    Francois, Danier, 41-53
    Germond, Jean-Claude, 240-246
    Goertz, R.C., 257-262
    Guilbaud, Jean-Pierre, 240-246
    Hamel, W.R., 3-18
    Heer, Ewald, 30-37
    Herndon, J.N., 59-66
    Hill, K.J., 220-229
    Honma, Norikazu, 247-253
    Howe, Michael J., 153-162
    Ishino, Yoshitane, 247-253
    Janett, Albin, 67-69
    Jelatis, D.G., 59-66
    Jennrich, C.E., 59-66
    Kohler, Gerhard Wolfgang, 197-219
    Krsnak, Henry G., 153-162
    Kuban, D.P., 70-78
    Leary, Robert H., 163-174
    Marchal, Paul, 41-53
    Martin, H.L., 3-18, 59-66, 70-78
    McNair, James M., 163-174
    Naruse, Toshihisa, 247-253
    Petit, Michel, 41-53
    Riche, Francois, 240-246
    Salaske, Manfred, 197-219
    Satterlee, P.E., Jr., 59-66
    Sawano, Toshiyuki, 247-253
    Selig, Manfred, 197-219
    Taylor, H.J., 142-150
    Vertut, Jean, 41-53, 89-100, 240-246
    Wagner, Eyke, 67-69
    Walters, Sam, 119-130
    Wernli, Robert L., 81-88, 101-107
    Witham, Carl, 23-29
Automatic
    modes, 191
    pan and tilt, 87
Automation, 121, 153, 177, 184
Automation, Robotics, and Machine Intelligence Systems, 121
Auxiliary
    guides, 207
    vehicles, 203

B

Backdrive, 148
Backlash, 44-45, 48, 60, 73, 77
    gear train, 148
Bilateral
    servo-electronics, 42
    servo-loop principles, 46
    servomechanisms, 7
Bipolar devices, 180

Boom
    control loops, 135
    elements, 133
    launches, 138
    units, 133
Booting assemblies, 51
Brakes, 225
Buoyancy, 103

C

Cable, 24, 28, 90-91, 212, 225, 236, 240-241, 251
    controlled systems, 89-100, 146
    cutting, underwater, 103
    handling systems, 89-92
    harness, 149, 224
    problems, 226
    routings, 26
    strength, 91
    tethered, 251
Cable Controlled Underwater Recovery Vehcile, 102
Cameras, 28, 60, 68, 87-88, 93, 95, 98, 105, 109-110, 121, 124, 137, 143, 200-201, 203, 212, 216, 247, 250-251
    boxes, 249-250
    imaging, 137
    lander, 138
    stereo, 68
    surveillance, 68
    systems, 106, 237
    television, 240-241, 244, 250
Canada, 13, 142
Cells, 189, 220, 223
    handling, 227
    hot, 178, 205, 211, 215
    shielded, 175
Chassis, 212, 214, 218, 242
    geometry, 217
Chemical munitions, 163-174
CID, See: Control Input Device
Circuits, 47, 258
Clean-up systems, undersea, 108-115
Clear sight devices, 250
Closed loop control, 85-87
Coal mining, 26
Commands, surface sampler, 135
Communications, 126, 130, 240-241, 244
    link, 173
    systems, 123, 203
Compact design, 55
Components, 21, 27, 83, 182, 187, 223
Computers, 84, 235-237, 239, 244, 251
    activities, lander, 135
    control systems, 83-84, 157, 163, 166, 171
    integration, 84
    mechanisms, 259
    mini, 166, 171, 173
    optimization processes, 147
    speed, 243
    steering, 243
Consoles
    control, 88, 184, 186, 225
    operating, 198
Contact stresses, 148
Contamination, 178, 198
Control, 126, 240
    algorithms, 144
    board, 244
    box, 235-236
    computer, 84
    desks, 205
    digital, 63
    equipment, undersea, 110
    handles, 244, 246, 257, 261-262
    logic, 135-136
    loops, 135
    methods, 4
    modes, 85
    modification, 180
    panels, 56, 225, 247
    process, 171-172
    remote, 257
    systems, 7, 84, 144, 157, 163, 171
    terminals, 172
    units, undersea, 110
Control consoles, 88, 184, 186, 225
    concept, 85
    undersea, 110
Control Input Device, 20-21
Controllability, 231, 259-260
Controllers, 88, 126, 172, 208, 213, 222
    arm design, 87
    interface units, 143
    master, 4, 6, 84
    position, 84, 86
    robotic, 173-175, 184
    software, 173
Controls, 58, 62, 86, 88, 130, 225, 257
Cooling systems, 108
Corrosion, 25
Cost and costs, 223
    analysis, 23
    demilitarization, 163
    development, 121
    effectiveness, 153, 177
    efficiency, 84
    labor, 8, 153
    manufacturing, 221
    operational, 121, 163
    production, 8-9
    reductions, 9, 14, 230
    robot, 177
    safety, 121
    savings, 23, 123, 223
    space, 121
Counterbalance, 57
Crack instability, 147
Cross-coupling, 74
Cryobaths, 165, 167
    sensors, 172
Cryo-fracture, 164
    machines, 170
    process flow, 165
    production facility, 165
    robot, 170
Cryogenic
    bath servicing tasks, 168
    embrittlement, 164
    treatment, 165

D

Data
    acquisition, 16
    analog, 203
    digital, 203
    handling, 135

manipulator, 218
position, 88
transmissions, 90, 93
vehicle, 218
Dead-weight compensation, 206, 211, 216
Debris capture, 127
Deceleration, 20
Decision-making capabilities, 159
Deflection, 47-48, 50
Degrees-of-freedom, 20, 55, 166
Depth perception, 124, 158
Design, 104, 192, 212, 216, 221, 230, 237, 258-259
    characteristics, 105
    configuration, 231
    constraints, 54
    criteria, 163
    development models, 134
    electrical, 149
    electronic, 235
    engineers, 84
    features, 8
    gear, 147
    inertia, 8
    integration, 235
    mechanical, 235
    modular, 182
    objectives, 8-9, 231
    performance, 42
    slave, 23
    spacecraft, 119
    system, 84
    thermal, 149
Development
    costs, 121
    manipulator, 257-262
    ocean, 247
Dexterity, 19, 23-24, 26-28, 54, 70, 97, 103-104, 121, 129, 157, 259-260
Diagnostics, 63, 173
Digging machines, 252
Digital
    control systems, 63
    electronics, 63
    integrated circuits, 180-181
    microelectronics, 71
    multiple unit, 48
    multiplexing, 244
    recording units, 50
Dimensions, 223
Displacement, 258, 260
Displays, 88, 130, 172
Docking, 128
    devices, 250
    operations, 19
    probe, 125
Drilling machines, 252
Drive
    arrangements, 55
    motors, 213, 228
    systems, 251
    tape, 71
    units, 207
Durability, 26-27

E

Elbows, 57, 223
    action, 228-229
    pivot, 224
Electric
    dead-weight compensation, 205
    manipulator, 260
    master-salve manipulators, 205
    motors, 75, 206, 211, 216
    power transmission, 208
    robots, 184
    supply, 203
Electrical
    design, 149
    manipulators, 259
    technicians, 158
    tools, 68, 200
Electromechanical
    arms, 96
    components, 134
Electronic
    amplifiers, 155
    controls, 76
    design, 235
Electrophoresis, 127
Emergency operations, 240
Encoders, 7, 149
End effectors, 24, 26, 28, 121, 125, 143-145, 156, 166, 172, 178, 182, 189
Engin de Reconnaissance et d'Intervention a Cable, 89, 92-93
    agility, 98
    characteristics, 94
    description, 93
Engineering models, 144
Engineers
    design, 84
    remote handling systems, 177
Environmental tests, 134
Equipment, 199, 216, 220
    disassembly, 190
    handling, 190
    process, 71
    selection, 183
    undersea, 110
    working, 253
ERIC II, See: Engin de Reconnaissance et d'Intervention a Cable
Error
    derivatives, 86
    recovery functions, 174
Europe, 13, 119
EVA, See: Extravehicular activity
Excavators, 252
Explosive handling systems, 27
Extravehicular activity, 121

F

Factories, orbital, 121
Fatigue, 85
    tests, 48, 50
Federal Republic of Germany, 197-219
Feedback, 98, 121, 185, 214, 240
    audio, 184
    aural, 98
    devices, 184
    manipulator, 244
    position, 83, 144
    postural, 98
    sensors, 74, 95

video, 121
vision, 97
visual, 98, 174
Fiber optics, 98, 103
Flexibility, 14-15, 28, 166, 211, 235, 237, 240
Flight models, 134
Flotilla, 119
    construction, 122
Fluids, 24-25
Force-feedback, 83-84, 125, 156
Forces, 214, 240, 260
    interaction, 9
    modulation, 9
    ratios, 63, 211
    sensors, 126, 189
Force reflection, 4-5, 175, 208, 241, 261-262
    arms, 59
    servomanipulators, 70-78
Forearm, 209
    rotation, 224, 228-229
Forging, 159-160
FORTRAN, 87, 173
Fracture control, 147
France, 13, 41, 89
    Navy, 89
Friction, 50, 57, 60, 73, 77
    torque, 75
Fuel reprocessing, 59, 70
Functionoid, 230-239
Future, 88, 111, 121
    research, 14
    space operations, 130

G

Gear, 208
    box, 221, 224
    design, 147
    group, low-speed, 148
    mechanism, 226
    ratios, 74, 148
    reductions, 74, 148
    shoulder joint, 144, 146-147
    stresses, 148
    train backlash, 148
Geometry, 10, 42, 243
    chassis, 217
    comparisons, 75
    manipulator, 87
    servomotor, 75
    space station, 120
    speed, 244
    steering, 245
    variable, 240, 242
    vehicle, 244
Glove boxes, 185, 189-190
Graphics systems, 15, 173-174
Gravity, 20
Gravity-assist flows, 133
Grip force, 203
Gripper action, 209, 228-229
Grippers, 26, 104-105, 160, 189, 198, 223, 258, 260
Guidance and controls, 135
Guide vehicle, 202, 204

H

Handling, 33
    devices, 212, 257
    equipment, 190
    processes, 154
Handling systems
    cable, 91
    explosive, 27
    payload, 142
    remote, 177
Hardware, 88, 154
    robot, 184
Hazardous materials, 164
Hierarchical systems, 16
History, 6
    servomanipulator, 41
Hot
    cells, 205, 211, 215
    laboratories, 240
Hydraulic
    cylinders, 248, 250
    power, 5, 24
    systems, 55, 184
    technicians, 158

I

Imaging collectors, 127
Indicator lights, 184-185
Industrial
    automation, 153
    manipulators, 153-162
    robots, 177, 182
Industry
    nuclear, 4, 121, 164, 221
    space, 35, 121, 129
Inertia, 50, 56-57, 73
    cargo, 20
    criteria, 20
    design, 8
    kinesthetic feedback, 98
    reference system, 237
Inspection, 82-84, 230, 247-248
Integration, 82, 87
    computer, 84
    design, 235
    evaluations, 134
Investigation, 89
Ireland, 102
Irradiation
    facilities, 240
    materials, 227
    tests, 208

J

Japan, 13, 31
Jaws, 198, 207, 261
    action, 223
    gripper, 189

Joints, 25
    axis cross coupling, 74
    performance, 144
    power conditioner, 149
    shoulder, 144-147
    yaw, 146
Joystick control, 4, 86

K

Kinematics, 26, 55, 70, 192
    configurations, 24, 73
    equations, 29
    improvements, 15
Kinesthetics, 98

L

Laboratories, 120, 240
Labor costs, 8, 153
Lander
    cameras, 138
    computer activities, 135
Landers, 135-138
Language, 87, 187
    programming, 184
    robot, 166, 173
Legs, 239, 248, 250-251, 253
    telescopic, 247
Lifting capacities, 21
Lights, indicator, 184-185
Limit switches, 225
Line monitoring, 173
Liquid nitrogen, 164, 168, 240
Load
    capacities, 73
    hook, 205
    sustaining, 90
Loads, 223
Logic, control, 135-136
Logistics module, 120
Lubrication systems, 149

M

MA 23 characteristics, 97
Machine intelligence systems, 121
Machines
    digging, 252
    drilling, 252
Maintainability, 55, 182, 237
Maintenance, 64, 71, 158, 163-164, 175, 178, 220, 230-231, 252
    areas, 165
    nuclear, 28
    orbital, 129
    remote, 59, 64, 70, 76, 164, 171
    support, 122
    systems, 63, 65
    tasks, 59
    techniques, 59
    technology, 70
    undersea, 108

Man amplifier, 161
    manipulator systems, 154
Man and machines, 121
Maneuverability, robot, 110
Manipulation, 229
    methods, 14
    types, 4
Manipulator, 222, 225
    applications, 158
    arms, 124, 142
    attributes, 8
    configuration, 155
    control modes, 85
    controller interface unit, 143
    crane vehicle, 197-198
    data, 218
    electric, 259-260
    feedback, 244
    flow charts, 86
    functions, 258
    geometry, 87
    joint performance, 144
    mini, 21
    motion, 88, 201
    movements, 228
    operators, 158
    relationships, 84
    requirements, 83
    techniques, 20
    units, 67
    vehicles, 197-219, 204, 211
Manipulators, 20, 103-105, 121, 130, 241-242, 246
    advantages, 158
    basic requirements, 257
    development, 257-262
    heavy-duty, 197
    industrial, 153-162
    light, 211
    manned, 158
    medium-sized, 198
    modes of operation, 43
    philosophy, 257-262
    power, 175-176
    shuttle, 19-21
    slave, 61
Manipulator systems, 175
    reliability, 70
    underground, 26
    undersea, 24
Man-like tasks, 60
Man-machine interface, 4, 23, 119-130
Manpower reduction, 163, 230
Manufacturing, 154
    costs, 221
    productivity, 8
Marine
    cleaning robots, 109
    growth, 108
Mars, 131-141
    surface operations, 136
Master
    arms, 4, 49, 62, 154, 209, 213
    controllers, 4, 6, 84
    manipulators, 61
    stations, 59-61, 65
Master/slave
    manipulators, 54-58, 70, 121, 205, 216, 240, 260
    mechanical, 175-176
    modes, 41, 50, 63

systems, 5
teleoperators, 9
Materials, 121
   irradiated, 227
   radioactive, 3, 67, 257
   thermal, 149
Materials handling, 26-27, 153
   operations, 164
   tasks, 153
Materials processing, 35, 121-127
Measuring
   profile, 248
   radiation, 178
   systems, 251
Mechanical arm, 150
   complexity, 84
   master/slaves, 175-176
   removal methods, undersea, 108
   subsystem, 144
Mechanical Master/Slave Manipulators, 70
Microcomputers, 187
Microelectronics, 71
Microgravity vehicles, 123
Microphones, 200-201
Microprocessor systems, 86, 105
Military, 231
Mine Neutralization Vehicle, 103
Mines and Mining, 103, 231
Mini
   arms, 21
   computers, 166, 171, 173
   manipulators, 21
Mirrors, 138
MNV, See, Mine Neutralization Vehicle
Mobile
   platforms, 221
   robotics, 230
Mobility, 10, 54-55, 98, 206, 220, 247
   base, 130
   head, 95
   problems, 97
Models
   engineering, 144
   flight, 134
Modes, 41
   automatic, 191
   control, 85
   of operation, 103
   simulation, 174
   walking, 235
Modular
   designs, 182
   imaging collectors, 127
   robot costs, 183
Modularity, 15, 76
Modules, 119, 127, 192
   logistics, 120
   motor, 76-77
   replacement schemes, 76
   servomanipulator, 76
   tong, 77
   voice, 184
   wrist, 77
Monitors, 211, 251
   radiation, 185
Motion, 214, 224, 235, 241, 257-262
   algorithms, 175
   axes, 62, 134
   commands, 87
   ranges, 55-57, 72-73

rates, 20, 87
responses, 175
Motors, 207-208, 221, 224, 235
   drive, 213, 228
   electric, 75, 206, 211, 216
   modules, 76-77
   roll positioner, 160
   sizes, 74
   submersible, 251
Movement, 205-206, 211, 222, 225, 228, 244, 248, 257
MSMs, See: Mechanical Master/Slave Manipulators, 70
Multiplex communication system, 244
Munitions, chemical, 163-174

N

NASA, See: National Aeronautics and Space Administration
National Aeronautics and Space Administration, 103-104, 119, 121, 128-142, 230
NEV, See: Nuclear Emergency Vehicle
New technologies, 105
No-Go, 136, 138
Nuclear
   activities, 231
   applications, 175-193
   facilities, 240
   fields, 10
   fuel reprocessing, 59, 70
   hot cells, 178
   industry, 4, 121, 164, 221
   maintenance manipulator systems, 28
   reactors, 204, 257
   service manipulator concept, 28
   technology, 9
   workcells, 175, 182
Nuclear Emergency Brigade, 197-219
Nuclear Emergency Vehicle, 104-105

O

Object
   definition, 88
   recovery, 103
Observatories, 123
Obstacle crossing, 243
Ocean development, 247
Odex, 12, 230-239
   advantages, 237
   profiles, 232
Omni-directionality, 231
OMV, See: Orbital Maneuvering Vehicle
Open loop control, 85
Operating
   console, 198
   costs, 121, 163
   speeds, 8
Operational
   analysis, 129
   control, 6
   time distribution, 105
Operations, 121
   automated, 184
Operators, 85-88, 157, 247, 258
   consoles, 64
   support, 173
   viewing, 60

Optimization processes, 147
Orbit, 119, 142
Orbital
    applications, 123
    factories, 121
    maintenance, 129
    photography, 192
    servicing system, 129
    shuttles, 19
    subsystems, 123
    systems, 123
Orbital Maneuvering Vehicle, 119, 122-123, 129
Orbiters, 135
Output signals, 185

P

PAGODE, characteristics, 92
Pallet configurations, 168
Pan-and-tilt systems, 87-88
Payload, 119, 142-143
    capability, 166
    capacity, 233
    handling system, 142
    placement, 122
Performance, 57-58, 250
    parameters, 233
    range, 233
    specifications, 231
Photography, 248, 250
    orbital, 132
Pick-and-place, 32
    operations, 180
    routines, 184
PID controller, 86
Pilot disorientation, 97
Pipelines, 111
Pitch, 160, 166
Platform drive, 225-226
Position
    controllers, 84, 86
    data, 88
    encoder, 149
    errors, 86, 175
    feedback, 83, 144
    loops, 6
    measuring systems, 251
    sensors, 83, 86, 106, 109
    transmitter, 208, 213
Potentiometer, 155
Power
    consumption, 105, 233
    efficiency, 231
    generating stations, 108
    manipulators, 175-176
    station systems, 113
    transmission, 207-208
Powering technique, 24
Process
    control, 171-172
    equipment, 71
    line monitoring, 173
    nuclear, 183
    status information, 173
    technology, 163
Production
    lines, 160
    output, 8

Productivity, 153
    manufacturing, 8
Profile measuring, 248
Programmability, 33
Programmed control, 87
Programming languages, 87, 184, 187
Projectiles, 168
Propulsion, 93, 103, 243
Proximity sensors, 187-188, 239

Q

Quality controls, 230

R

Radiation, 28, 187, 203
    applications, 182
    considerations, 180
    effects, 181
    exposure, 29
    levels, 198
    measurements, 178
    monitors, 185
Radioactive
    environments, 10
    materials, 3, 67, 257
Radio control, 203, 240
Ranges of motion, 55, 57, 72-73
Ratcheting, 87
Rate controls, 28, 85
Rate of motion, 20, 87
Reactors, 257
    nuclear, 204
Real-time
    diagnostics, 63
    interaction, 4
Recovery
    claws, 102, 105
    procedures, 173
    systems, 83, 101-107
    tasks, 82, 106
ReCUS underwater surveyor, 247
Reference systems, 237
Relays, 222
Reliability, 28, 70-71, 182, 230-231, 237
Remote
    access, 230
    applications, 59
    control, 257
    handling, 177
    maintenance, 59, 64, 70, 76, 164, 171
    manipulation, 23
    observation, 89
    operations, 164, 247
    operators, 239
    orbital servicing system, 129
    presence, 121
    radio control, 203, 240
    servicing, 124
    technology selection, 164
Remote Manipulator System, 3
Remote Unmanned Work System, 103, 105
Remotely Operated Mobile Manipulation, 220
Remotely Operated Vehicles, 81
Removal methods, undersea, 108

Repair, 121, 124
Repeatability, 9, 177
Rescue vehicles, 240
Research, 15
Resupply, 128
Reversibility, 262
RMS, See: Remote Manipulator System
Robot and robotics, 3-18, 121, 153-154, 164, 166-167, 172, 175-193
  aisle, 187
  arms, 181, 188
  benefits, 177
  bodies, 248
  bottom-travelling, 247
  cleaner units, 109, 114
  controllers, 173-175, 184
  control systems, 7
  costs, 177
  cryo-fracture, 170
  definition, 30, 82
  electric, 184
  flexibility, 166
  general purpose, 31
  handling, 252
  hardware, 184
  industrial, 177, 182
  languages, 166, 173
  maneuverability, 110
  marine cleaning, 109
  material handling, 166
  mobile, 230
  rubble leveling, 251
  sensory, 31
  servo-controlled, 31
  Soviet, 32
  special purpose, 31
  submersible, 247
  systems, 166, 176, 183, 248
  tasks, 178
  telepresence, 128
  underwater, 112
  vendors, 184
  walking, 247-253
Robotic systems, 176, 183
  availability, 182
  unloading, 169
Robot Institute of America, 30
Rocket boosters, 103
Roll positioner motions, 160
ROMAN, See: Remotely Operated Mobile Manipulation
Rotary brushes/scrapers, 110, 115
Rotation, 60, 209-211, 223-224, 228-229, 240-241, 244, 258
ROVs, See: Remotely Operated Vehicles
RUWS, See: Remote Unmanned Work System

S

Safety, 26-27, 163, 173, 184, 230, 257
  tasks, 205
Sampling
  coordinates, 137
  sequences, Viking, 138
Satellites, 3, 122, 125
Scaling characteristics, 21
SCARAB, See: Submersible Craft Assisting Repair and Burial
SCAT, See: Submersible Cable-Actuated Teleoperator

Sea
  floor, 89
  mines, 103
Searchlights, 200, 212, 218
Sensors, 74, 106, 110, 126, 128-130, 153, 171-173, 183, 201, 239
  contact, 237
  cryo-bath, 172
  devices, 182
  feedback, 74, 95
  fine-guidance, 129
  force, 126, 183, 189
  moment, 183, 189
  polling programs, 173
  position, 83, 86, 106, 109
  proximity, 187-188, 239
  submarine, 110
  temperature, 138, 199, 203
  visual, 128
Servicing, 121
  arms, 177
  remote, 124
  system, 122, 129
  tasks, 128
Servo
  control systems, 144
  drives, 212, 244
  electronics, 42
  loops, 6, 9, 46, 155
  systems, 55
  valves, 86
Servomanipulator
  advanced, 71
  arms, 60
  concepts, 71
  force-reflecting, 70-78
  history, 41
  modules, 76
  performance, 63
  slaves, 71
Servo master/slaves, 175-176
  manipulators, 54-58
Servomechanisms, 262
  bilateral, 7
Servomotors
  drive units, 60, 62
  geometries, 75
  parameters, 75
Servopower amplifier, 149
Shakeout areas, 161
Shielded cells, 175
  transfer systems, 187
Shoulder, 223
  action, 228-229
  braces, 147
  differential, 57
  drives, 57
  joints, 144-147
Shroud units, 133
Shuttle Attached Manipulator, 19
Shuttle Remote Manipulator System, 142
Simulation
  facility, 149
  mode, 174
Slave, 25, 55, 155
  arms, 42, 56, 60, 62, 206-208, 212-213, 218
  design, 23
  manipulators, 61
  packages, 59-60, 64

servodrive unit, 212
standard, 49
systems, 70
SM-229, 54
Software, 15, 63, 171, 173-174, 190, 235-236
Solenoids, 135
Soviet robots, 32
Space
  arenas, 11
  assembly, 121
  costs, 121
  exploration, 19, 34
  industrialization, 35, 121, 129
  operations, 121
  operations, future, 130
  program, 121
  strategy, 119
  systems, 34
  telescope, 121
Spacecraft, 30, 34, 131
  design, 119
  docking, 125
  repair, 121
  retrieval, 122
  services, 124
  servicing, 121
  transmitters, 135
Spacelab, 119
Space shuttle, 3, 13, 102, 128, 142
  manipulators, 19-21
  payload handling, 142
Space station, 119, 121, 124
  architecture, 120
  geometries, 120
Spain, 102
Speed, 223, 233, 246
  cleaning, 111
  compatibility, 8
  computer, 243
  operating, 8
  rotational, 244
  vehicle, 244
SRMS, See: Shuttle Remote Manipulator System
Stability, 231, 247-248
Steering, 242-244
  computer, 243
  geometry, 245
  ratios, 215
Stereo
  cameras, 68
  loudspeakers, 68
  vision, 124
Stereo-optic vision systems, 124
Stereoscopic viewing, 69
Strength-to-weight ratios, 231, 237
Stresses
  contact, 148
  gear, 148
Submarine sensors, 110
Submergence teleoperator systems, 89-100
Submersible Cable-Actuated Teleoperator, 101-102, 105
Submersible Craft Assisting Repair and Burial, 24
Submersibles, 101, 103, 105
  motors, 251
  robots, 247
Surface
  imaging, 138

operations, Mars, 136
Surface sampler
  activities, 139
  commands, 135
  components, 131-132
  controls, 135
  operations, 139
  performance, 140
  problems, 141
  sequences, 136
  subsystem, 131-141
Surveillance, 28, 98, 230
  cameras, 68
Surveying, 247-248
Suspension systems, 203
Switches, 4, 85, 88, 225, 261
Switzerland, 67
Synchronism, 58
  positional, 260
System and Systems
  arm, 15
  autonomy, 130
  classification, 82
  clean-up, 108
  communications, 203
  configurations, 83-84
  control, 84, 87, 144, 157, 163, 171
  cooling, 108
  design, 84
  dexterity, 129
  drive, 251
  explosives handling systems, 27
  failures, 71
  forging, 159
  graphics, 174
  hierarchical, 16
  hydraulic, 55, 184
  inertial reference, 237
  inspection, 82-83
  lubrication, 149
  machine intelligence, 121
  man-amplifier manipulator, 154
  manipulator, 175
  microprocessor, 86, 105
  nuclear, 28
  operators, 159
  optical, 127
  position measuring, 251
  recovery, 83, 101-107
  reference, 237
  robotic, 166, 176, 183, 248
  servicing, 122, 129
  servocontrol, 144
  solid lubrication, 149
  space, 34
  submergence teleoperator, 89-100
  suspension, 203
  telerobot, 179, 190-191
  television camera, 237
  test rigs, 150
  undersea, 36, 108
  upgrades, 130
  viewing, 68, 257
  vision, 28, 124, 130, 173, 261
  waste removal, 185
  weights, 24
  work, 82-83, 104

T

Tables, command, 135
Tape
    drives, 71
    recordings, 50
Task
    analysis, 128, 177
    capabilities, 84
    classification, 84
    efficiency, 4
    functions, 173
    identification, 177
    sequences, 169
    structuring, 178
    times, 106
    variability, 26
Tasks, 128, 175
    appliance handling, 157
    automated, 177
    recovery, 82, 106
    robots, 178
    safety, 205
    servicing, 128
    teleoperated, 179
Technicians, 158
Technology
    new, 105
    process, 163
Teleoperated tasks, 179
Teleoperation definition, 3-18, 30
Telepresence
    definition, 121
    project applications, 127
    robot, 128
    technology, 124
    teleoperator comparison, 130
Telerobot, 179
    systems, 190-191
Telescopes, 127
    retrieval, 126
    space, 121
Telesymbiotics, 94, 98
Television, 261
    cameras, 87, 101, 103, 199-200, 212, 218, 237, 240-241, 244, 250
    monitors, 174, 184-185, 205
    viewing, 70
Temperature
    profile data, 138
    sensors, 138, 199, 203
Terrain
    exploration, 231
    rough, 240
Test
    model hardware, 134
    vehicles, 240
Testing, 106, 149
Tethers, 93, 103, 251
Thermal
    control materials, 149
    design, 149
Three-dimensional imaging, 124
Time
    bases, 134
    efficiency, 241
    -invariant scaling, 21
Tongs, 77, 200, 207-208, 211, 214, 220, 258, 261

Tool and Tools, 97, 201, 211, 214
    carriers, 200
    center point, 191
    electrical, 68, 200
    mechanical, 200
    special, 68, 166
Topographic
    features, 250
    profile maps, 137
Torque, 75, 148
Tracks, 216, 221, 240
    servowelding, 189
Training, 157, 259
Trajectory planning, 29
Transmissions, 93, 207-208
Transmitters, 203
    position, 208, 213
    spacecraft, 135
Trigonometric functions, 87
Troubleshooting, 158

U

Underground systems, 26
Undersea, 101-107
    cleaning systems, 108, 114
    construction, 247-253
    exploration, 3
    lighting, 251
    maintenance, 108
    manipulator systems, 24, 101
    operations, 247, 251-252
    robot equipment, 109
    robots, 112, 247
    systems, 36
    vehicles, 30, 81, 83
    vehicles, problems, 88
    work systems, 81
    work tasks, 83
United Kingdom, 13
United States Army, 27
United States Navy, 11, 101
Universal manipulator controllers, 15

V

Variable
    geometry, 242
    rate control, 25
Vehicle, 241
    assembly, 119
    auxiliary, 203
    data, 218
    descriptions, 242
    geometry, 244
    guide, 202, 204
    inspection, 82
    logistics, 119
    manipulator crane, 197-198
    microgravity, 123
    pilots, 88
    propulsion, 244
    rescue, 240
    speed, 244

test, 240
  undersea, 30, 81, 83
Velocity, 21, 144, 260
  angular, 57
  loops, 7
Vendors, 180
  robot, 184
Versatility, 14, 67, 85, 235, 237, 239
Video
  coordinators, 88
  displays, 124
  feedback, 121
  information, 123
Viewing, 241
  stereoscopic, 69
  systems, 68, 257
Viking flights, 134
VIRGULE, 240-246
  vehicle, 41-42
Vision, 32, 172,
  feedback, 97
  research, 16
  systems, 124, 130, 173
Visual
  displays, 88
  feedback, 98, 174
  sensors, 128
  systems, 28, 261
Voice modules, 184

W

Waist, 223
  rotation, 228-229
Walking
  robots, 247-253
  sequences, 237, 249
Waste removal system, 185
Weight, 24, 28, 55-56, 233, 242
  constraints, 25
West Germany, 12
Wheels, 243-245
Work cells, 83, 179
  nuclear, 175, 182
Working
  envelopes, 166
  equipment, 253
Worksite, 121
Work systems, 82-83
  design, 104
  operators, 88
  remote, 103
  undersea, 81
Work Systems Package, 83-84, 103-106
Wrist, 189, 208-209
  action, 228-229
  modules, 77
  motions, 73
  pivot, 224
  rotation, 223-224, 228-229
WSP, See: Work Systems Package

X

X-ray astrophysics, 127

Z

Zero gravity, 19

TJ 211 .T44 1985

Teleoperated robotics in hostile environments